A Fragrant Introduction to Terpenoid Chemistry

To Hilary,

Thank you.

A Fragrant Introduction to Terpenoid Chemistry

Charles S Sell
Quest International, Ashford, Kent, UK

RS•C
advancing the chemical sciences

The jacket illustration depicts a molecule of longifolene superimposed on a sprig of *Pinus longifolia*. Longifolene is a major component of the leaf oil of *P. longifolia* and its intriguing structure and often surprising reactions (see Chapter 7) typify the excitement of terpenoid chemistry.

ISBN 0-85404-681-X

A catalogue record for this book is available from the British Library

Published by The Royal Society of Chemistry,
Thomas Graham House, Science Park, Milton Road,
Cambridge CB4 0WF, UK

Registered Charity Number 207890

For further information see our web site at www.rsc.org

Typeset by Alden Bookset, Northampton, UK
Printed by TJ International Ltd, Padstow, Cornwall, UK

Preface

The mind is a fire to be kindled, not a vessel to be filled.
Plutarch

The book is aimed primarily at university undergraduates, post-graduates and professional chemists who wish to build up their knowledge of terpenoid chemistry. It is intended to serve as a general introduction to the exciting field of terpenoid chemistry. Terpenoids play an important part in all our lives, from perfumes through insect pest control to pharmaceuticals such as steroid hormones and the anti-cancer drug paclitaxel. The subject therefore also serves to illustrate the importance of chemistry in everyday life.

In the interests of length and also of the author's expertise, we will concentrate on the mono- and sesquiterpenoids and primarily those of interest as fragrance ingredients. Higher terpenoids will be mentioned and the reader will be able to extrapolate the basic principles of terpenoid chemistry from the more detailed examples using lower terpenoids to these higher homologues.

Chemistry is a multi-faceted discipline and each part is interconnected with every other. It is also the central natural science, lying between physics and biology. To understand chemistry we must understand something of physics. Equally, since living organisms function through chemistry, we must understand that chemistry in order to fully understand them. I have therefore included some elements of biochemistry and molecular biology in order to illustrate the key role which terpenoids play in the processes of life and the senses of sight and smell in particular.

Terpenoid chemistry touches on all aspects of stereochemistry and mechanism. However, one prominent feature of terpenoid chemistry is that of carbocation reactions and the fundamental research which forms the basis of our understanding of this area, was carried out on fragrant terpenoids. Some of the most elegant of all total syntheses involve sesquiterpenoid targets. The book will therefore also serve as a refresher course on mechanism, stereochemistry and synthetic methodology. Where appropriate, basic principles are discussed in order to prepare for their application to terpenoids. For example, the elements of stereochemistry are reviewed in Chapter 4 before showing how

v

important they are in understanding the chemistry of carvone and menthol. I recommend that any student readers of this book acquire a set of molecular models. These will be particularly helpful in understanding stereochemistry and carbocation rearrangements in cyclic molecules.

The first two chapters are designed to excite by showing the diversity of terpenoids and their roles in living organisms. Also amazing is that such a diversity can be produced from one simple feedstock and a handful of chemical reactions. Students going through the book from the beginning should not be put off by the apparently complex chemistry described, especially in Chapter 2. The basic principles of the chemistry are covered in detail in later chapters.

There is a selection of problems involving terpenoid chemistry and this is followed by worked solutions. As always, problems are a good way of testing one's understanding of a subject and this is one of the reasons for including a number in this book. However, some of them serve a dual purpose and are almost integral parts of the text since they explain some points which are, deliberately, passed over rather superficially in the main body of the text. If the reader finds something which appears to have been glossed over, then it would be useful to check the problems section to see if the explanation lies there.

There is a bibliography which will serve to direct those who wish to know more to some of the key sources of information. These are arranged by subject in order to make it easy to use. There are also specific references which are cited in the main body of the text. These are mostly to original research papers and are designed to encourage the students to test the excitement of exploring the original literature. There is a small degree of overlap between the references and the bibliography. I felt it better to accept this than to create a complex system of cross-referencing which would reduce accessibility.

I believe that science and art should not be separated but should be taken together since each helps in our understanding and appreciation of the other. Many great scientists were also accomplished in the arts. For example, Albert Einstein played the violin and Alexander Borodin, besides being a professor of chemistry at a medical school in Saint Petersburg and a leading figure in research into alkaloids, was one of the greatest Russian composers of his day. Perfumery is clearly a blend of creative art and chemical science. I have therefore tried to develop a link to philosophy and the arts through the use of appropriate quotations at the start of each chapter and by the use of perfumery as an example of discovery chemistry.

Table of Contents

This chapter explains the definitions and classification of terpenoids. It also describes where and why terpenoids occur in nature and how they are extracted from natural sources.

This chapter includes a brief introduction to the processes used in biogenesis. It explains how nature constructs the basic 5-carbon building blocks used for terpenoid biogenesis and how these hemi-terpenoid units are connected together to form chains of 10, 15, 20, *etc.* carbon atoms. It includes a brief overview of how these chains can be cyclised and modified to produce the staggering array of terpenoids which are present in nature.

Chapter 3   Linear and Monocyclic Monoterpenoids   43

This chapter gives a short introduction to the methods of structural
determination adopted before spectroscopy came into existence. Confirmation of proposed structures by synthesis provides an introduction
to synthetic strategy. Myrcene and citral are used as examples of these
disciplines and the chemistry of linalool and terpineol serve as a gentle
introduction to carbocation chemistry.

Chapter 4   Menthol and Carvone   65

These two key monocyclic monoterpenoids provide an excellent
illustration of isomerism: structural, geometrical and stereoisomerism.
The chapter demonstrates the importance of isomeric purity in biological
processes involving molecular recognition, the interaction of different

forms of isomerism and the implications of stereochemistry for synthesis.

Chapter 5 Bicyclic Monoterpenoids   97

The chemistry of pinanes, camphanes and bornanes introduces the subject of carbocation chemistry. The basic principles governing the reactivity of carbocations and the factors which determine the selectivity of processes involving them, are therefore to be covered in detail in this chapter. The dramatic changes in chemical structure which can result from simple cation rearrangements are illustrated in these relatively easy to visualise molecules.

Chapter 6   Precious Woods   **135**

Sandalwood and cedarwood constituents demonstrate the increasing complexity of carbocation reactions as the molecular size increases from 10 to 15 carbon atoms. In addition, Cedarwood chemistry demonstrates how changes to conditions can radically affect the outcome of carbocation reactions. Total synthesis of sandalwood materials introduces the Wittig reaction as a means of delivering geometric selectivity in synthesis. At this point, a revision of the basics of carbanion chemistry including the stereochemistry of the aldol reaction is appropriate. Examples from the chemistry of cedrene and selinene remind us that nothing can be taken for granted in terpenoid chemistry.

Chapter 7   Other Woody Odorants   **177**

Vetiver, patchouli, *Pinus longifolia*, cloves and hops provide us with
examples of further increases in complexity in rearrangements, including
the santonin rearrangement, and the corresponding increase in the
challenges of total synthesis. The vetivones take us back to the use of
degradation as a tool for structural elucidation and then, through the
attendant need for total synthesis, forward to the sheer elegance of
Stork's synthesis of β-vetivone. Longifolene shows how the course of a
reaction can be dramatically changed by the presence of a strategically
placed neighbouring atom.

Chapter 8   Degradation Products   **229**

Nature is never static and this is demonstrated by chemical degradation of higher terpenoids in organisms and in the environment. The emphasis will be on the degradations which yield desirable products such as ambergris, the ionones, damascones, irones and theaspirones. The use of carotenoid derived pigments in vision introduces us to receptor proteins and the senses by which we perceive the universe around us.

Chapter 9   Commercial Production of Terpenoids   **269**

In this chapter, the two main reasons for organic synthesis are compared and contrasted. In previous chapters, synthesis was a tool for structural elucidation and the key driving force was the unambiguous nature of the product's structure. Similar thinking is necessary for discovery chemistry. However, for commercial production, the key factors are safety, cost and security. Effluent, including unwanted by-products, is given prominence as part of the cost factor. For complex structures related to natural products, the most cost-efficient and secure starting materials might well be other natural products and this introduces the issue of sustainability. The often complex interplay between all of these parameters is illustrated through appropriate examples.

Chapter 10   Discovery and Design of Novel Molecules   309

The basic underlying principles of discovery chemistry are the same as those in discovery of novel molecules for all applications including pharmaceuticals, flavour ingredients, adhesives, lubricants and so on. The first part of the chapter outlines these basic principles and so should serve as an introduction for any area of discovery chemistry. The second part of the chapter uses fragrance ingredients as an example of the discovery process.

Perfumery is a blend of science and art. The language and artistic elements of perfumery will be discussed together with the technical aspects of commercial perfumery. The role of structure/property correlations (QSPRs) in the discovery of new ingredients will be covered. This will lead on through discussion of the scope and limitations of QSPRs to a summary of the current state of knowledge of the process of olfaction. The contribution of discovery chemistry to the history of perfumery is outlined and illustrated by specific examples.

Acknowledgements

I would like to thank Professor William Motherwell and Dr Anton van der Weerdt for their support and encouragement for this book. Without Professor Motherwell, the work would never have been started. I am also indebted to the late Professor Arthur Birch FAA, FRS for introducing me to the fascinating world of terpenoid chemistry. Professor Giovanni Appendino has also been an inspiration through his infectious enthusiasm and love of chemistry, botany, the arts, history and the interplay between them. In practical terms, I am indebted to various colleagues for their comments on specific sections of the book; to Laurence Payne for valuable input on experimental design, David McNulty also on experimental design, Neil Vincer on safety, Paul Hawkins on environment and Chris Furniss who bravely undertook the task of reading the entire draft and making useful comments on the contents. My wife Hilary deserves special thanks for her patience, good humour, understanding, love and support, all of which helped to sustain me through the process of writing and editing the book.

© Image 100/Royalty Free/CORBIS

Carotene, the orange terpeniod pigment in carrots, is used as a source of vitamin A which is essential for sight

CHAPTER 1

Background

I do not know what I may appear to the outside world, but to myself I seem to have been like a boy playing on the sea-shore, and diverting myself in now and then finding a smoother pebble or a prettier shell than ordinary, whilst the great ocean of truth lay all undiscovered before me.

Isaac Newton

Using a simple five carbon building block, nature creates an array of terpenoid chemicals with an infinite variety of structural variation and vast range of biological functions. Such a cornucopia cannot but leave the terpene chemist feeling as Newton did.

KEY POINTS

The simple isoprene unit is the basis of an enormous range and a variety of chemical structures which we know as terpenoids.

In nature, terpenoids serve a variety of purposes including defence, signalling and as key agents in metabolic processes.

Terpenoids have been used in perfumery, cosmetics and medicine for thousands of years and are still extracted from natural sources for these uses.

1.1 DEFINITIONS AND CLASSIFICATION

Plants and animals produce an amazingly diverse range of chemicals. Most of these are based on carbon and so the chemistry of carbon came to be known as organic chemistry, *i.e.* the chemistry of living organisms, the chemistry of life. These chemical products of plants and animals can be classified into primary and secondary metabolites. Primary metabolites are those which are common to all species and can be sub-divided into proteins, carbohydrates, lipids and nucleic acids. These four groups of materials are defined according to the chemical structures of their members. The secondary metabolites are often referred to as "natural products". These can be sub-divided into terpenoids, alkaloids, shikimates and polyketides. This classification is based on the means by which the materials were made. These synthesis routes are referred to as biosynthetic or biogenetic pathways.

Individual secondary metabolites may be common to a number of species or may be produced by only one organism. Related species often have related patterns of secondary metabolite production and so a species can be classified according to the secondary metabolites they produce. Such a classification is known as chemical taxonomy. Occasionally, two plants are found to have identical physical aspects which botanists use for classification, but differ in the secondary metabolites they produce. For example, two flowers may look identical but one is odourless whilst the other possesses a strong scent due to the production of a fragrant terpenoid chemical. Such different strains are known as chemotypes.

Terpenoids are defined as materials with molecular structures containing carbon backbones made up of isoprene (2-methylbuta-1,3-diene) units. Isoprene contains five carbon atoms and therefore, the number of carbon atoms in any terpenoid is a multiple of five. Degradation products of terpenoids in which carbon atoms have been lost through chemical or biochemical processes may contain different numbers of carbon atoms, but their overall structure will indicate their terpenoid origin and they will still be considered as terpenoids.

The generic name "terpene" was originally applied to the hydrocarbons found in turpentine, the suffix "ene" indicating the presence of olefinic bonds. Each of these materials contain two isoprene units, hence ten carbon atoms. Related materials containing 20 carbon atoms are named as diterpenes. The relationship to isoprene was discovered later, by which time the terms monoterpene and diterpene were well established. Hence the most basic members of the family, *i.e.* those containing only one isoprene unit, came to be known as hemiterpenoids. Table 1.1 shows various sub-divisions of the terpenoid family based on this classification. It also shows two specific sub-groups of terpenoid materials, namely, the carotenoids and the steroids. Steroids and carotenoids are sub-groups of the triterpenoids and tetraterpenoids, respectively, as will be explained later.

Occasionally, the word terpene is used to indicate any terpenoid. In this book, the word terpene will be restricted to its original meaning. Similarly, the term "isoprenoid" is often used in place of "terpenoid."

Table 1.1 *Classification of Terpenoids*

Name	No. of isoprene units	No. of carbon atoms
Hemiterpenoids	1	5
Monoterpenoids	2	10
Sesquiterpenoids	3	15
Diterpenoids	4	20
Sesterterpenoids	5	25
Triterpenoids	6	30
Tetraterpenoids	8	40
Polyisoprenoids	>8	>40
Steroids triterpenoids which produce Diels's hydrocarbon when distilled from zinc dust		
Carotenoids	8	40

1.2 THE ISOPRENE RULE

The isoprene rule, proposed by Wallach in 1887, defines terpenoids as chemicals containing a carbon skeleton formed by the joining together of isoprene units. Isoprene, the "building block" of terpenoids, is 2-methylbuta-1,3-diene. If we look at the parent 2-methylbutane, we could consider the molecule to resemble a nanoscalar tadpole with a "head" at the branched end of the molecule, the other end therefore constituting the "tail." Thus, in principle, two isoprene units could be joined head-to-head, tail-to-tail or head-to-tail. By far the commonest fusion is head-to-tail. Figure 1.1 shows two isoprene units being joined head-to-tail to produce a monoterpenoid backbone. Occasionally, a tail-to-tail coupling occurs. This is a characteristic feature of steroids and carotenoids. In both of these classes, there is a tail-to-tail fusion exactly in the centre of the backbone, the other joins being head-to-tail type. The hypothetical head-to-head fusion does not occur.

After formation of the basic C_{5n} skeleton, the chain may be folded to produce rings and functionalised by the introduction of oxygen or other

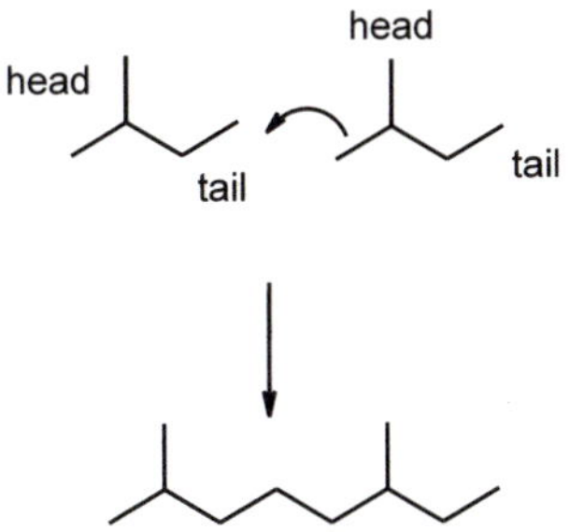

Figure 1.1

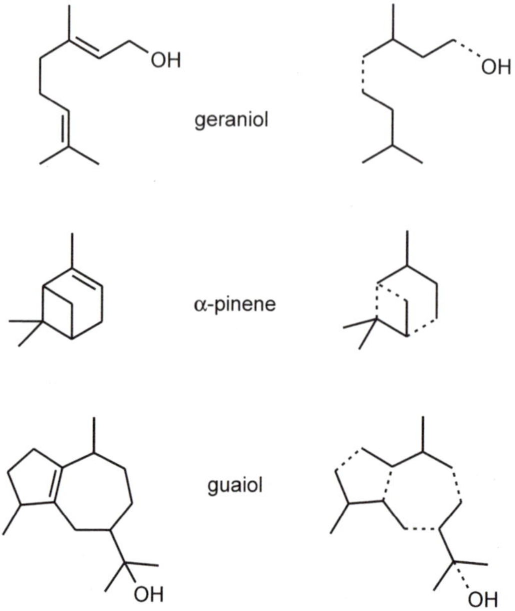

geraniol

α-pinene

guaiol

Figure 1.2

heteroatoms. Figure 1.2 shows how the isoprene units and the original backbone can be traced out in three simple terpenoids. Occasionally, skeletal rearrangements occur which make this process more difficult and fragmentation or degradation reactions can reduce the number of carbon atoms so that the empirical formula does not contain a simple multiple of five carbons. Nonetheless, the natural product chemist will still quickly recognise the characteristic terpene framework of the structure. Sometimes molecules contain both terpenoid fragments and fragments from other biogenetic classes.

1.3 TERPENOID NOMENCLATURE

The terpenoids are divided into groups and sub-groups according to the pathway by which nature synthesised them and hence, by their skeletal structures since these arise directly from the biosynthesis. As described above, the first basis for classification is the number of isoprene units which make up the terpenoid. The names for these groups are shown in Table 1.1. The next classification depends on whether the skeleta remain as open chains or have been cyclised giving one, two or more rings. Families of terpenoids possessing the same skeleton are named after a principal member of that family, usually either the most common or the first to have been discovered. Charts of these names are given in Devon and Scott's dictionary.[1.1] To name an individual terpenoid, it is customary to use the IUPAC or CAS systems of nomenclature. However, it is often more convenient to use either a trivial name or a semi-systematic name derived from the terpenoid structural family to which the material in question belongs. The trivial names often relate to a natural source in which the terpenoid occurs.

As an example of the co-existence of systematic, semi-systematic and trivial names, we could look at the monoterpenoid ketone, carvone. Carvone occurs in both enantiomeric forms in nature, the *laevo*-form in spearmint and the *dextro*-form in caraway. The trivial name carvone is derived from the Latin name for caraway, *Carum carvi*. The basic carbon skeleton is that of 1-isopropyl-4-methylcyclohexane. This skeleton is very common in nature and is particularly important in the genus *Mentha*, which includes various types of mint, since it forms the backbone of most of the important components of mint oils. The skeleton has therefore been given the name *p*-menthane and the numbering system used for it is shown in Figure 1.3. Therefore, any of the following names may be used to describe the same molecule: carvone, *p*-mentha-1,8-dien-6-one and 1-methyl-4-(1-methylethenyl)cyclohex-1-ene-6-one. To classify it, we could say it was an unsaturated ketone of the *p*-menthane family of monoterpenoids.

Greek letters are used in various ways to distinguish between isomeric terpenoids. They may indicate the order in which the isomers were discovered or their relative abundance in the oil. For instance, α-pinene is the most significant component of turpentine, usually comprising almost three quarters of the oil by weight. The next most significant component is β-pinene. These structures are shown in Figure 1.4.

In the case of cyclic terpenoids, the letters α, β and γ often refer to the location of the double bond in isomeric olefins. In these cases, the letter α indicates an endocyclic trisubstituted double bond, β refers to

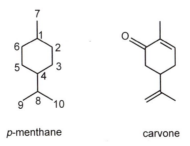

p-menthane   carvone

Figure 1.3

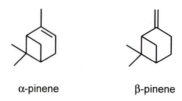

α-pinene   β-pinene

Figure 1.4

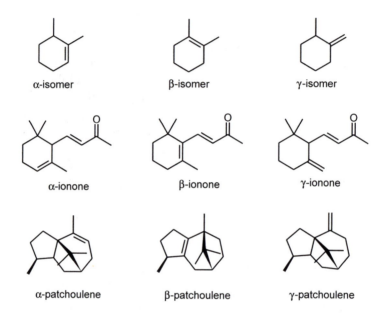

α-isomer   β-isomer   γ-isomer

α-ionone   β-ionone   γ-ionone

α-patchoulene   β-patchoulene   γ-patchoulene

Figure 1.5

a tetrasubstituted olefinic bond and γ to an exocyclic methylene function. Generic structures are shown in Figure 1.5 along with the examples of the isomeric ionones and patchoulenes.

1.4 THE ROLE OF TERPENOIDS IN NATURE

Terpenoids are produced by a wide variety of plants, animals and micro-organisms. As for all metabolites, the synthesis of terpenoids places a metabolic load on the organism which produces them and so, almost invariably, there is a role which the material plays and for which it is synthesised. The roles which the terpenoids play in living organisms can be grouped into three classes: functional, defence and communication.

Figure 1.6 shows some examples of what is meant by functional terpenoids, *i.e.* those that play a key part in the metabolic processes of the organism in which they are produced. Vitamin A, or retinol, is the precursor for the pigment in eyes which detects light and is therefore responsible for the sense of sight. Vitamin E, or tocopherol, is an important antioxidant which prevents oxidative damage to cells. Vitamin D_2, also known as calciferol, regulates calcium metabolism in the body and is therefore vital for the building and maintenance of bone. Chlorophyll-a is a green pigment found, for example, in plant leaves and is a key factor of photosynthesis through which atmospheric carbon dioxide is converted to glucose.

There are a number of ways in which plants and animals use terpenoid chemicals to protect themselves. Probably the two commonest methods are the production of resins by plants which have been damaged and the production of materials which will render a plant or animal unattractive to predators.

Many plants, when damaged, exude resinous materials as a defence mechanism. Rosin is produced as a physical barrier to infectious organisms, by pine trees when the bark is damaged. Similarly, rubber is

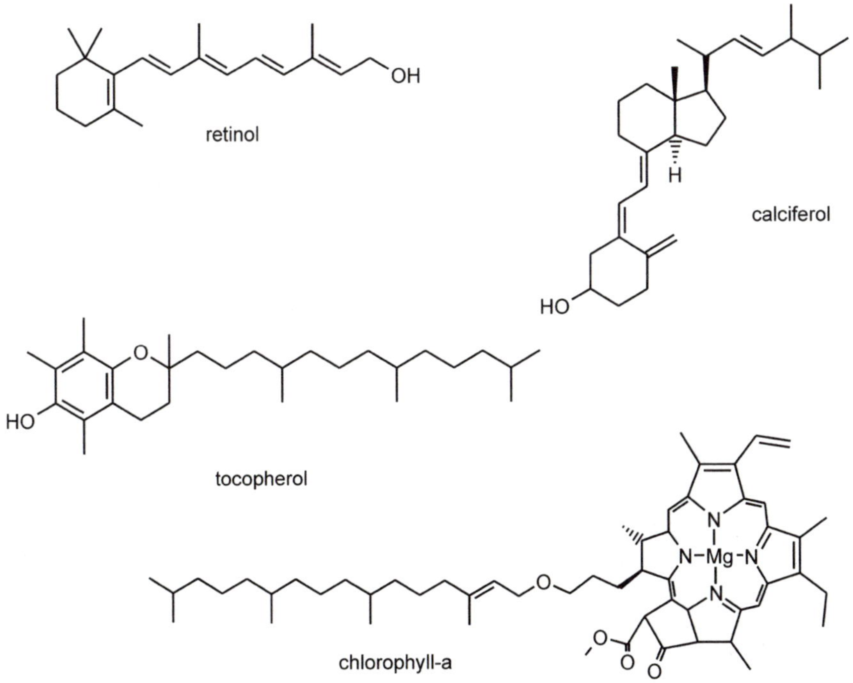

Figure 1.6

a defensive secretion. The shrub *Commiphora abyssinica* produces a resin which contains a number of antibacterial and antifungal compounds. One of these is the eudesmane derivative (1.1), which is shown in Figure 1.7. The role of the resin is to seal the wound and prevent bacteria and fungi from entering and damaging the plant. The resin has a pleasant odour and so was put to use by man as a perfume ingredient. It is known as myrrh. Because of its antimicrobial properties, myrrh was also used as an antiseptic and preservative material, for instance, in the embalming of corpses. Frankincense, derived from trees of the genus *Boswellia*, is another such resin and has been used in religious rites for thousands of years. Thus, two of the three gifts brought to the Christ Child by the magi were perfume ingredients containing terpenoids. Knowledge of terpenoids thus helps us to understand the symbolism involved; gold, frankincense and myrrh represent, respectively, king, priest and sacrifice.

Bufotalin, also shown in Figure 1.7, is a cardiac glycoside which functions as a heart stimulant. It is produced by toads in order to prevent other animals from preying on them; would be predators soon learn that toads do not make good food. Similarly, many plants produce terpenoids making them unpalatable to insects which would otherwise eat their foliage. Two examples are shown in Figure 1.7. The first is azadirachtin which is produced by *Melia azadirachta* and also by the Indian neem tree, *Azadirachta indica*. The other is warburganal which is produced by plants

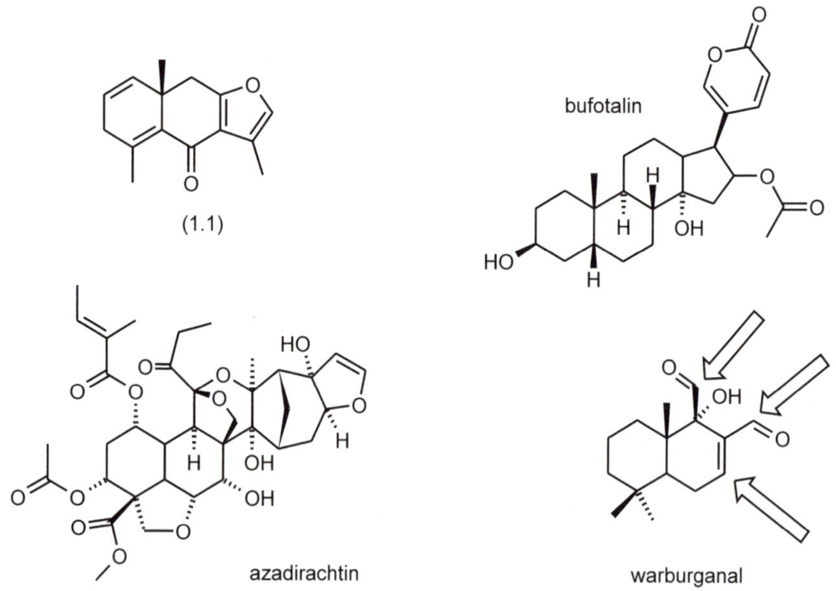

(1.1)

bufotalin

azadirachtin

warburganal

Figure 1.7

of the genus *Warburgia*. Warburganal contains two aldehyde functions and one of these is α,β-unsaturated. Thus, it is capable of undergoing triple alkylation of nucleophilic materials such as the nitrogen atoms of proteins and nucleic acids. This property makes it a skin sensitiser (*i.e.* a material which can induce an allergic reaction in some subjects upon repeated exposure) and carcinogen. It is therefore a doubly effective deterrent because of its unpleasant taste and high toxicity. In the figure, the arrows indicate the three potential sites of nucleophilic attack.

Terpenoids are also used as chemical messengers. If the communication is between different parts of the same organism, the messenger is referred to as a hormone. Examples of hormones are shown in Figure 1.8. Giberellic acid is a hormone used by plants to control their rate of growth. Testosterone and oestrone are mammalian sex hormones. Testosterone is a male hormone and oestrone, a female.

Chemicals that carry signals from one organism to another are known as semiochemicals. These can be grouped into two main classes. If the signal is between two members of the same species, the messenger is called a pheromone. Pheromones carry different types of information. Not all species use pheromones. In those which do, some may use only one or two pheromones while others, in particular the social insects such as bees, ants and termites, use an array of chemical signals to organise most aspects of their lives.

Sex pheromones are among the most widespread. Male moths can detect females by smell at a range of many miles. Some terpenoid pheromones are shown in Figure 1.9. Androst-16-en-3-ol is a porcine sex pheromone and the compound which produces "boar taint" in pork.

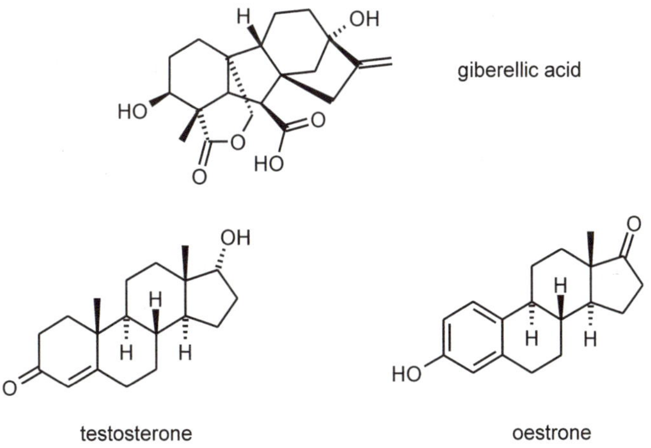

giberellic acid

testosterone

oestrone

Figure 1.8

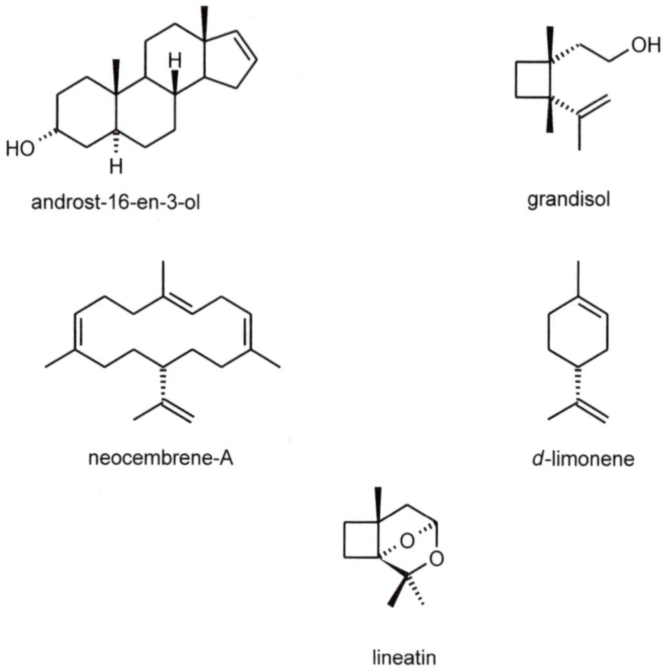

androst-16-en-3-ol

grandisol

neocembrene-A

d-limonene

lineatin

Figure 1.9

(Boar taint is a flavour found in the meat of boars but not of sows.) It is produced by boars and is released in a fine aerosol when the boar salivates and champs his jaws. When the sow detects the pheromone in air, she immediately adopts what is known as "the mating stance" in readiness for the boar. Grandisol is a sex attractant for the male boll weevil, a serious pest for cotton growers.

Ants and termites use trail pheromones to mark a path between the nest and a food source. This explains why ants are often seen walking in single file over long distances. One such trail pheromone is neocembrene-A which is produced and used by termites of the Australian species *Nasutitermes exitiosus*. The social insects also use alarm, aggregation, dispersal and social pheromones to warn of danger and to control group behaviour. For example, d-limonene is an alarm pheromone of some Australian termites and lineatin is the aggregation pheromone of *Trypodendron lineatum*. So, exposure to these two terpenoids will produce opposite reactions in their target species, repulsion in the first case and attraction in the second.

Chemicals which carry messages between members of different species are known as allelochemicals. Within this group, allomones benefit the sender of the signal, kairomones its receiver and with synomones both the sender and receiver benefit. Examples are shown in Figure 1.10.

Camphor and *d*-limonene are allomones in that the trees which produce them are protected from insect attack by their presence. For instance, Arthur Birch, one of the great terpene chemists of the twentieth century, reported finding *d*-limonene in the latex exuded by trees of the species *Araucaria bidwilli*.[1,2] These trees are protected from termite attack because the *d*-limonene they produce is an alarm pheromone for termites that live in the same area. Similarly, antifeedants could be considered to be allomones since the signal generator, the plant, receives the benefit of not being eaten. Myrcene is a kairomone, in that it is produced by the ponderosa pine and its presence attracts the females of the bark beetle, *Dendroctonous brevicomis*. Geraniol is found in the scent of many flowers such as the rose. Its presence attracts insects to the flower and it can be classified as a synomone since the attracted insect finds nectar and the plant obtains a pollinator.

One terpenoid which has an unusual signalling property is nepetalactone. This is actually a mixture of two isomers, as shown in Figure 1.11, the major being the *trans,trans*-isomer (1.2) and the minor the *trans,cis*-isomer (1.3). Nepetalactone is the principal component of the oil of catnip or cat mint (*Nepeta cataria*), constituting 70–90% of the oil. It is an insect repellent, which is probably why the plant produces it. However, it has a surprising effect on all felines, from domestic cats to lions and tigers, in that it induces grooming and rolling behaviour in

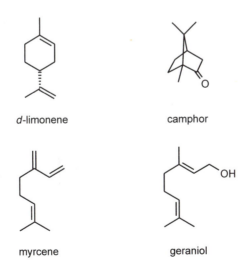

d-limonene   camphor

myrcene   geraniol

Figure 1.10

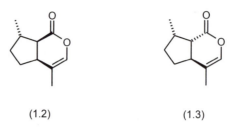

(1.2)   (1.3)

Figure 1.11

them. This is probably purely coincidental as it is hard to see what benefit this would be to the plant.

1.5 EXTRACTION AND USE OF TERPENOIDS

Many commercial uses of terpenoids reflect their natural uses. Those that are produced in nature because of their biological activity may well find commercial use as drugs or pest control agents. For example, α-santonin (see Chapter 7 for some of its interesting chemistry) is extracted from Levant wormseed, *Artemisia maritima*, for use as an anthelmintic. The poisonous nature of foxglove is due to the presence of terpenoid glycosides which have strong stimulant action on heart muscle. Digitoxin is a glycoside of digitoxigenin and is extracted from foxglove for use in treatment of certain heart conditions. The odorous terpenoids are, of course, used as fragrance ingredients in cosmetics, toiletries and household products. Cineole, extracted from various eucalyptus species, serves both purposes since it is used in perfumery as well as a nasal decongestant. (Figure 1.12)

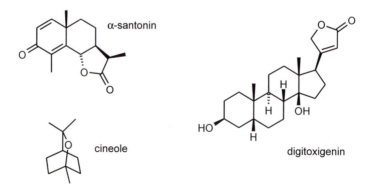

Figure 1.12

Terpenoids are also put to uses for which their physical or chemical properties suit them but which are not the uses for which nature originally intended them. Rubber is a polymer of isoprene which is produced in the rubber tree as a defensive secretion but is widely used by humans because of its elastic properties. Turpentine has a long history of use as a solvent, particularly for paints and, similarly, lac resin as varnish. Nowadays, turpentine is also used as a feedstock for the synthesis of other materials of commercial interest, in addition to a wide variety of fragrance ingredients.

The methods used to extract perfume ingredients from their natural sources have changed over time as technology in general has advanced. However, both old and new methods fall into four basic classes; tapping, expression, distillation and solvent extraction.

Many plants, particularly trees, exude resins when their bark is damaged. Deliberate damage and subsequent collection of the resin is known as tapping. This method is used to collect latex for rubber production and for gum turpentine. It is also used to produce frankincense (also known as olibanum), myrrh and other similar fragrance materials; although in these cases there is usually further processing of the resin after collection.

When oils are forced out of the natural source by physical pressure, the process is referred to as expression and the product is called an expressed oil. If you squeeze a piece of orange peel, you will see the oil bearing glands burst and eject a fine spray of orange oil. Many commercially available citrus oils, bergamot in particular, are prepared in this way.

Volatile terpenoids can be isolated from their natural sources by distillation. Since volatility and odour often go hand in hand, distillation is of major importance for the isolation of fragrance ingredients. Distillation of perfume ingredients from their natural sources can be done in three ways: dry (or empyrumatic) distillation, steam distillation or hydrodiffusion. Dry distillation involves high temperatures since heat, and in most cases this will be direct flame, is applied to the surface of the vessel containing the plant material. Usually this technique is reserved for the highest boiling of the oils, typically those derived from wood, because the high temperatures are necessary to vaporise their chemical components. Cade and Birch Tar are the major oils obtained by dry distillation. Cade and Birch Tar oils contain distinctive burnt, smoky notes as a result of pyrolysis of plant material. In steam distillation, water or steam is added to the still pot and the oils are co-distilled with the steam. The presence of water in the pot during steam distillation limits the temperature of the process to 100 °C. Thus much less degradation occurs in this process than in dry distillation. However, some degradation does occur. For example, tertiary alcohols, particularly the higher boiling sesquiterpenoid and diterpenoid alcohols, present in the plant often dehydrate in the pot and distil as the corresponding hydrocarbons. The steam distilled oil is separated from the water by means of a Florentine flask which separates them based on their differing densities. The aqueous distillate is sometimes referred to as the waters of cohobation. Figure 1.13 shows a simple schematic representation of a Florentine flask.

Hydrodiffusion is a relatively new technique. It is essentially a form of steam distillation. However, it is steam distillation carried out upside down since the steam is introduced at the top of the pot and the water and oil taken off as liquids at the bottom. The plant materials diffuse through the cell membranes into the steam and are carried to the bottom of the still by the descending flow of condensate. This technique therefore saves energy because it is not necessary to vaporise the oil.

For a comparison of the various distillation techniques, the reader is referred to the paper by Boelens *et al.*[1.3]

Figure 1.14 is a simple schematic representation of the various distillation processes. Materials obtained in this way are referred to as essential oils. Thus, for example, the oil obtained by steam distillation of lavender is known as the essential oil of lavender or lavender oil. The term essential oil arises from the Aristotelian theory that matter is continuous and composed of four elements, *viz.* fire, air, earth and water. The fifth element, or quintessence, was considered to be spirit. Distillation was believed to separate the spirit of the plant from its physical matter; hence the terms spirits or essential (short for quintessential) oils were applied to distillates. Occasionally, the monoterpene hydrocarbons

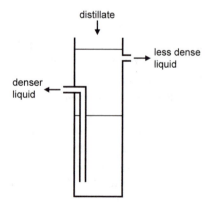

Figure 1.13

Distillation

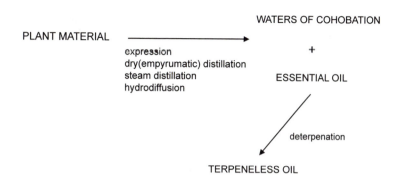

Figure 1.14

are removed from the oils by distillation or solvent extraction to give a finer odour in the product. The process is known as deterpenation and the product is referred to as a terpeneless oil.

Terpenoids can also be extracted from natural sources by solvent extraction. Many perfume ingredients have been obtained in this way for thousands of years and a language has grown up to describe the various processes and products thus obtained. These are summarised in Figure 1.15. The processes are written in cursives and the technical names for the various products in capitals.

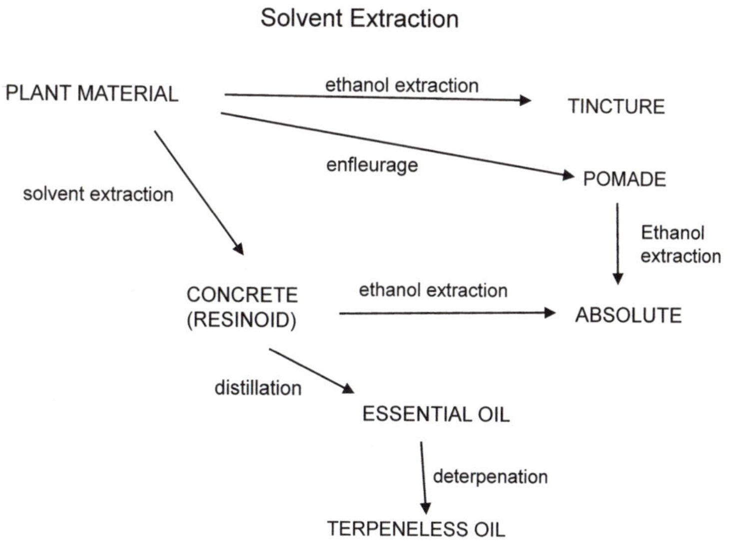

Solvent Extraction

PLANT MATERIAL —— *ethanol extraction* → TINCTURE

solvent extraction

enfleurage → POMADE

Ethanol extraction

CONCRETE (RESINOID) —— *ethanol extraction* → ABSOLUTE

distillation → ESSENTIAL OIL

deterpenation

TERPENELESS OIL

Figure 1.15

Ethanolic extraction is seldom used for plant materials because of the high proportion of water compared to oil in the plant. It is more important with materials such as ambergris. The sperm whale produces a triterpenoid known as ambreine in its intestinal tract. This is excreted into the sea and, on exposure to salt water, air and sunlight, undergoes a complex series of degradative reactions to produce the material known as ambergris. (More detail of this chemistry will be given in Chapter 8.) This waxy substance can be found floating in the sea or washed up on beaches. Extraction of it with ethanol produces tincture of ambergris.

Enfleurage was used by ancient Egyptians to extract perfume ingredients from plant material and exudates. Its use continued up to the early twentieth century but is now of no commercial significance. In enfleurage, the natural material is brought into intimate contact with purified fat. For flowers, for example, the petals were pressed into a thin bed of fat. The perfume oils diffuse into the fat over time and then the fat can be melted and the whole mixture filtered to remove solid matter. On cooling, the fat forms a pomade. Although the pomade contains the odorous principles of the plant, this is not a very convenient form to have. The concentration is relatively low and the fat is not the easiest or most pleasant material to handle, besides which, it will eventually turn rancid. Ancient Egyptians used to apply pomade directly to their heads, but in more recent times, it was usual to extract the fat with ethanol. The odorous oils are soluble in alcohol because of their degree of oxygenation. The fat used in the extraction and any fats and waxes extracted from the plant along with the oil, are insoluble in ethanol and so are separated from the oil. Removal of the ethanol by distillation produces what is known as an absolute.

The most important extraction technique nowadays is simple solvent extraction. The traditional solvent for extraction was benzene, but this has been superseded by other solvents due to concern over possible toxic effects of benzene on those working with it. Petroleum ether, acetone, hexane and ethyl acetate, together with various combinations of these, are typical solvents for extraction. Recently, there has been a great deal of interest in the use of liquid carbon dioxide as an extraction solvent. The pressure required to liquefy carbon dioxide at ambient temperature is considerable and thus the necessary equipment is expensive. This is reflected in the cost of the oils produced; however, carbon dioxide has the advantage that it is easily removed and there are no concerns about residual solvent levels.

The product of such extractions is called a concrete or resinoid. It can be extracted with ethanol to yield an absolute or distilled to give an essential oil. The oil can then be deterpenated, *i.e.* freed of monoterpene hydrocarbons.

1.6 NATURAL INSPIRATION

The chemistry of terpenoids is rich and varied, and attempts to understand it have, on many occasions, contributed fundamentally to our total understanding of their chemistry. One example is the work of Wagner and Meerwein whose studies in terpene chemistry led, amongst many discoveries, to the elucidation of the rearrangement that bears their names. This work made a very significant contribution to our fundamental understanding of the properties and reactions of carbocations, as will be seen in Chapter 5.

Occasionally, the level of a terpenoid in its natural source is too low to make commercial extraction feasible. An example is paclitaxel (also known under the trade name, Taxol®) (Figure 1.16) which is produced by the Pacific Yew, *Taxus brevifolia*. Paclitaxel has been shown to be an effective agent for the treatment of breast and ovarian cancer. However, the level of paclitaxel in the bark of the tree is so low that it would require three mature trees to produce sufficient paclitaxel to treat one patient. Mature, in this case, means several hundred years old. Thus, there are insufficient trees in the world to treat even a small percentage of women with these diseases. Therefore there has been much activity in attempts to produce total or partial (*i.e.* starting from an available analogue) synthesis of paclitaxel. Similarly, there is currently a great deal of research activity in the search for analogues of paclitaxel which would be easier to produce and would retain the anticancer activity.

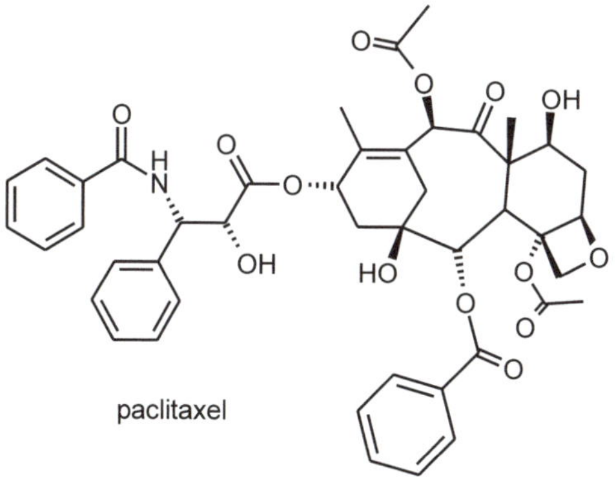

paclitaxel

Figure 1.16

Nature remains a great source of inspiration for chemists and many of today's perfumery ingredients, drugs and agricultural chemicals are replicates (nature identical) of natural terpenoids or close analogues of them.

REFERENCES

1.1.   T.K. Devon and A.I. Scott, *Handbook of Naturally Occurring Compounds*, Vol. 2, *The Terpenes*, Academic Press, London, 1972.

1.2.   A.J. Birch, *J. Proc. R. Soc. New South Wales*, 1938, **71**, 259.

1.3.   M.H. Boelens, F. Valverde, L. Sequeiros and R. Jimenez, *Perfumer and Flavorist*, 1990, **15**, 11.

© David Buffington/CORBIS

CHAPTER 2
Biosynthesis

The infinite variety of terpenoids produced
in growing plants are made from two key
feedstocks and a handful of reaction types

I praise you for I am fearfully and wonderfully made

Psalm 139

When a poet, an artist or a perfumer looks closely at an individual flower, he will find inspiration for his work from the beauty of its shape, colour and the scent it emits. When the natural products chemist delves deeper into the detail, he sees the diversity and intricacy of the multitudinous molecular structures of which the flower is made and the amazing, complex and very elegant processes which nature uses to produce them. I believe that this can only increase the awe with which we view the wonderful universe in which we live.

KEY POINTS

Isopentenyl pyrophosphate is produced from glucose and can be isomerised to prenyl pyrophosphate.

Head-to-tail coupling of these two gives geranyl pyrophosphate.

Further sequential addition of isopentenyl pyrophosphate units to geranyl pyrophosphate then builds chains containing multiples of five carbons in which the individual "isoprene" units can be identified.

Using enzyme-mediated carbocation chemistry, the C_{5n} materials can be cyclised and rearranged into an almost infinite variety of skeletal forms.

Tail-to-tail coupling is also possible and this leads to, for example, steroids and carotenoids.

2.1   INTRODUCTION

The process by which nature produces the chemicals it needs, is known as biosynthesis. By studying biosynthesis, we not only learn about nature's chemistry – but also generate knowledge which is also useful in many ways. For example, by understanding nature's chemistry and how natural products are constructed, the patterns of nature can be understood and used to assist in identifying the structures of newly isolated materials and in manufacturing new compounds with similar properties. What follows is a very brief summary of biosynthesis. Further information can be found in the books by Bu'Lock, Mann *et al.*, Torssell, Rawn and Matthews, and van Holde, details of which can be found in the Bibliography.

2.2   ENZYMES AND COENZYMES

The chemical reactions that are observed in biosynthesis (or biogenesis) are essentially the same as those the synthetic organic chemist uses to produce materials in the reaction flask. The key difference between natural and synthetic chemistry, lies in the catalytic systems found in nature. The catalysts which drive biochemical reactions are known as enzymes. These are globular proteins, *i.e.* proteins that prefer to adopt globular shapes, rather than those that remain linear or form into sheets or helices. The role of enzymes is to make biochemical reactions faster and much more selective.

Somewhere within the globular structure of an enzyme, there exists a location where the chemical reaction occurs. This location is known as the active site. Examination of active sites reveals how an enzyme can improve the rate and selectivity of the reaction which it catalyses. Firstly, the active site serves to hold together all the necessary components of the reaction in the correct configuration for the reaction to proceed. This is achieved through molecular recognition. The active site contains docking points for each of the reaction components and these are arranged in space in such a way as to facilitate the formation of a transition state. Selectivity is achieved by restrictions around the docking points so that only certain substrates will be accepted and also, where appropriate, through preferential formation of one transition state over another. This organisation of the reacting species by the enzyme reduces entropic barriers to reaction. The enzyme can also reduce enthalpic barriers to reaction through what is known as an allosteric effect. This phenomenon results from energy that is stored through folding the protein molecule. For example, the following sequence of events might occur during the course of an enzyme-catalysed reaction. The reaction components dock at the active site and, in doing so, induce strain at a remote part of the protein structure. To relieve this strain, the protein changes its tertiary structure and this in turn changes the newly formed transition state/active-site complex from one in which the transition state resembles the starting materials to one in which the transition state resembles the products. After the release of the products from the active site, the protein returns to its original state in readiness for the next reaction cycle. Thus, the energy needed to drive the reaction through the crucial transition has come from forces holding the protein chain in its tertiary structure. Since the protein molecule eventually returns to its original shape, the overall reaction is energy neutral with respect to the protein. The ultimate energy required was derived from the Gibbs Free Energy of the reaction the enzyme catalysed.

The course of reactions in terpenoid biogenesis is heavily controlled by the enzymes that plants or animals possess. This is particularly the case in reactions where a pyrophosphate intermediate undergoes heterolytic cleavage to give the pyrophosphate anion and a carbocation. Usually, there is a variety of reaction pathways that the carbocation can follow, but the enzyme directs it specifically along just one of them. Thus, from a common precursor and the same initial heterolysis, one organism will produce one metabolite and another organism will produce something quite different. The availability of enzymes in an organism is controlled by its DNA. Thus, related organisms with similar DNA will have similar enzymes at their disposal and therefore they will produce similar terpenoid structures. On the other hand, two organisms that differ more widely in their genetic make-up, are more likely to produce quite different terpenoids. The classification of plants by means of the metabolites they produce is known as chemotaxonomy. Sometimes, two plant species appear identical from the point of view of size, leaf shape, flowers and so on, but they produce different secondary metabolites. Such species are referred to as chemotypes.

Occasionally, enzymes require coenzymes, also known as co-factors, to take part in a reaction. For example, in a reduction or oxidation reaction, the coenzyme may provide the reducing or oxidising power to drive the reaction. The coenzyme is then recycled by being re-oxidised or re-reduced, as appropriate, in a subsequent cycle or even in a different enzyme system. There are three coenzymes that are particularly important in the biosynthesis of terpenoids and it is worth looking at them in a little more detail before we move on to the main topic.

2.2.1   Adenosine Triphosphate (ATP)

Adenosine triphosphate, as can be seen from Figure 2.1, is formed from the nucleotide adenine, the sugar ribose and three phosphate units. Having three phosphate units in line is energetically unfavourable

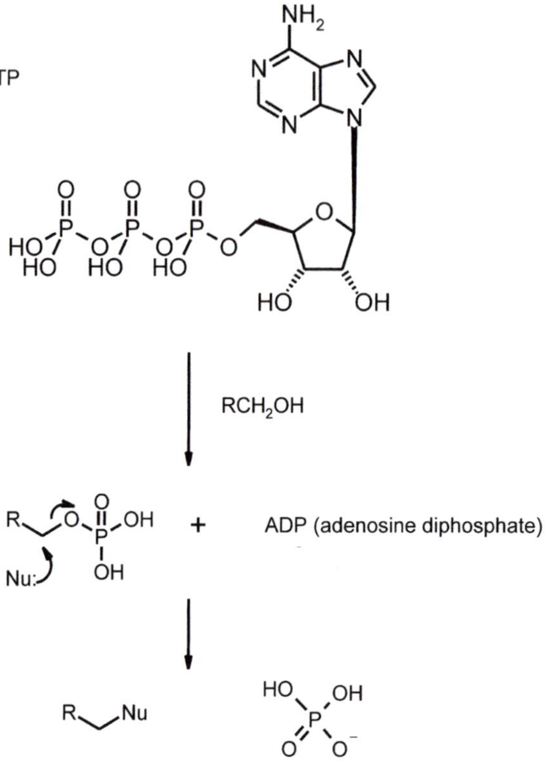

ATP

Figure 2.1

and thus, there is a driving force to lose one of these units and form the corresponding diphosphate. In doing so, ATP donates one phosphate unit to another molecule. If this other molecule is an alcohol, the resultant product is a phosphate ester. Since phosphate is a much better leaving group than hydroxide, the phosphate ester is much more susceptible to attack by a nucleophile than was the starting alcohol. Thus, as shown in Figure 2.1, ATP serves to provide the energy to drive nucleophilic substitution reactions.

2.2.2   Nicotinamide Adenine Dinucleotide Phosphate (NADP/NADPH)

The structure of this coenzyme is shown in Figure 2.2. It is made up of nicotinamide, ribose, phosphate and the nucleotide adenine. It exists in two forms, as shown. These differ in their oxidation state. NADP contains a pyridinium ring while NADPH contains a dihydropyridine. They can be interconverted by transfer of a hydride anion. The related coenzymes NAD and NADH are the same as NADP and NADPH except that they lack the phosphate ester at the 2′ position (*i.e.* the

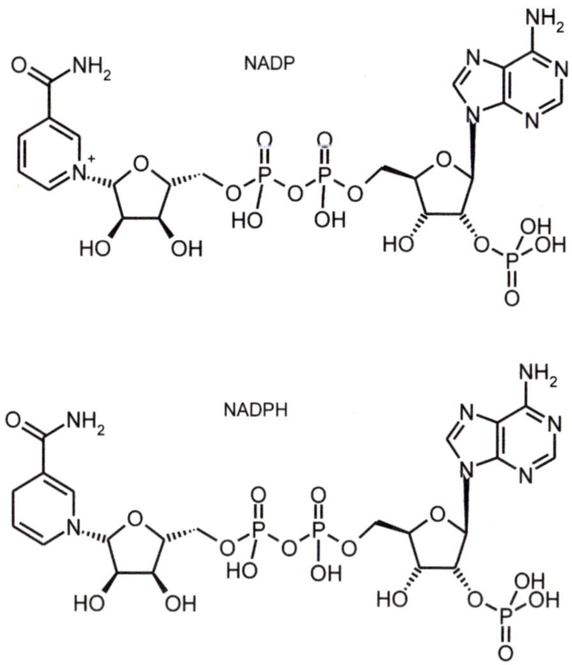

Figure 2.2

2-carbon of the ribose attached to the adenine). Although both pairs of coenzymes carry out the same chemical role, NADP and NADPH are usually found in anabolic (*i.e.* biosynthetic) processes, whereas NAD and NADH are found in catabolic (*i.e.* degradative) processes.

Since NADPH can be converted to NADP by donation of a hydride anion, it is nature's equivalent of reagents such as LiAlH$_4$, NaBH$_4$, *etc*. Similarly, NADP can accept a hydride ion and is therefore nature's equivalent of oxidative reagents such as those of Jones, Collins, Swern and Oppenauer. Examples of these two processes are shown in Figure 2.3. The upper half of the figure shows a ketone being reduced to an alcohol by NADPH in the presence of an acid. The NADPH is oxidised in the process to give NADP. In the lower half of Figure 2.3, an alcohol is oxidised to a ketone by NADP in the presence of a base and the NADP is simultaneously reduced to NADPH.

2.2.3 Coenzyme A (CoA)

This coenzyme is composed of adenine, ribose, phosphate and a chain formed by amide links between one hydroxy acid, one amino acid and thioethanolamine. The chemical role of CoA is derived from the thiol

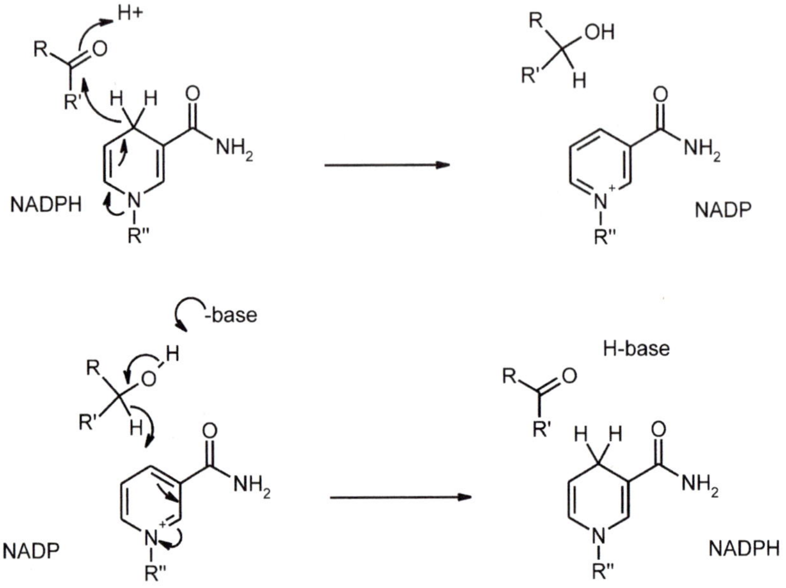

Figure 2.3

group at the end of the chain. This thiol function can form thioesters with carboxylic acids. Having done so, it modifies the acid unit in two ways. Firstly, thiolate anions are much better leaving groups than hydroxide and so the carbonyl carbon of the thiolester is much more susceptible to nucleophilic attack than it was when in the form of an acid. Secondly, because of the electron-withdrawing properties of the sulfur atom, the hydrogen atoms α-to the carbonyl group are much more acidic than those in the free acid. This means that it is easier to form the carbanion and initiate aldol- and Claisen-type chemistry. The structure of CoA and its effects in biosynthetic chemistry are shown in Figure 2.4.

2.2.4 (Co)enzymes in Summary

In all of these coenzymes, much of the molecule is composed of adenine, ribose and phosphate moieties. At first sight, these seem rather unnecessary as they do not play an active role in the chemistry taking place. Their function is, however, an important one. These components are recognised by the enzyme and serve as the ties which lock the reacting species together at the active site of the enzyme. An example of this impressive degree of organisation is shown in Figure 2.5. Here we see the enzyme lactate dehydrogenase about to reduce a pyruvate anion to the corresponding lactate anion using NADH as the reductant.

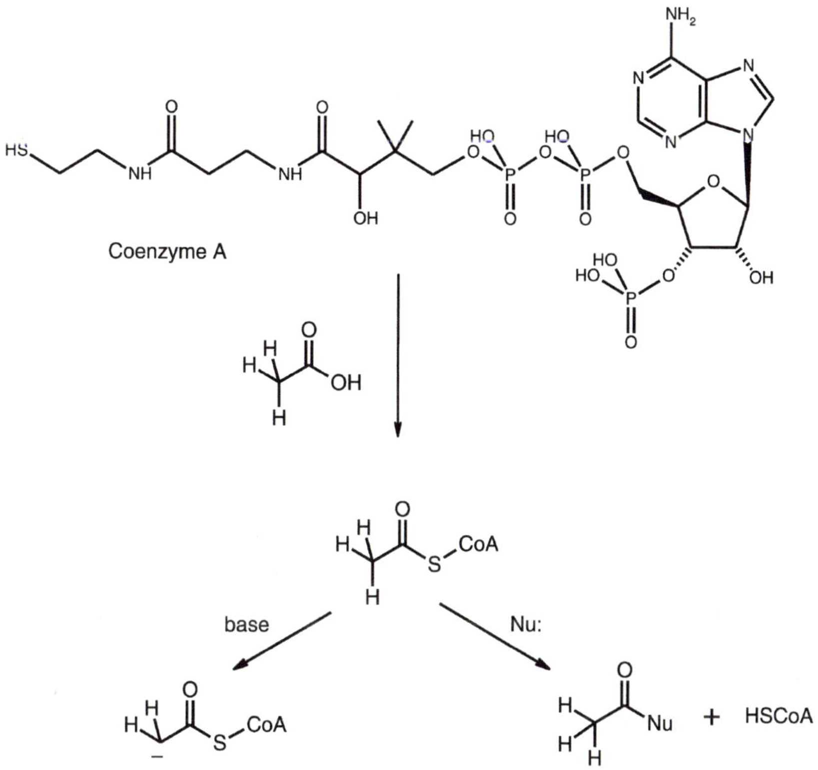

Coenzyme A

Figure 2.4

The coenzyme and the substrate are held in the correct orientation for reaction by no less than ten hydrogen bonds to the protein structure. The three-letter codes are those conventionally used for the amino acid constituents of the proteins and the subscript numbers indicate the position of that specific amino acid in the total amino acid sequence of the protein backbone. The range of numbers quoted and the gaps between them show how the active site is formed from amino acids from all along the peptide chain and that they are brought together by the folding in the tertiary structure of the protein.

2.3   BIOSYNTHESIS OF C₅ PYROPHOSPHATES

Green plants and algae synthesise glucose from carbon dioxide and water. This energetically unfavourable process is known as photosynthesis since sunlight is used as the energy source to drive the reaction. Glucose in turn can be broken down, in a process known as glycolysis, to give the phosphate ester of the enol form of pyruvic acid, phosphoenol pyruvate.

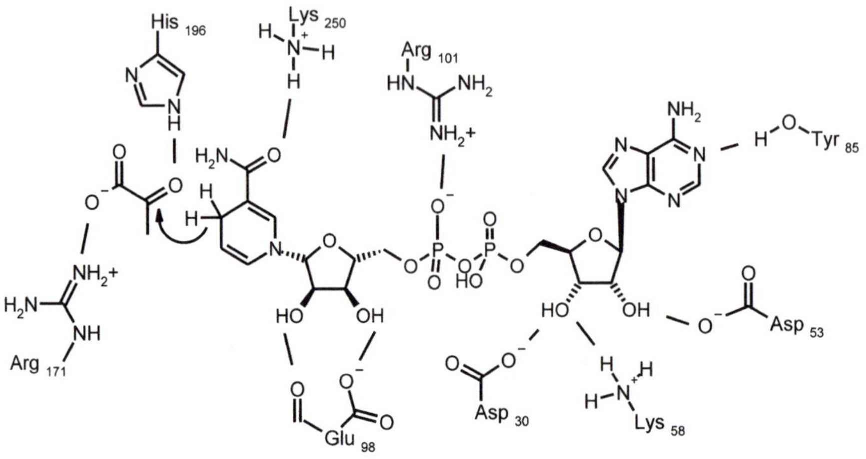

Figure 2.5

This process can take place either in the plant that made the glucose or in another organism that consumed the original plant. Hydrolysis of phosphoenol pyruvate gives free pyruvate, which is subjected to nucleophilic attack at the ketone carbon by the thiol group of coenzyme A, followed by decarboxylation, to give acetyl coenzyme A. This process is shown in Figure 2.6.

The process by which acetyl coenzyme A is converted into the two isomeric building blocks of the terpenoids is shown in Figure 2.7. One molecule of acetyl coenzyme A (2.1) undergoes nucleophilic attack by the anion (2.2) of another, with loss of the anion of one unit of coenzyme A,

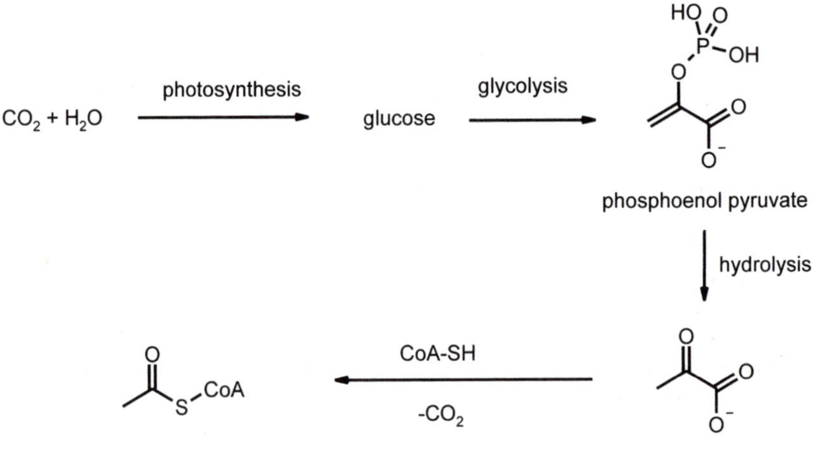

Figure 2.6

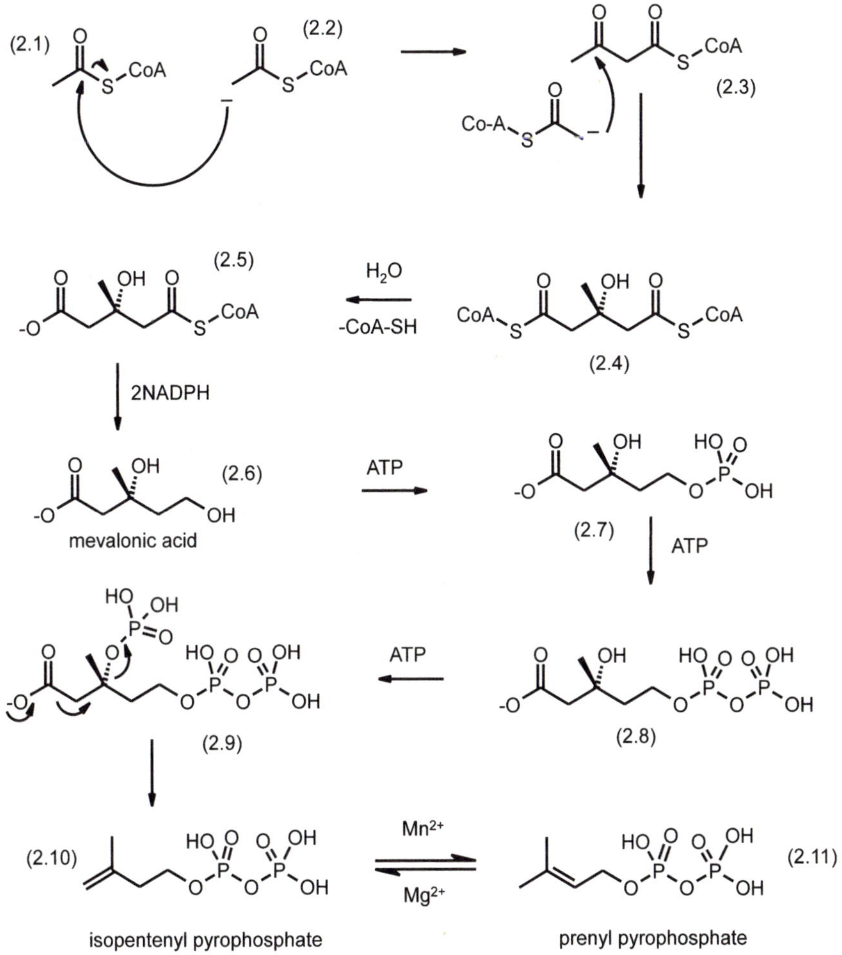

Figure 2.7

to give acetoacetyl coenzyme A (2.3). Addition of the anion of a third acetyl coenzyme A to the ketone carbon gives the hydroxy dithioester (2.4). This is followed by hydrolytic loss of a coenzyme A unit to give the mono-thioester of 3-hydroxy-3-methylpentandioic acid (2.5). The thioester function is then reduced by two equivalents of NADPH to give the dihydroxy acid known as mevalonic acid (2.6). Mevalonic acid is then phosphorylated by three equivalents of ATP. The first produces the phosphate of the primary alcohol (2.7); the second converts this to the pyrophosphate (2.8); and the third phosphorylates the tertiary alcohol to give the triphosphate (2.9). The tertiary phosphate ester is prone to elimination and, in this case, this tendency is enhanced by the concomitant decarboxylation to give isopentenyl pyrophosphate (2.10).

The double bond of isopentenyl pyrophosphate can be isomerised into the thermodynamically more favourable trisubstituted position by manganese ions, giving prenyl pyrophosphate (2.11). The reverse isomerisation can be achieved by magnesium ions. Both isomerisations are, of course, directed by enzymes.

2.4   LINEAR TERPENOIDS VIA HEAD-TO-TAIL COUPLING

The reaction schemes in Figures 2.6 and 2.7 allow us to see how the basic five-carbon unit is formed in nature. By examining the way in which these building blocks are assembled, we can now gain an understanding of the patterns that are visible in the skeletons of terpenoid materials. As discussed in Chapter 1, by far the commonest linkage between two of these five carbon units is head-to-tail fusion. This is brought about by the mechanism shown in Figure 2.8. The basic starting point is one molecule of prenyl pyrophosphate (2.11). Hydrolysis of the pyrophosphate group at this point would produce a hemiterpenoid. Alternatively, in the presence of base, a molecule of isopentenyl pyrophosphate (2.10) can be added to give a ten-carbon species. An allylic methylene hydrogen is removed from the isopentenyl pyrophosphate (2.10) by the base. The resulting allylic anion, via its terminal carbon atom, attacks the allylic pyrophosphate of the prenyl pyrophosphate (2.11) in an S_N2 reaction. This produces geranyl pyrophosphate (2.12), the starting point for all monoterpenoids. The selectivity of the entire reaction is based on the inherent chemical reactivity of the reagents but, of course, enhanced by enzymic control. Thus, the oxygenated end of one unit is coupled to the opposite end of another giving the head-to-tail fusion. Hydrolysis at this point leads to monoterpenoids. As before, the alternative is to add another unit of isopentenyl pyrophosphate (2.10), since geranyl pyrophosphate (2.12) contains an exactly analogous reactive site to that of prenyl pyrophosphate (2.11). This addition, as shown in Figure 2.8, produces farnesyl pyrophosphate (2.13), the precursor to all sesquiterpenoids. Once again, either hydrolysis or addition of a further isopentenyl pyrophosphate (2.10) unit can occur. The first will produce sesquiterpenoids and the second, geranylgeranyl pyrophosphate (2.14), the precursor of the diterpenoids. The addition of five carbon units can continue in the same way to give higher terpenoids and, eventually, polyisoprenoids.

2.5   CYCLIC TERPENOIDS THROUGH CARBOCATION
CHEMISTRY

As we look at the reactions shown in Figures 2.9–2.16, we will see the typical reactions of carbocations. The chemical parameters controlling

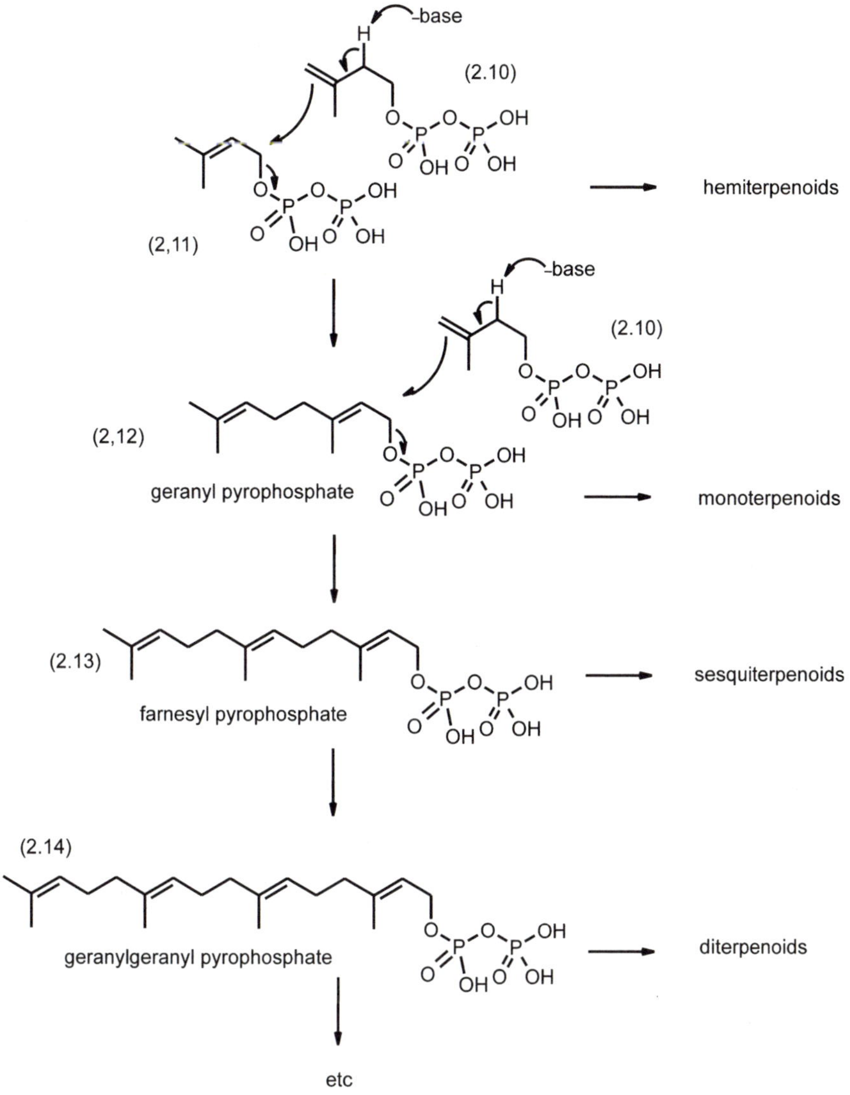

Figure 2.8

carbocation reactions are discussed in detail in Chapter 5. During biosynthesis, the basic chemical reactivity is enhanced and controlled by the enzymes that catalyse the various reactions. When chemistry alone presents alternative reactions, the enzymic catalysts involved in biosynthesis will usually direct the reaction path specifically toward one of these. Since different organisms contain different enzymes, the terpenoids they produce from the same precursor will often be different.

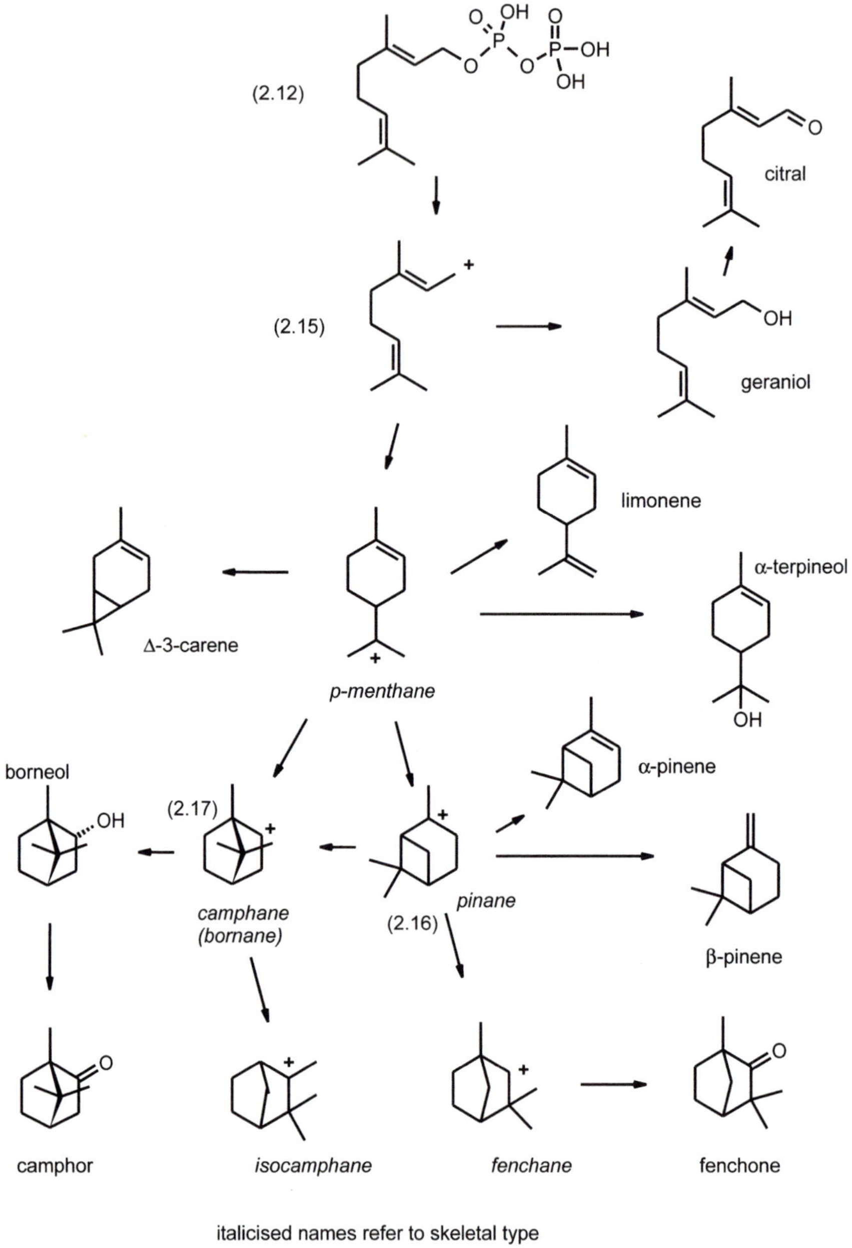

Figure 2.9

This results in a unique pattern of terpenoid components in each of these different organisms and hence our ability to distinguish and classify species according to the terpenoids that they produce.

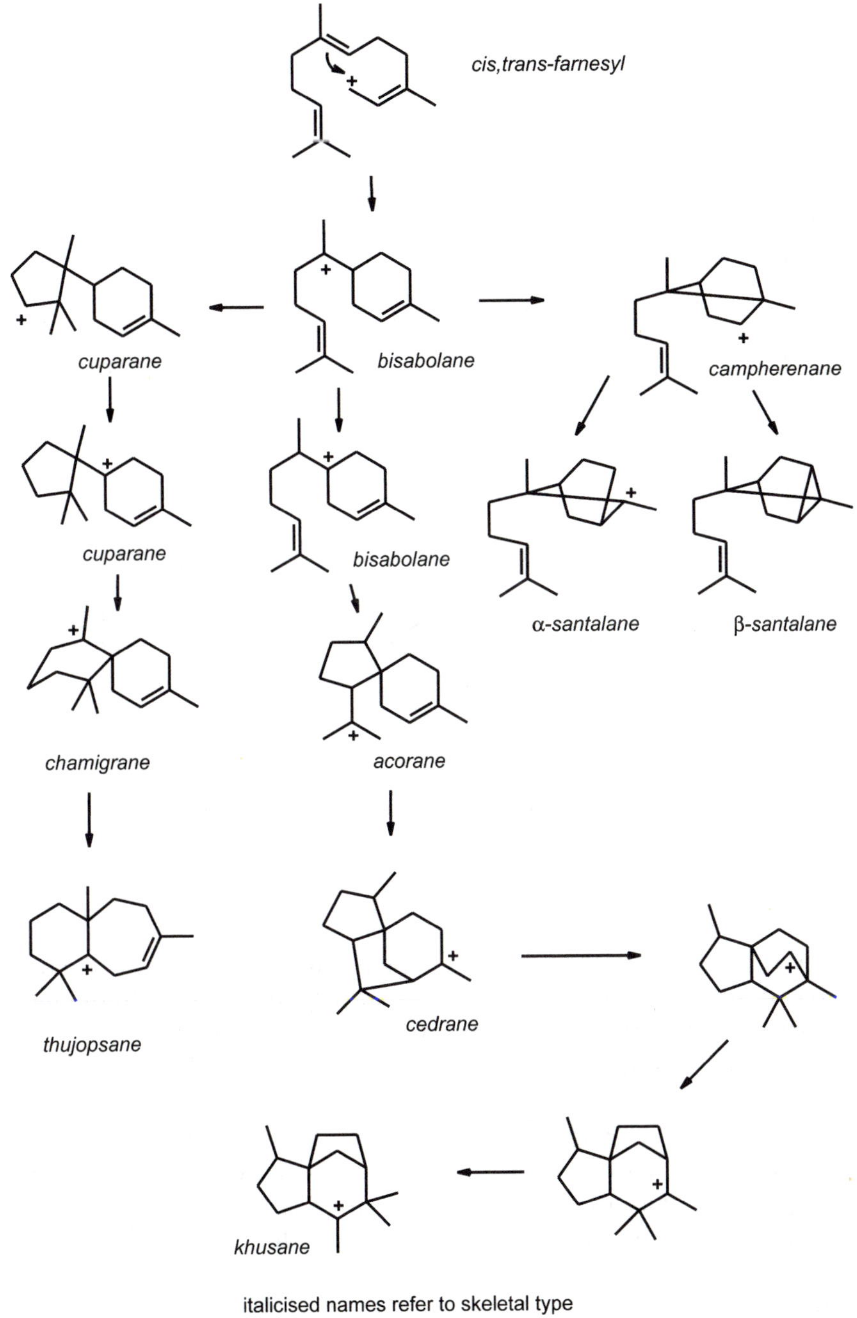

cis,trans-farnesyl

cuparane

bisabolane

campherenane

cuparane

bisabolane

α-santalane

β-santalane

chamigrane

acorane

thujopsane

cedrane

khusane

italicised names refer to skeletal type

Figure 2.10

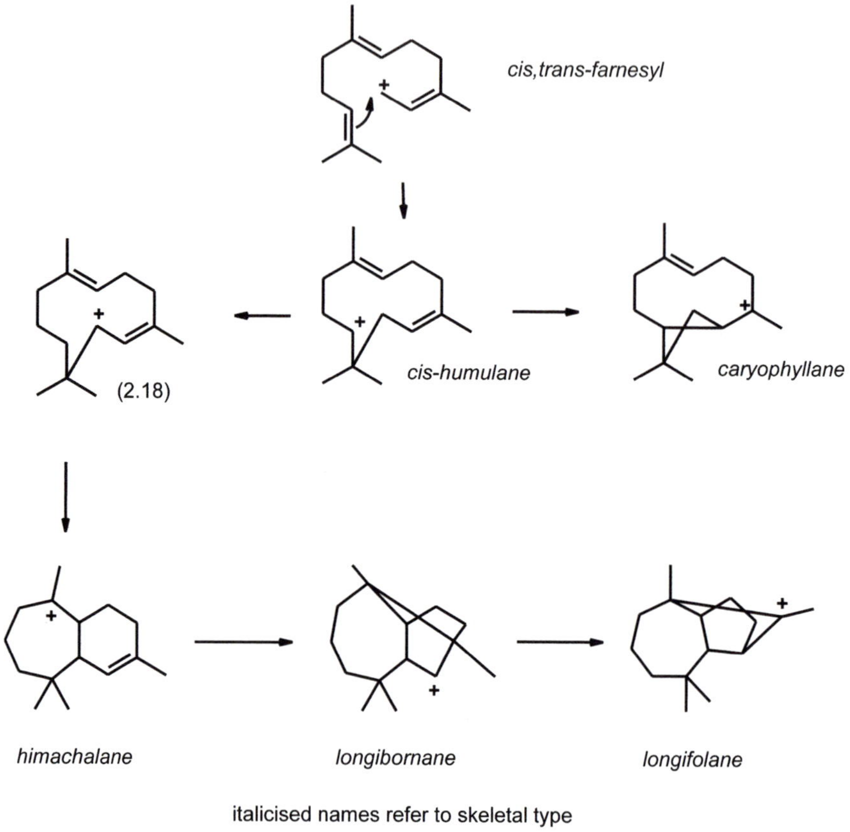

cis,trans-farnesyl

cis-humulane

caryophyllane

(2.18)

himachalane

longibornane

longifolane

italicised names refer to skeletal type

Figure 2.11

The reactions described in this chapter concern solely the formation of the basic skeletons of the various terpenoid families. Subsequent biochemical conversions, such as oxidation reactions, further increase diversity and, again, the patterns of such reactions will be characteristic of the organisms in which they occur.

The aim of this chapter is to demonstrate how, using only a few precursors and a few types of cationic reactions, it is possible to construct a staggering variety of different terpenoids. The reader should not be overwhelmed by the carbocation chemistry in this chapter. It might be advantageous to return to this chapter after reading Chapter 5 and thereby gaining an increased understanding of the reactions involved. Indeed, some reactions are discussed at the end of Chapter 5.

In Figures 2.9–2.16, normal typeface is used for names of specific terpenoids, *e.g.* limonene. Italicised text is used to denote a typical

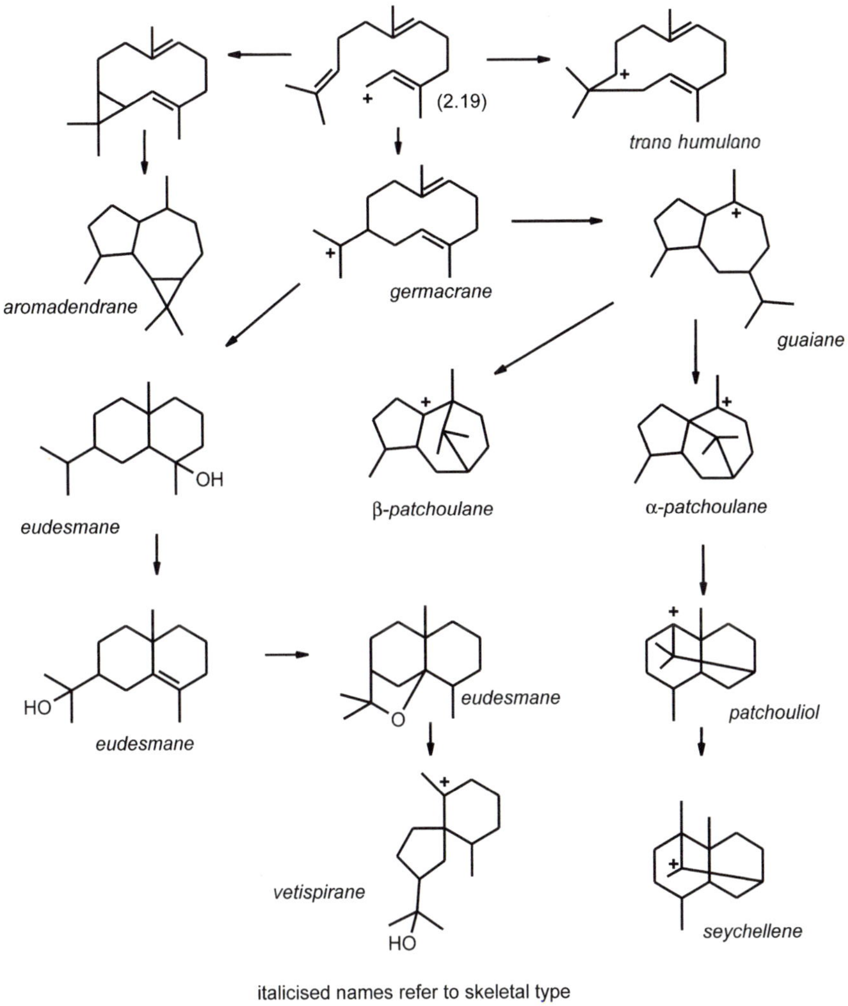

aromadendrane

trano humulano

germacrane

guaiane

eudesmane

β-*patchoulane*

α-*patchoulane*

eudesmane

eudesmane

patchouliol

vetispirane

seychellene

italicised names refer to skeletal type

Figure 2.12

member of a structural class of materials. For example, in Figure 2.9, the word *menthane* is placed alongside the cation from which limonene is formed. This indicates that the basic ring structure, *i.e.* 4-isopropyl-1-methylcyclohexane, is given the generic name menthane.

The following discussion concentrates on those terpenoid skeleta which will be discussed in subsequent chapters. For a more comprehensive treatment of ring formations and interconversions, the reader is referred to the book by Devon and Scott.[2.1]

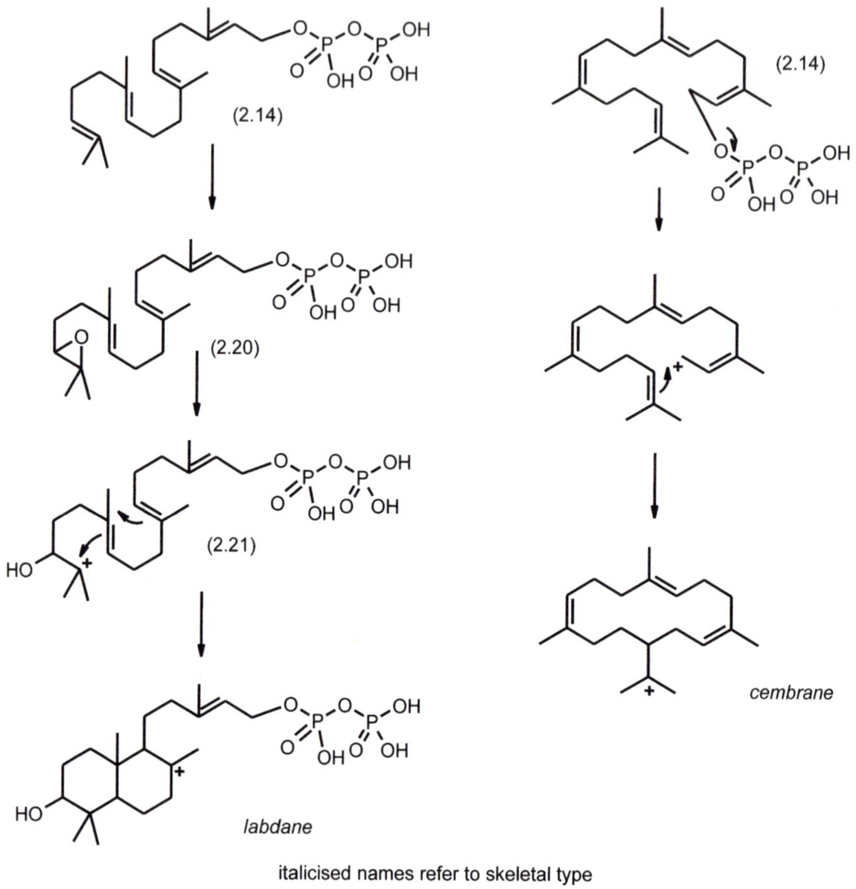
(2.14)
(2.20)
(2.21)
HO
labdane
cembrane
italicised names refer to skeletal type

Figure 2.13

2.6 MONOTERPENOIDS FROM GERANYL PYROPHOSPHATE

Figure 2.9 shows how the commonest monoterpenoid families are formed from geranyl pyrophosphate (2.12). Heterolysis of the carbon–oxygen bond in geranyl pyrophosphate produces the geranyl carbocation (2.15). Addition of water to this cation will produce geraniol, an important terpenoid alcohol which occurs in many flowers, such as roses, and other natural sources. Oxidation of geraniol to the corresponding aldehyde gives citral, the characteristic flavour component of lemons.

Carbocations are, of course, electron-deficient species and will therefore seek out electron-rich centres with which to react. In the

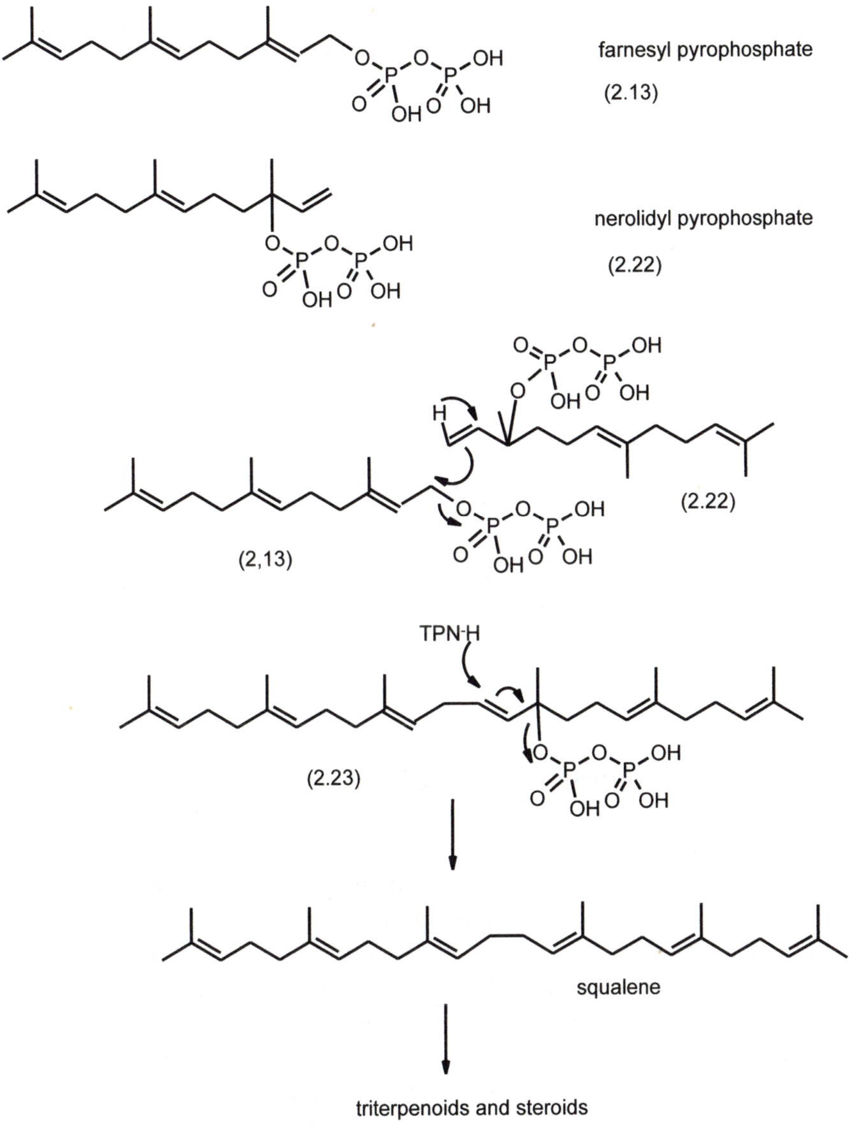

farnesyl pyrophosphate
(2.13)

nerolidyl pyrophosphate
(2.22)

(2.22)

(2,13)

TPN-H

(2.23)

squalene

triterpenoids and steroids

Figure 2.14

formation of geraniol, the electron-rich reagent is the oxygen atom of a water molecule. The olefinic bond at the other end of the geranyl carbocation (2.15), is another electron-rich species. Furthermore, it is placed six atoms away from the carbon carrying the initial positive charge and is therefore ideally placed to form a 6-membered ring. Intramolecular reactions that form 5- or 6-membered rings are favoured

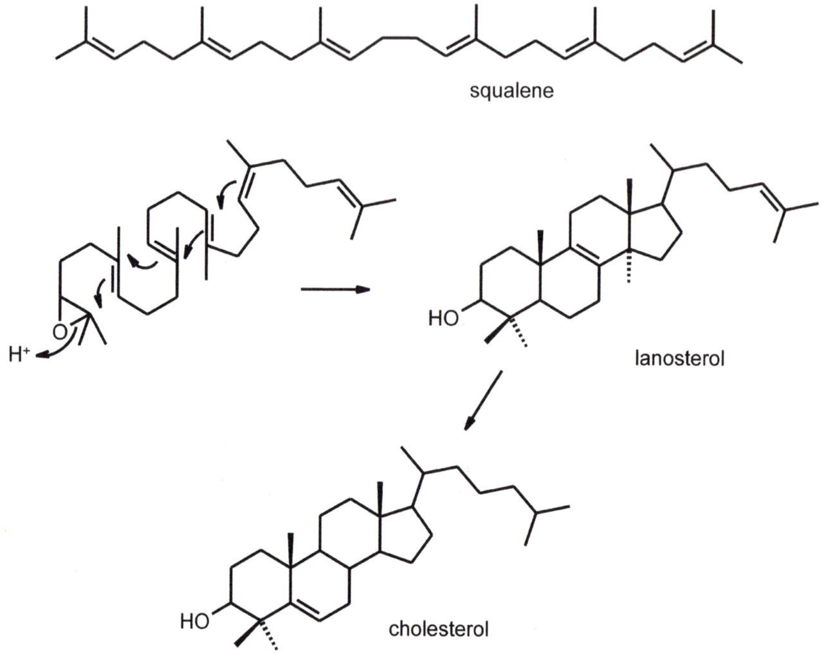

Figure 2.15

entropically and so these reactions are often difficult to prevent. In the case of the geranyl carbocation, the formation of the 6-membered ring leads to the menthane skeleton. This is often referred to as *p*-menthane to indicate the 1,4 relationship of the two substituents on the cyclohexane ring. Elimination of a proton from this carbocation gives limonene, the major component of orange oil. Trapping the carbocation by water gives α-terpineol, an important component of, *inter alia*, lilac flower oil.

This initial menthyl carbocation can add back across the ring to the other double bond. If it adds to the olefinic carbon carrying the methyl group, then the product is a bicyclic material of the camphane or bornane family. If the addition is to the other end of the olefin, the result is a bicyclic material of the pinane family. It will be noticed that this structure contains a strained 4-membered ring. This is the preferred regiochemistry for the addition, but there is an energy penalty in forming the strained ring system. The ring strain in the pinane skeleton is, as would be expected, an important factor in shaping the chemistry of the pinanes. Elimination of a proton from the pinyl cation (2.16) can take place in either of two directions giving the two isomeric pinenes as shown in Figure 2.9.

(2.14) geranylgeranyl pyrophosphate

(2.24) geranyllinalyl pyrophosphate

phytoene

carotenoids

β-carotene

astaxanthin

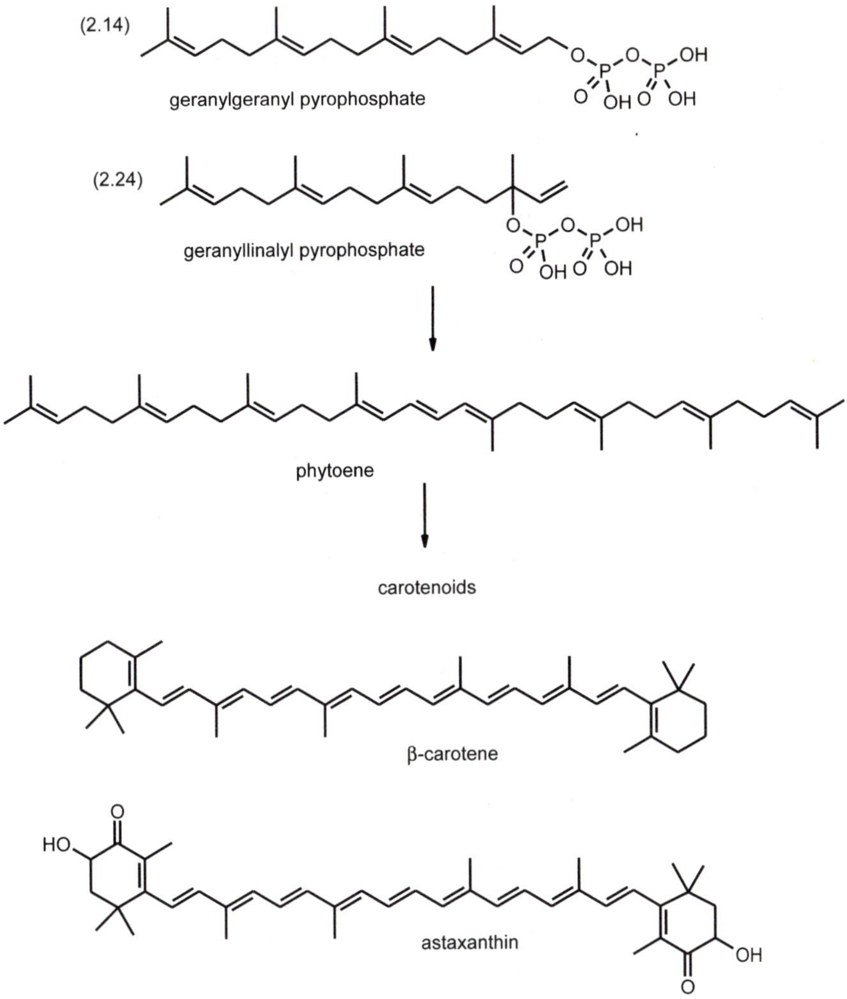

Figure 2.16

The ring strain at this point in the biosynthesis can also be relieved by means of a Wagner–Meerwein rearrangement (see Chapter 5 for details of this reaction) to produce the much less strained fenchane system. Addition of water, followed by oxidation, gives the ketone, fenchone.

Similarly, the bornyl carbocation (2.17) can react with water to give borneol and this can be oxidised to camphor, the characteristic odorant of camphor wood. The isocamphane skeleton is formed by a Wagner–Meerwein rearrangement of the bornyl carbocation.

2.7 SESQUITERPENOIDS FROM *CIS,TRANS*-FARNESYL PYROPHOSPHATE WITH INITIAL CLOSURE AT THE 6,7-DOUBLE BOND

When considering sesquiterpenoids, the number of possible reaction pathways increases significantly over those available in the monoterpenoid series. This results from the increased size of the precursor, farnesyl pyrophosphate, and the fact that it contains three double bonds. Both the 2,3- and 6,7-double bonds are initially formed in the *trans* configuration, but the 2,3-bond is part of an allylic carbocation system and therefore it can isomerise to the *cis* form. Initial ring closure can occur at either of the two remote double bonds and with the 2,3-bond in either of the *cis* or *trans* configuration.

Figure 2.10 shows some of the pathways that are possible with the 2,3-bond in the *cis* form and when the initial closure occurs at the 6,7-bond. This ring closure resembles the cyclisation of the geranyl cation to the *p*-menthane skeleton, and the product, the bisabolane skeleton, similarly resembles a *p*-menthane skeleton, which has been extended by addition of an isoprene unit to its isopropyl tail. Elimination of a proton from this bisabolyl carbocation, or its trapping by a nucleophile, will lead to a sesquiterpenoid of the bisabolane family. However, there is a number of possibilities for further skeletal rearrangement, three of which are shown in Figure 2.10.

Firstly, taking the path shown on the left of Figure 2.10, the electrons of the terminal double bond can trap the carbocation to give the cuparane skeleton. The initial carbocation formed in this series is secondary and therefore less stable that the ion from which it was formed. Abstraction of a hydrogen from the nearby tertiary carbon of the 6-membered ring leads to a more stable tertiary carbocation. A 1,2-carbon shift can then occur to give the two 6-membered rings of the chamigrane family. A further 1,2-carbon shift then produces the thujopsane skeleton.

Alternatively, as shown on the right of Figure 2.10, the initial bisabolane carbocation can be trapped by the double bond in the ring. This leads to the campherenane skeleton. This name draws attention to the fact that the skeleton resembles that of the camphane series of monoterpenoids where an extra isoprene unit is attached to one of the methyl groups on the one-carbon bridge. A Wagner–Meerwein rearrangement of this carbocation gives the α-santalane skeleton, whilst *trans*-annular elimination of a proton, leads to the β-santalane family.

The third possibility shown in Figure 2.10 is that for a 1,2-hydrogen shift to give a bisabolane with the cationic centre in the 6-membered ring.

This carbocation can be trapped by the terminal olefin to give the acorane skeleton. Subsequent trapping of that carbocation by the remaining double bond leads to the cedrane skeleton. Three successive 1,2-carbon shifts then give the framework of the khusane family of sesquiterpenoids.

2.8 SESQUITERPENOIDS FROM *CIS,TRANS*-FARNESYL PYROPHOSPHATE WITH INITIAL CLOSURE AT THE 10,11-DOUBLE BOND

When the carbocation formed by heterolysis of *cis,trans*-farnesyl pyrophosphate reacts with the 10,11-double bond, an 11-membered ring is formed. This gives the *cis*-humulane family as shown in Figure 2.11. Trapping of the humulane carbocation by the 2,3-double bond gives the caryophyllane skeleton. This is a very interesting bicyclic system since it contains two strained rings, a 4-membered one and a 9-membered one, fused together. Furthermore, the 9-membered ring contains a *trans* double bond. Construction of a molecular model of this will give a vivid demonstration of the amount of ring strain experienced by members of the caryophyllene family. This strain has a significant influence on their chemistry as will be seen in Chapter 7. The general principles concerning ring strain are discussed in Chapter 5.

Alternatively, the initial *cis*-humulane carbocation can undergo a 1,3-hydrogen shift to give a carbocation (2.18), which can then be trapped by the 6,7-double bond to give the himachalane skeleton. Trapping this carbocation by the remaining double bond gives the longibornane skeleton and a subsequent 1,2-carbon shift produces the longifolane series.

2.9 SESQUITERPENOIDS FROM *TRANS,TRANS*-FARNESYL PYROPHOSPHATE

The farnesyl carbocation (2.19) in which both double bonds are in the *trans* configuration, can add to the 10-11-double bond, either at the 11-carbon to give *trans*-humulane or the 10-carbon to give the germacrane skeleton, as shown in Figure 2.12. A third possibility is to insert into the 10,11-double bond to give a cyclopropane ring. In this case, the ring strain in the resultant material can be relieved by *trans*-annular cyclisation to give the aromadendrane skeleton.

The most chemically obvious cyclisation is that which forms the germacrane skeleton. However, this system also contains a strained 10-membered ring. The ring strain can be relieved by either of two cyclicsations. One gives the 5,7-system of the guaiane skeleton and the other, the 6,6-system of the eudesmanes. The guaiane skeleton can undergo further *trans*-annular cyclisation to give either the α- or β-patchoulane skeletons. The first of these can then undergo further 1,2-carbon shifts to give, firstly, the patchouliol, and then the seychellene ring systems.

Figure 2.12 also shows how various oxygenation patterns can arise in the eudesmane series and eventually, with the correct pattern, the eudesmane ring structure can rearrange to the vetispirane skeleton.

2.10 DITERPENOIDS

The key precursor to the diterpenoids is geranylgeranyl pyrophosphate (2.14) as shown in Figure 2.13. This molecule contains 20 carbon atoms and 4 double bonds and so the variety of possible diterpenoids is orders of magnitude greater than those of sesquiterpenoids. It would be impossible to discuss all of them in a book of this size and so Figure 2.13 shows only the two diterpenoid families most closely associated with fragrance. Closure of the geranylgeranyl carbocation at the 14-carbon, gives the cembrane skeleton containing a 14-membered ring. This biosynthetic pathway is initiated by heterolysis of the carbon–oxygen bond, which links the terpenoid unit to the pyrophosphate group. Thus, it is analogous to the type of ring-forming reaction that we have seen with monoterpenoids and sesquiterpenoids where the initial cation is formed at the "tail" of the molecule. In the diterpenoid series, we also see a second method of initiating ring closure, this time starting at the "head" of the molecule. In this type of reaction the double bond at the remote end of the molecule from the pyrophosphate group is functionalised so that a carbocation can be generated at the 15th carbon atom. For example, the double bond between the 14th and 15th carbon atoms can be epoxidised. Protonation of the epoxide (2.20) will form an oxonium ion. This ion will undergo ring opening, preferably to place the resultant alcohol group on the 14th carbon and the positive charge on the 15th. This gives a tertiary carbocation that is more stable than the alternative secondary ion, which would be formed if the ring were to open in the opposite direction. The product (2.21) with the secondary alcohol and tertiary carbocation is shown in Figure 2.13. There then follows a concerted reaction involving two more double bonds with the formation of the decalin ring system, which is found in the labdane skeleton, also shown in Figure 2.13.

2.11 TAIL-TO-TAIL COUPLING – TRITERPENOIDS AND STEROIDS

So far, we have seen terpenoid chains built using only head-to-tail links. In the triterpenoids and tetraterpenoids, we also see chains containing a tail-to-tail coupling. These are brought about as shown in Figure 2.14. Allylic rearrangement of farnesyl pyrophosphate (2.13) (the precursor to the sesquiterpenoids) gives nerolidyl pyrophosphate (2.22) in which the oxygenated function is now on the third carbon and the double bond has moved to the terminal position. Nucleophilic displacement of the pyrophosphate group of farnesyl pyrophosphate by the electrons of the terminal double bond of nerolidyl pyrophosphate results in the formation of a bond between the "tail" carbons of two C_{15} units. The reaction involves a concomitant 1,2-hydride shift in the nerolidyl chain, as shown in Figure 2.14.

The allylic pyrophosphate (2.23), formed in the above reaction, then reacts with TPN-H. This is a coenzyme which is similar to NADPH and, like NADPH, serves as a hydride anion donor. The hydride ion adds to the newly formed double bond, displacing it in the direction of the pyrophosphate group, which is then lost by cleavage of the carbon–oxygen bond to produce squalene. Squalene is a triterpenoid hydrocarbon containing six double bonds and a tail-to-tail fusion in the centre of the molecule. It occurs widely in nature, shark liver oil being a particularly rich source.

Squalene is very important in nature as it is the precursor of steroids, a family of molecules that serve as hormones in numerous organisms. The formation of the steroid framework from the mono-epoxide of squalene is shown in Figure 2.15.

Squalene is epoxidised selectively, under enzymic control to give a single epoxide at one end of the molecule. Protonation of the epoxide sets in train a series of synchronous reactions, known as a cascade reaction, involving cyclisation across four of its five remaining double bonds. This results in the characteristic 6,6,6,5-ring system of the steroid skeleton. The reacting centres in the cyclisation reaction all sit in a *trans-anti-*periplanar arrangement (see Chapter 5) and this defines the stereochemistry of the steroid system. Figure 2.15 shows lanosterol as one of the early products of steroid biosynthesis. Lanosterol, as its name suggests, occurs in lanolin, the oil in sheep's wool. Further bioconversions lead to other steroids such as cholesterol, a steroid notorious for its role in heart disease.

2.12 TETRATERPENOIDS AND CAROTENOIDS

Just as farnesyl pyrophosphate can be isomerised to nerolidyl pyrophosphate, so too can geranylgeranyl pyrophosphate (2.14) be isomerised to geranyllinalyl pyrophosphate (2.24). Coupling of geranylgeranyl pyrophosphate with geranyllinalyl pyrophosphate, in a reaction exactly analogous to that between farnesyl pyrophosphate and nerolidyl pyrophosphate, gives a 40-carbon structure with a tail-to-tail link at the centre. In squalene, the central bond remains unsaturated but, in the tetraterpenoids, it is dehydrogenated to give phytoene as shown in Figure 2.16.

Cyclisation to give a 6-membered ring at each end of the tetraterpenoid chain results in the carotenoid skeleton. The carotenoids form an important group of natural terpenoids. The most striking feature of the carotenoids is the number of double bonds they contain and the extended conjugation of these gives rise to absorption spectra well into the visible region. Probably the most important carotenoid is retinol (vitamin A, see Figure 1.6 for the structure), which is the precursor for the pigment responsible for vision, as described in Chapter 8. Figure 2.16 shows two carotenoid pigments. The first is the hydrocarbon β-carotene. This pigment is responsible for the colour of carrots and it occurs in many other natural sources such as grass. Cleavage at the central bond can give two molecules of retinol and so, β-carotene is used in biological systems to store the more delicate retinol. Astaxanthin is an oxygenated carotenoid. Its delicate pink colour is responsible for the characteristic red hue of the flesh of salmon.

REFERENCE

2.1. T.K. Devon and A.I. Scott, *Handbook of Naturally Occurring Compounds*, Vol. 2, *The Terpenes*, Academic Press, London, 1972.

CHAPTER 3

Linear and Monocyclic Monoterpenoids

© CORBIS

Roses rely heavily on terpenoids for their
beautiful scent

*But earthlier happy is the rose distilled than that which withering on the virgin
thorn, grows, lives and dies in single blessedness.*

William Shakespeare

Shakespeare was well acquainted with flowers and herbs and makes
many references to them in his writings. This quotation from
"A Midsummer Night's Dream" shows that he could see value in
chemical technology and its ability, through distillation, to preserve the
scent of a rose well beyond its natural life expectancy.

KEY POINTS

Before spectroscopic methods of structural determination became
available, the structure of natural products was determined by degradation of the molecule to give known fragments.
The fragments would be pieced together rather like a jigsaw
puzzle and the suggested structure would be proved or disproved
by synthesis.

When a synthesis is undertaken for the purpose of confirmation
of structure, the methods used must give unambiguous bond
formation. Different criteria apply to synthetic routes which are
intended to provide a means of commercial production.

Even in the chemistry of these relatively simple terpenoids, we begin
to learn about the principles of carbocation chemistry such as
rearrangement of carbocations and the *trans-anti*-periplanar rule.

Much of the fundamental work on understanding the basic
principles of organic chemistry was carried out on terpenoids.
In this and the following five chapters, we will see how work on

terpenoid chemistry led to a much deeper understanding of all of organic chemistry. Terpenoid chemistry contributed especially to our understanding of biogenesis, structural determination, carbocation chemistry, stereochemistry and conformational analysis.

3.1   STRUCTURAL DETERMINATION

A structural determination of a molecule such as myrcene would nowadays be done in less than an hour using NMR spectroscopy. A century ago, when modern spectroscopic methods were not available, it would have taken months of painstaking and accurate experimental work and a great deal of thought. This took the form of a chemical detective work using degradation to produce pieces of the molecular jigsaw which then had to be put together again. The proposed structures were then confirmed by synthesis and the discipline of total synthesis was found. The intellectual achievement of such early work is considerable. It must be admired and never under-rated.

If you enjoy solving puzzles, you might like to read through the two accounts of structural determination (myrcene and citral) below, without looking at the figures and work out the structures for yourself from the evidence as it unfolds.

3.2   MYRCENE

Myrcene is a hydrocarbon which occurs in many essential oils such as those obtained from parsley leaf, rosemary, celery leaf, hops, lemongrass, cardamom seeds and blackcurrant buds. It is also obtained when β-pinene, a major constituent of turpentine, is heated to a high temperature. It has a pleasant odour for a hydrocarbon, being described as sweet, balsamic, herbal and refreshing.

When confronted with a material of unknown structure, chemists in the late nineteenth and early twentieth centuries would resort firstly to elemental analysis. A small amount of the material would be burnt in dry air. The weights of carbon dioxide and water produced from a known weight of sample would enable them to calculate the carbon and hydrogen content of the material. Similarly, the percentage content of other elements such as nitrogen, sulfur and halogens, could be determined by appropriate gravimetric methods. Oxygen would be assumed by difference since there was no simple gravimetric determination for it. Thus, the full empirical formula of an unknown material could be determined through this technique known as elemental analysis.

In the case of myrcene, the result would be that the empirical formula is $C_{10}H_{16}$. The fact that the molecule contains 10 carbon atoms suggests that it may be a monoterpenoid. Nowadays, thanks to the isoprene rule, this would immediately suggest a few skeletal types such as 2,6-dimethyloctane or *p*-menthane. However, in the days when the structure of myrcene was elucidated, a much wider range of possibilities would have been considered.

Saturated acyclic hydrocarbons containing *n* carbon atoms, possess $2n + 2$ hydrogen atoms. Thus a saturated acyclic hydrocarbon with 10 carbon atoms would have 22 hydrogen atoms in the molecule. Myrcene has only 16 and is therefore lacking 6. This means that its structure contains three double bonds and/or rings.

In order to distinguish between double bonds and rings, the simplest technique was to hydrogenate the molecule. Each double bond would take up two atoms of hydrogen whereas rings would remain unaffected. Hydrogenation of myrcene gives a compound with an elemental composition of $C_{10}H_{22}$. Each molecule of myrcene has taken up six hydrogen atoms and therefore must have contained three double bonds.

In 1929, Diels found that, when myrcene is heated with maleic anhydride, it forms a 1:1 adduct by means of the Diels–Alder reaction. Therefore, at least two of myrcene's three double bonds are conjugated.

Ozone is known to cleave double bonds cleanly to give, under appropriate work-up conditions, aldehydes and ketones. The two new C=O bonds are formed where the C=C bond was in the original molecule and so, by identifying the aldehyde and ketone fragments, it is possible to mentally reconstruct the original hydrocarbon.

Ozonolysis of 1 mole of myrcene gives 1 mole of acetone, 2 moles of formaldehyde and 1 mole of a keto-dialdehyde. The acetone indicates the presence of one $(CH_3)_2C$=group and the formaldehyde, two H_2C= groups. When treated with chromic acid, 1 mole of the keto-dialdehyde gives 1 mole of succinic acid and 1 mole of carbon dioxide. It must, therefore be 2-ketoglutaraldehyde (pentan-2-one-1,5-dial).

Ozonolysis of 1 mole of the Diels–Alder adduct between myrcene and maleic anhydride gives 1 mole of acetone, among other products. This shows that the double bond carrying the two methyl groups is not part of the conjugated diene system.

Taking all of the above into consideration, there is only one possible structure for myrcene. Figure 3.1 shows the reaction scheme which can be pieced together from the experimental observations.

The ability of myrcene to take part in Diels–Alder reactions leads to some interesting chemistry and some interesting products.

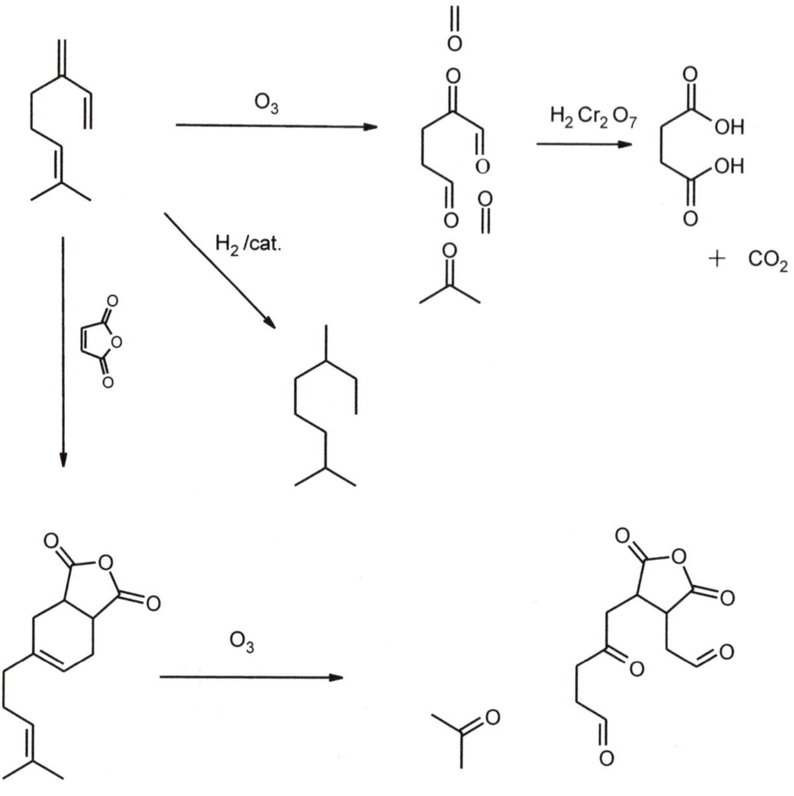

Figure 3.1

For example, the addition of acrolein to myrcene gives a product known by trade names such as Myrac Aldehyde® or Empetal®. The product is a mixture of isomers because the dienophile can line up in two orientations relative to the diene. For simplicity, only the major isomer, in which the two substituents on the cyclohexane ring are 1,4 relative to each other, is shown. Hydration of the acyclic double bond of Empetal® gives a hydroxy-aldehyde, known as Hydroxyempetal® or Lyral®. Lyral® possesses a fine odour reminiscent of the delicate odour of the flowers of lily of the valley and is widely used in perfumery.

The direct hydration of Empetal® gives only a very poor yield of the desired product because of competing side reactions. These illustrate an important principle in organic chemistry, *viz.* if an intramolecular reaction can occur to form a 5- or 6-membered ring, then it will usually take precedence over intermolecular reactions. An example in this instance is the intermolecular Prins reaction shown in Figure 3.2.

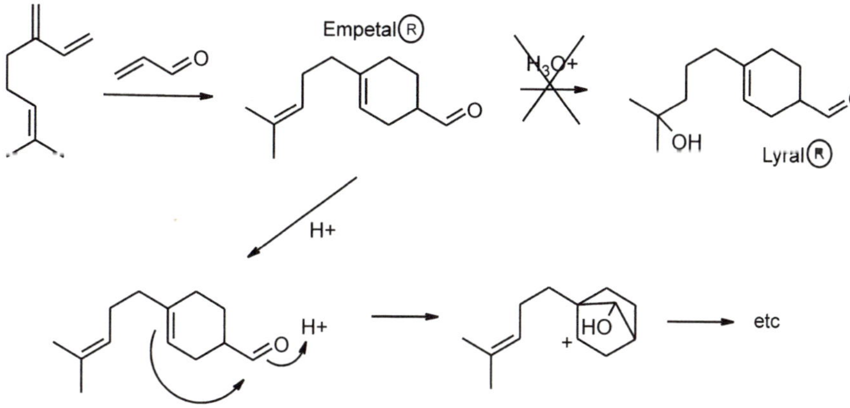

Figure 3.2

In order to prepare Lyral,[R] therefore, one approach would be to protect the aldehyde function, for example as an acetal, through acid catalysed addition of an alcohol or diol. If the double bond in the tail could then be hydrated in preference to that in the ring, deprotection of the aldehyde would lead to the desired product. However, there is a more elegant solution to the problem and this is shown in Figure 3.3.

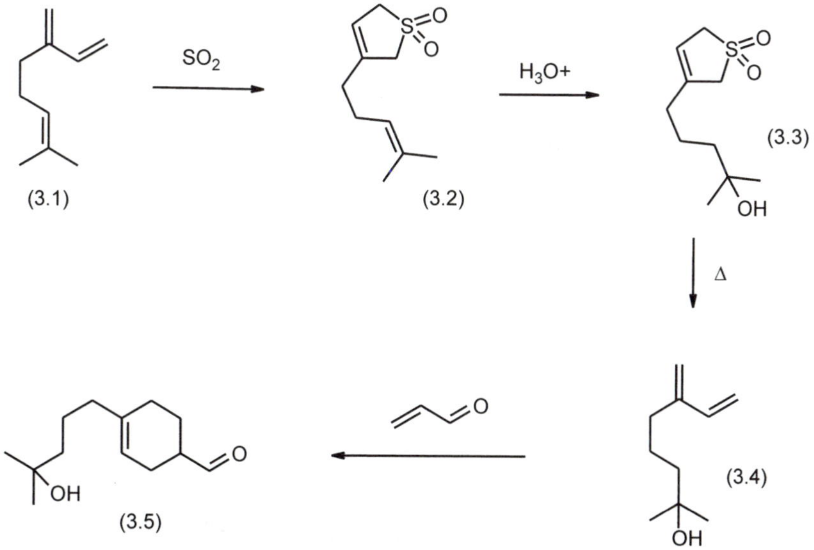

Figure 3.3

Sulfur dioxide reacts with dienes to form a cyclic sulfone in a reaction somewhat reminiscent of the Diels–Alder reaction. In the case of Myrcene (3.1), the adduct produced is (3.2) and its remote double bond is selectively hydrated by mineral acid. Pyrolysis of the product (3.3) under neutral conditions leads to cleavage of the dihydrothiophene ring to give the hydrated diene (3.4) and Diels–Alder reaction of this product then produces Lyral® (3.5).

3.3 OTHER MONOTERPENES

Some other common monoterpenoid hydrocarbons are shown in Figure 3.4. Citronellene is also known as dihydromyrcene. It does not occur in nature but is prepared by pyrolytic cracking of pinane. Hydration of the trisubstituted bond of citronellene produces dihydro-myrcenol which is used to provide fresh notes in perfumery. Further details of both of these processes are given in Chapter 9. α-Ocimene occurs in various essential oils such as those of basil, catnip and magnolia, while the β-isomer occurs in the leaf oils of lemon, orange and grapefruit and also in narcissus flowers. Allo-ocimene occurs in the leaves of sequoia trees and is the thermodynamically most stable of the three ocimenes since the double bonds are all in conjugation and as heavily substituted as possible. Double bonds can be isomerised under

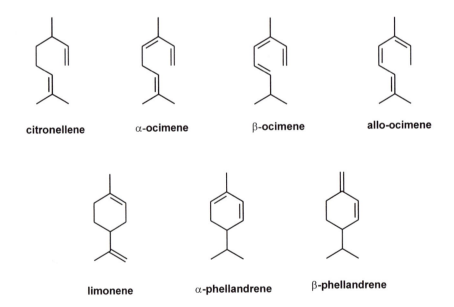

citronellene   α-ocimene   β-ocimene   allo-ocimene

limonene   α-phellandrene   β-phellandrene

Figure 3.4

acidic conditions through protonation to give a carbocation and then deprotonation of that cation to give an isomeric olefin. Double bonds prefer to have as much alkyl substitution as possible and so treatment with acid will, eventually, result in the olefinic bonds moving to the most substituted position. Frequently, some acid is formed by degradation or hydrolysis of plant components during the extraction process. This means that traces of acid are often present during the distillation of a natural extract. Therefore, the presence of allo-ocimene could be an indication of isomerisation during distillation rather than of natural occurrence in the oil. Similarly, the presence of hydrocarbons in the oil could be, partially at least, the result of elimination of water from alcohols present in the natural source. Modern techniques of headspace analysis (*i.e.* direct analysis of the air above the plant material) will avoid the possibility of such artefact formation. Limonene is perhaps the most widely occurring of the monoterpenes. It is found in a very wide range of essential oils but the richest sources are the citrus oils. For example, orange peel oil contains about 80% of *d*-limonene. The two isomers of phellandrene are also very widespread in nature. Both are often found in eucalyptus oils, the α-isomer also is found in angelica, dill and parsley and the β- in dill and fennel.

3.4 CITRAL

Citral occurs in many essential oils such as those of hops, rose, ginger, orange and basil. Its importance stems from its occurrence in lemon and lemongrass, the oils of which both depend heavily on citral for their characteristic odours. It can comprise as much as 85% of lemongrass oil.

Elemental analysis of citral showed that it has an empirical formula of $C_{10}H_{16}O$ and therefore contains three double bonds or rings. Hydrogenation demonstrated that it contains three double bonds and therefore is acyclic. When treated with hydroxylamine, it formed an oxime. Reduction with sodium in ethanol gave an alcohol and oxidation with Tollens' reagent gave an acid. These last three pieces of evidence leave little doubt that citral is an aldehyde. In 1891, Semmler found that, on pyrolysis, citral produced *p*-cymene and he therefore proposed that the skeleton was that of 2,6-dimethyloctane. The positions of the two carbon–carbon double bonds and the aldehyde group were demonstrated in 1895 by Tiemann and Semmler. They oxidised citral in a two-stage process using firstly alkaline permanganate and then chromic acid. This treatment is known to oxidize aldehydes to carboxylic acids and to cleave double bonds, converting the olefinic carbon atoms to carboxylic acids or ketones depending on the substitution pattern on the olefin. From this oxidation of citral, they obtained 1 mole of acetone, one of oxalic acid and one of 4-keto-pentanoic acid from each mole of citral. Taking this into consideration with the previous evidence, they deduced, correctly, that citral is 3,7-dimethylocta-2,6-dienal. Their conclusion was supported by the work of Verley in 1897, who degraded citral to acetaldehyde and 2-methylhept-2-en-6-one by means of a retro-aldol reaction using potassium carbonate as base. Figure 3.5 shows a summary of all of these results.

Having deduced the structure of a compound from degradation experiments, the next task for the natural products chemist is to confirm

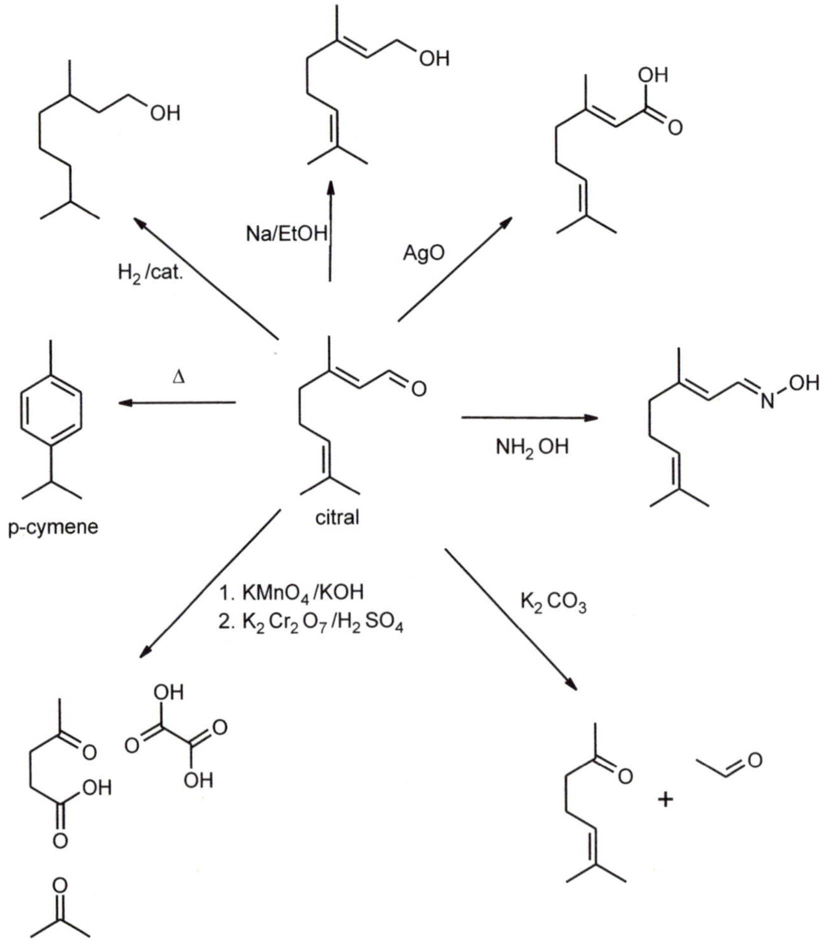

Figure 3.5

the structure by synthesis. Syntheses are described as total or partial. Partial syntheses are those which start from a compound which already has much of the structure in place. Total synthesis, which starts from basic building blocks is more rigorous, and more difficult. Since the object of the synthesis is solely to confirm the structure, the most important thing is to use synthetic methods which will lead to unambiguous bond formation. Using the best methods available to them at the time, the first total synthesis of citral was started by Barbier and Bouveault in 1896 and completed by Tiemann in 1898. This synthesis is shown in Figure 3.6.

The French team started with the available compound, 2,4-dibromo-2-methylbutane (3.6). They reacted this with acetylacetone (pentan-2,4-dione) in the presence of base. There is no ambiguity in this reaction. The most acidic proton is that between the carbonyl groups of acetylacetone

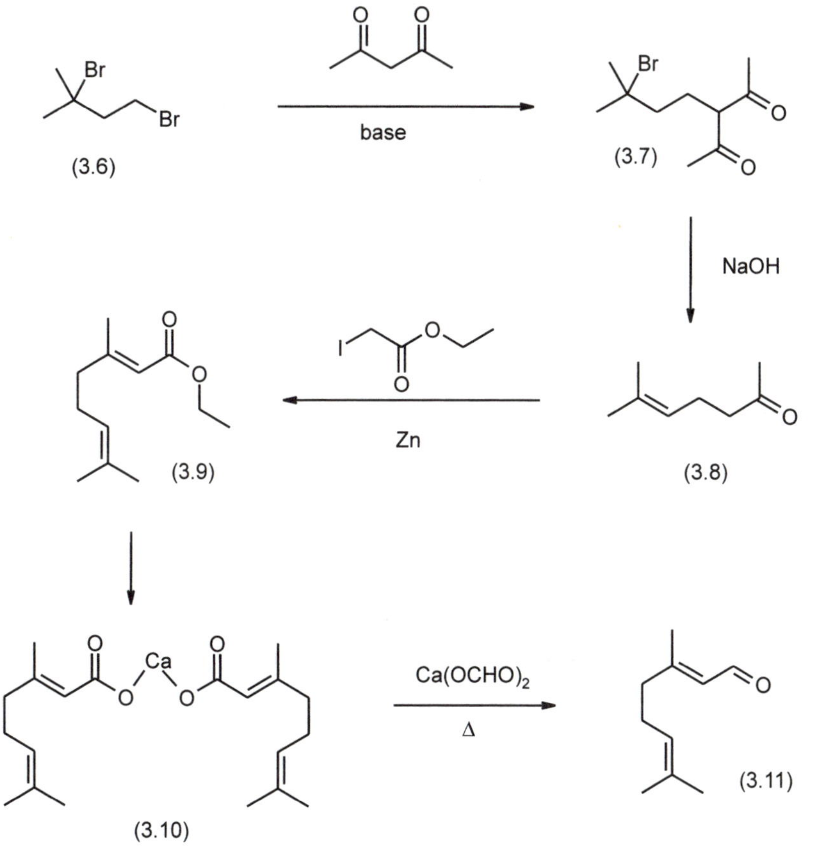

Figure 3.6

and the primary bromide will react very much faster than the tertiary as a result of the steric and electronic effects of the two methyl groups of the latter. The sole product is therefore the bromodiketone (3.7) shown in the figure. Treatment of this with sodium hydroxide achieves two things. First, hydrogen bromide is eliminated. Under the conditions employed, the resultant double bond will end up in the thermodynamically most stable position, that is trisubstituted. The other reaction which takes place is the reverse aldol reaction of the diketone. The product of this treatment is therefore 2-methylhept-2-en-6-one (3.8). This is, of course, the ketone which Verley obtained by retro-aldol reaction from citral. The reverse of Verley's procedure, *i.e.* the forward aldol reaction between methylheptenone and acetaldehyde is fraught with problems. The greatest of these is the self-condensation of acetaldehyde. Of the various possible combinations of aldol reactions, this is the most favourable and would predominate over the desired crossed aldol reaction. To solve this problem, Barbier and Bouveault used the Reformatsky reaction between methylheptenone and ethyl iodoacetate. This is a clean reaction with an unambiguous product, the desired ester, ethyl geranate (3.9). This is the point in the synthesis where Tiemann took over. He hydrolysed the ethyl geranate and then pyrolysed its calcium salt (3.10) in the presence of calcium formate, a reaction shown by Bertagnini in 1848, to produce aliphatic aldehydes from the corresponding carboxylic acids. This gave synthetic citral (3.11) which was shown to be identical in all respects to the natural material and thus confirmed the structural elucidation of Semmler, Tiemann and Verley.

Citral is the key odoriferous principle of lemon and lemongrass oils and is therefore potentially useful in perfumes and flavours. More importantly, citral is a key intermediate in many synthetic routes to the ionones (a group of perfumery ingredients which will be described more fully in Chapter 8) and vitamins A, E and K. The synthesis of Barbier, Bouveault and Tiemann served its purpose as confirmation of the structure of citral. However, in terms of preparing large amounts of material for commercial use or even for use as an intermediate in further laboratory syntheses, it is somewhat lacking. This will be discussed in detail in Chapter 9; for the present suffice it to say that the need for an efficient synthesis of citral stimulated much synthetic effort over the years.

The synthesis of Arens and van Dorp is shown in Figure 3.7. This synthesis, completed in 1948, may not look much cleaner nor more effective than that of Barbier, Bouveault and Tiemann, but it is an important development since it laid the intellectual basis of the most efficient modern syntheses of citral, as will be seen in Chapter 9. The key feature of this synthesis is its use of acetylenic chemistry.

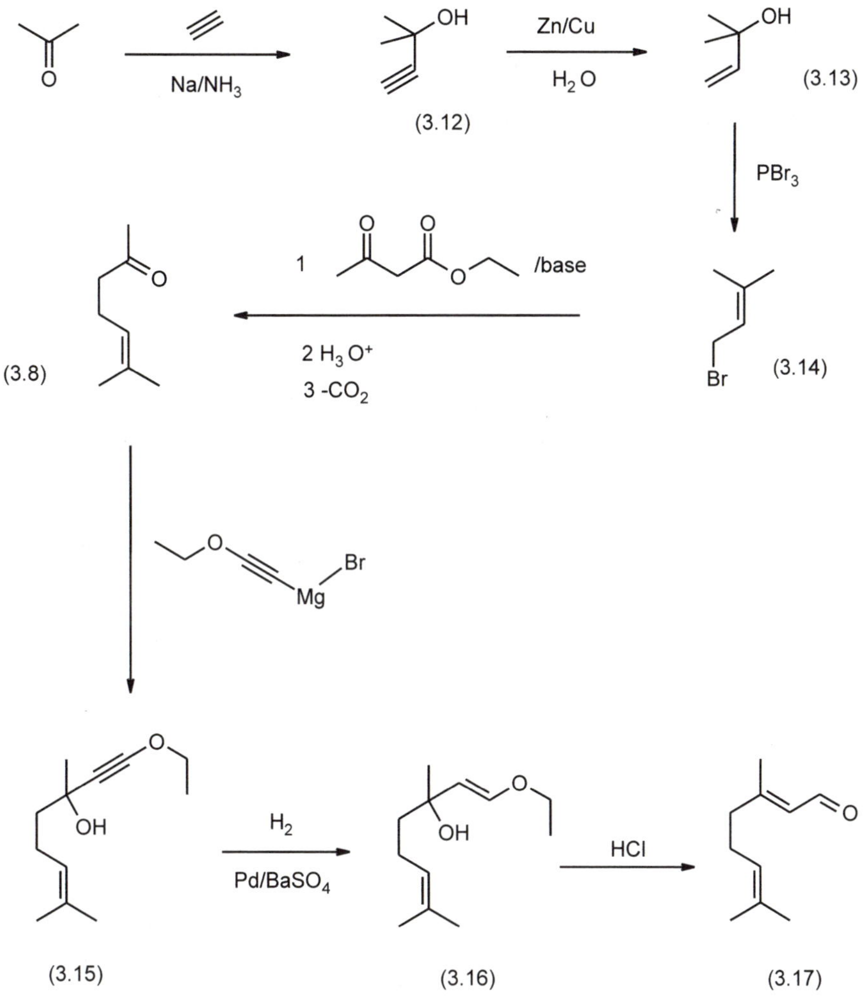

Figure 3.7

One nucleophilic addition of an acetylenic anion is used to construct the first half of the citral molecule and a second extends it to the full 10 carbon atoms.

Arens and van Dorp used acetone and acetylene as the starting point for their synthesis. Sodamide was generated *in situ* from sodium metal and liquid ammonia and used to remove the acidic proton of acetylene. The acetylene anion was then added to acetone to give methylbutynol (3.12). The triple bond of this product was reduced with a zinc/copper couple to methylbutenol (3.13). Treatment of this alcohol with phosphorus tribromide gave prenyl bromide (3.14). This type of allylic transposition is common in terpenoid synthesis. The intermediate allylic cation can be attacked at either end by the incoming nucleophile, in this case the bromide anion and the combination of steric factors and the production of the more thermodynamically stable product can be combined to give high selectivity for addition of the bromide to the terminal carbon. Prenyl bromide was then used to alkylate the anion of ethyl acetoacetate. The resultant keto-ester was then hydrolysed and the acid decarboxylated to give methylheptenone (3.8). Now came the second addition of an acetylenic unit. In this case, ethoxyacetylene was used in order to leave an oxygen function on the terminal carbon of the product (3.15), *i.e.* in the position where oxygenation is required in the eventual target, citral (3.11). As before, the acetylenic bond of the product was reduced to the corresponding olefin (3.16), this time by hydrogenation in the presence of Lindlar catalyst. The product of this hydrogenation was a vinyl ether, in other words a masked aldehyde function. It therefore remained only to hydrolyse this to the aldehyde and eliminate one molecule of water to produce citral (3.11).

Citral contains two trisubstituted double bonds. One of these, the one nearer to the aldehyde function, is asymmetrically substituted at both ends and so citral exists as two isomers. These were originally named citral-a and citral-b. Nowadays we refer to them as geranial and neral. These names are derived from the alcohols, geraniol and nerol, respectively, which are produced on reduction of the two. Their structures are shown in Figure 3.8. The figure also shows how the two isomers of citral can be interconverted through the enol form. In the enol form, the carbon–carbon bond which is a double bond in citral, is a single bond and can be rotated relatively freely. Re-ketonisation can therefore produce either isomer, depending on the orientation of this central bond as the protonation/hydration sequence proceeds. This cannot happen in the case of the alcohols and so geraniol and nerol are much less easily interconverted than are geranial and neral. Thus, the ratio of geraniol to nerol in an essential oil will be a good reflection of their relative proportions in the plant from which the oil was extracted. On the contrary, the ratio of geranial to neral in an essential oil will depend on the extraction process as well as the starting ratio in the plant.

3.5 GERANIOL

Geraniol (3.20) occurs in many essential oils including rose, geranium, citronella and palmarosa. In some instances, for example some sources

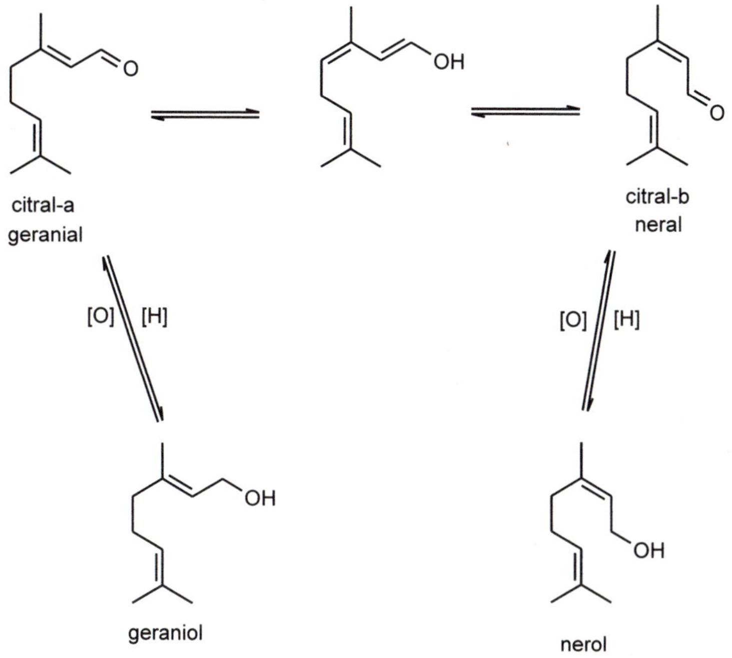

Figure 3.8

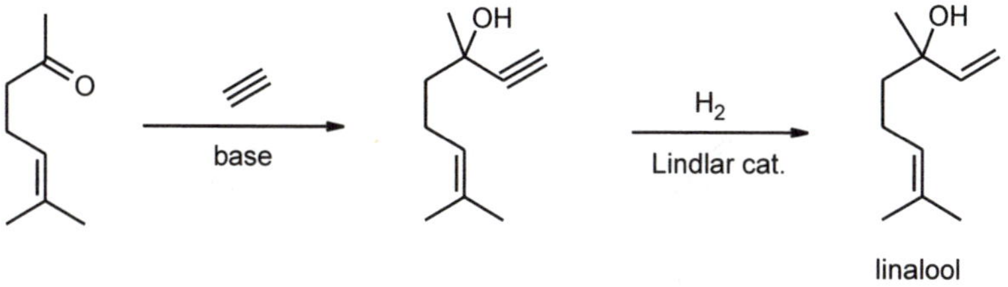

Figure 3.9

of palmarosa, it will comprise up to 90% of the oil. Nerol (3.19) is less widespread; rose and palmarosa are the most important oils containing it. Both alcohols possess floral scents reminiscent of rose and hence are often referred to, along with linalool and citronellol, as the rose alcohols. Linalool is an allylic isomer of geraniol and nerol. It is easily synthesised by addition of the acetylide anion to methylheptenone. This variation on Arens and van Dorp's citral synthesis is shown in Figure 3.9.

3.6 LINALOOL

Linalool (3.17) occurs even more widely than geraniol. The richest source is Ho leaf oil which can contain over 95% linalool. Rosewood contains 80–85% and freesia about 80% linalool. It also occurs at levels around 50% in lavender and in herbs such as coriander and basil. Citrus leaves and flowers also contain significant amounts of linalool. However, it takes its name from the oil of linaloe wood, of which it accounts for about 30%.

Linalyl acetate also occurs in many essential oils, significant levels being present in lavender, clary sage and the leaf oil of bitter orange. Early attempts to prepare linalyl acetate by esterification of linalool (3.17) using acetic anhydride gave an unexpected result. The acetate ester obtained was not that of linalool but that of geraniol. Conversely, when geranyl acetate (3.18) was treated with steam at 200 °C under pressure, the hydrolysis product was not geraniol but linalool. This reaction sequence is shown in Figure 3.10.

An explanation for these observations lies in Figures 3.11 and 3.12. In Figure 3.11, we see how protonation of the alcohol functions of nerol (3.19), geraniol (3.20) and linalool (3.17) lead to the corresponding allylic cations; (3.21), (3.22) and (3.23) respectively.

Figure 3.12 shows how the π-electrons of the double bond of an allylic cation can move towards the carbon atom bearing the positive charge. This generates an isomeric allylic cation, with the positive charge on the opposite end of the 3-atom system. In free allylic cations, this exchange is so rapid that the two isomers are indistinguishable. In fact, in molecular orbital theory, we consider the system to consist of a single set of orbitals which stretches across all three atoms. Since there is single bond character in each of the bonds, rotation is possible and the three cations, *viz.* geranyl (3.22), neryl (3.21) and linalyl (3.23), become equivalent. This is often represented as a smear of electrons as shown in structure (3.24) at the foot of Figure 3.12. Therefore in reactions such as those of

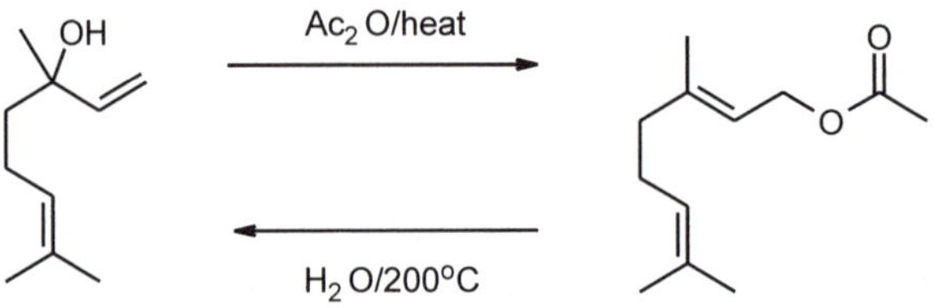

(3.17)   (3.18)

Figure 3.10

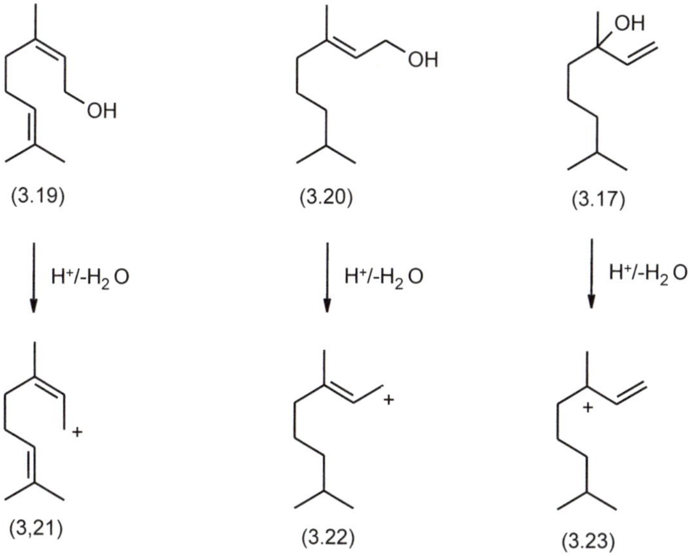

(3.19) (3.20) (3.17)

H⁺/-H₂O H⁺/-H₂O H⁺/-H₂O

(3,21) (3.22) (3.23)

Figure 3.11

Figure 3.10, once the carbon–oxygen bond of the starting material has been broken, the incoming nucleophile can add to either the same carbon atom which carried the original substituent, or to the one at the opposite end of the allylic system. The exact course of the reaction will be determined by the reagents and the reaction conditions.

We might therefore expect total scrambling of regiochemistry and stereochemistry to occur in such carbocation reactions. However, things

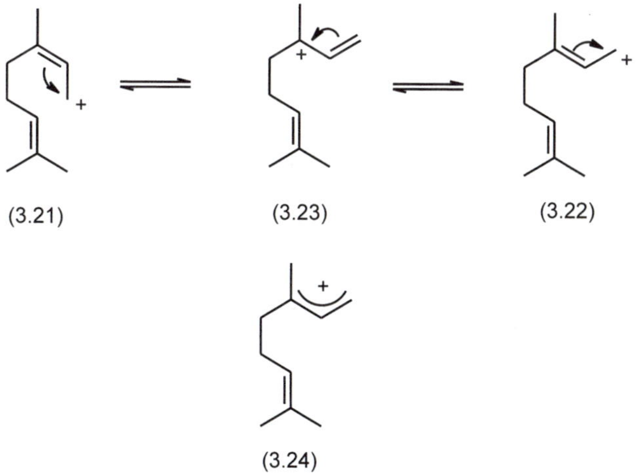

(3.21) (3.23) (3.22)

(3.24)

Figure 3.12

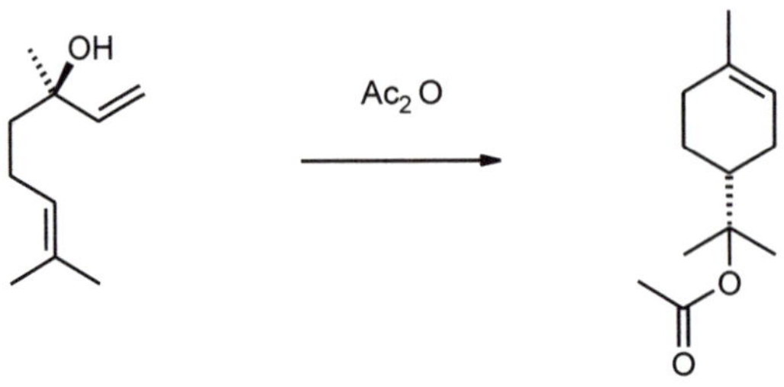

l-linalool
(3.25)

d-terpinyl acetate
(3.26)

Figure 3.13

are not so simple. In 1957, when Prelog treated *l*-linalool (3.25) with acetic anhydride, he obtained *d*-terpinyl acetate (3.26) as shown in Figure 3.13.

If this reaction had occurred through a simple loss of water from linalool to give a carbocation, followed by cyclisation and trapping of the newly formed carbocation by acetate, then a racemic product would have resulted. This sequence is shown in Figure 3.14. Loss of water from the protonated linalool would give the allylic cation. If this were a free

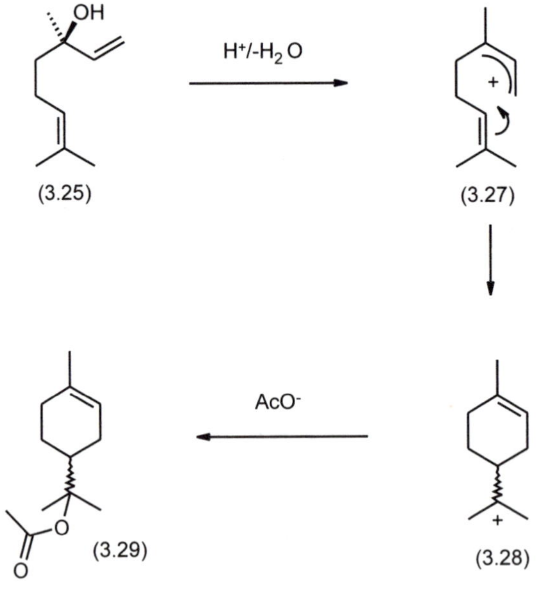

(3.25)

(3.27)

(3.29)

(3.28)

Figure 3.14

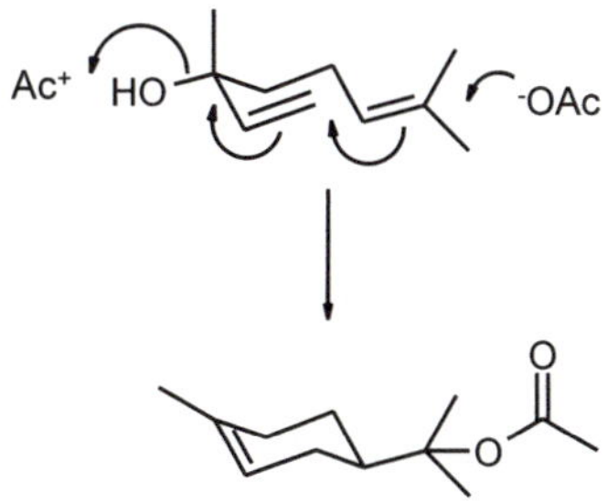

Figure 3.15

species as shown in Figure 3.14, then the cyclisation reaction with the double bond could occur equally from either face to give a 50:50 mixture of stereoisomers in the terpinyl cation (3.28). Trapping of this would then produce racemic terpinyl acetate (3.29). (If needed, a review of the principles of stereochemistry can be found in the opening section of Chapter 4.)

Since Prelog's product had retained the enantiomeric purity of the starting material, the mechanism must have involved some factor which constrained the reactive centres throughout the course of the reaction. The factor involved is known as the *trans-anti*-periplanar rule. This rule states that the bonds being formed or destroyed during a reaction prefer to exist in a geometry which places them in the same plane as each other or in a *trans-* or *anti-* relationship to each other. This arrangement allows for maximum interaction of the orbitals involved in the reaction, throughout its course. The role of the *trans-anti*-periplanar rule in Prelog's reaction is shown in Figure 3.15.

In the top half of Figure 3.15, we see the starting materials lined up ready to react. The bonds which will be broken are the oxygen–carbon bond of the alcohol and the two double bonds. The new bonds which will be formed are the bonds from the acylating species to the oxygen of the alcohol, the new carbon–carbon bond which will complete the ring and the bond from the incoming acetate anion to the olefinic carbon at the end of the linalool molecule. As can be seen from the figure, all of these bonds are lined up trans- to each other in a single plane. The stereochemistry of the carbon carrying the alcohol group in the starting material therefore determines the stereochemistry of the asymmetric ring carbon atom in the product. Since the starting material is homochiral (*i.e.* containing only one enantiomeric form), the product will also be homochiral. The reaction is essentially synchronous with all bonds being formed or broken simultaneously. The question therefore arises as to what is the nature of a carbocation. This will be discussed in more detail in Chapter 5.

3.7   CITRONELLOL AND CITRONELLAL

Figure 3.16 shows two other important acyclic monoterpenoids, citronellol and citronellal. Both occur in a variety of essential oils and are, chemically speaking, the reduced forms of geraniol/nerol and citral, respectively. Citronellol occurs in rose and geranium where the level can be up to 50% of the oil. The oil of the hedging rose, *Rosa rugosa*, contains about 60% citronellol. Levels of just under 10% can be found in ginger (*Zingiber officinale*) and juniper berries (*Juniperus communis*). The richest source of citronellal is a eucalypt, *Eucalyptus citriodora*, which has a content of up to 85% citronellal in its oil, depending on the exact subspecies and country of origin. Citronella oil (*Cymbopogon nardus*) contains about 40% citronellal and lemon balm (*Melissa officinalis*) about 35%. Citronellal is also found in herbs such as lovage (*Levisticum officinale*) and wormwood (*Artemisia asinthium*). Citronellal is a very effective insect repellent and so oils containing it are often used for this purpose. For example, Citronella oil is often added to candles which then provide both light and freedom from biting insects when lit on summer evenings.

As stated earlier, there is a general rule in organic chemistry that, if an intramolecular reaction can occur to give a 5- or 6-membered ring, it will be so favoured that it will predominate over intermolecular reactions. The cyclisation of linalool to terpinyl acetate shown in Figure 3.13 is another good example of this rule to add to that shown in Figure 3.2. Yet another excellent example is found in the chemistry of citronellal as shown in Figure 3.17.

When citronellal (3.30) is treated with a Bronsted acid, the oxygen atom is protonated and generates the oxonium ion (3.31) as shown in Figure 3.17. The positive charge is shared between the oxygen and the

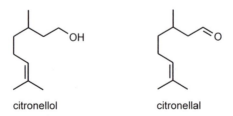

citronellol   citronellal

Figure 3.16

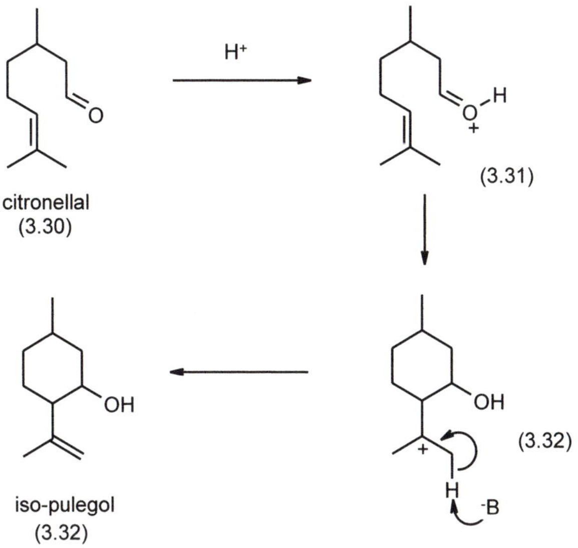

Figure 3.17

carbon atoms of the carbonyl bond. Trapping of the positive charge on the carbon atom by the π-electrons of the double bond at the other end of the molecule, leads to the formation of a 6-membered ring, as shown in structure (3.32). Elimination of a proton from one of the terminal methyl groups, then gives iso-pulegol as the product. This reaction is very favourable and is difficult to stop when there are even traces of acid in the presence of the citronellal. The stereochemistry of the reaction is interesting and will be discussed further in Chapter 4. The product of hydration of citronellal is known as hydroxycitronellal (3.34) and is of use in perfumery for its strong lily of the valley (muguet) odour. Any attempt to prepare it directly from citronellal will fail because of the competing cyclisation to give iso-pulegol which will always predominate. In order to prepare hydroxycitronellal, it is therefore necessary to prevent the alternative reaction from occurring. One way in which this can be achieved is through conversion of the aldehyde group to an oxazolidine as shown in Figure 3.18.

Reaction of citronellal (3.30) with diethanolamine produces the oxazolidine derivative (3.35). If this product is dissolved in 98% sulfuric acid, two things happen. Firstly, the nitrogen atom is protonated to form the hydrogen sulfate salt. This process no doubt helps the organic material to dissolve in the acid. Sulfuric acid then adds across the double bond at the far end of the molecule to give the sulfate ester (3.36) as

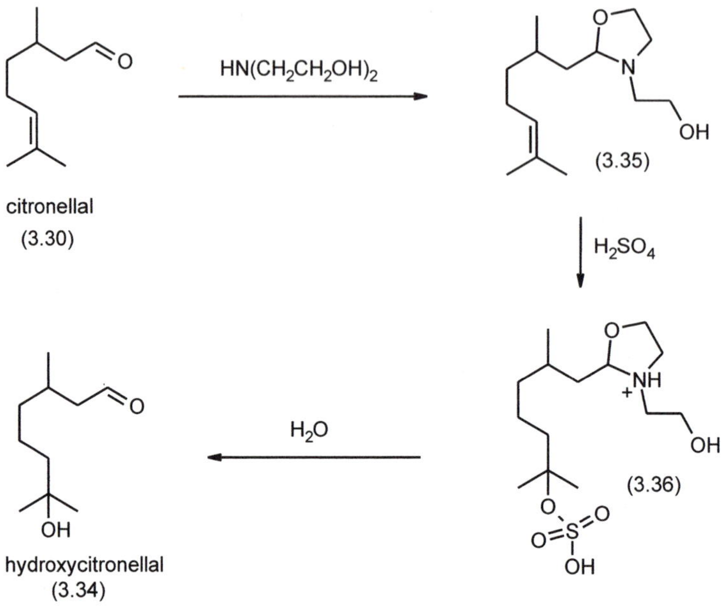

Figure 3.18

shown in Figure 3.18. If the sulfuric acid solution of this product is added to water, both the sulfate ester and the oxazolidine ring are hydrolysed under the acidic aqueous conditions to give hydroxycitronellal (3.34).

3.8   TERPINEOL

Terpineol, or more strictly, α-terpineol (3.38), is one of the most widespread of monocyclic monoterpenoid alcohols in nature. It is found in flowers such as narcissus and freesia; herbs such as sage, marjoram, oregano and rosemary; in the leaf oil of Ti-tree (*Melaleuca alternifolia*) and in the oil expressed from the peel of lemons. Reports of the level of terpineol in oils occasionally vary considerably and one wonders how much this is due to variations in the plants and to variations in the isolation process since terpineol could be an artefact. The layman will often describe the odour of terpineol as "pine disinfectant" since terpineol is, in fact a major component of pine disinfectant. This product is prepared by distillation of turpentine in the presence of acid which results in opening of the ring of α-pinene (3.37) to produce α-terpineol as shown in Figure 3.19.

The details of the mechanism of this reaction are shown in Figure 3.20. The driving force for the reaction comes from the release of the strain in

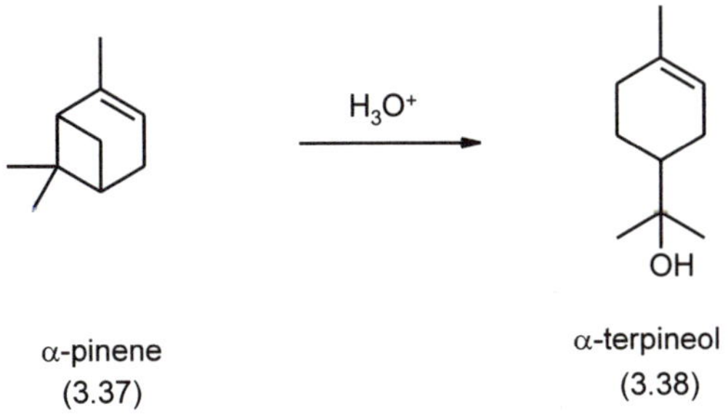

α-pinene
(3.37)

α-terpineol
(3.38)

Figure 3.19

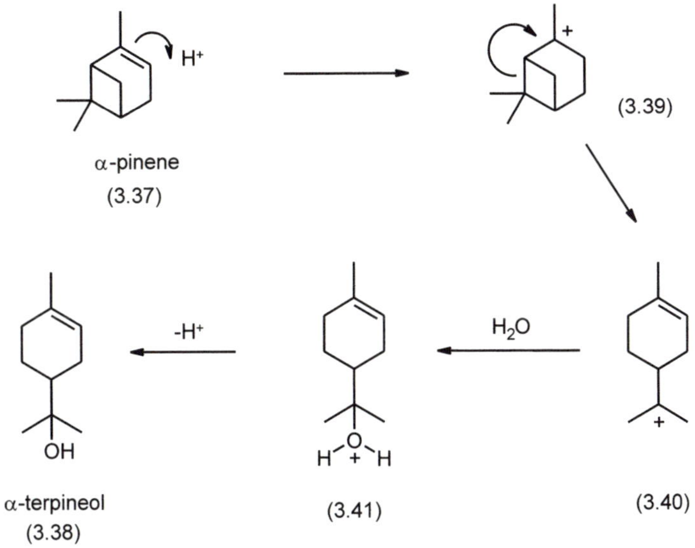

α-pinene
(3.37)

(3.39)

α-terpineol
(3.38)

(3.41)

(3.40)

Figure 3.20

the 4-membered ring in the α-pinene skeleton. The initial step in the process is the protonation of the double bond in α-pinene (3.37). This generates a carbocation centre adjacent to the strained cyclobutane ring as shown in structure (3.39). The ring strain can be released by movement of one pair of electrons from a single bond of the cyclobutane ring towards the positive charge with formation of a double bond to give (3.40). The resultant carbocation in the tail of the molecule, can then be quenched by addition of water followed by loss of a proton (from (3.41), giving α-terpineol (3.38)). This cationic rearrangement will serve as a gentle introduction to the subject of carbocation chemistry which will be discussed in much more detail in Chapter 5.

© S. Solum/Photolink

CHAPTER 4

Menthol and Carvone

The terpenoids of mint provide us with some interesting studies in stereochemistry

Perhaps looking glass milk isn't good to drink.

Lewis Carroll

Not many small girls would raise such an interesting question as this one of Alice's when confronted by a glass of milk in the world she discovered on the other side of the looking glass. Perhaps Lewis Carroll was thinking of Louis Pasteur's pioneering work on stereochemistry when he put these words into the mind of his fictional character. "Alice through the Looking Glass" was written 18 years after Pasteur reported the resolution of tartaric acid, and so the topic might have been a popular one in the Oxford academic circles in which Carroll moved. "Looking glass" molecules, enantiomers, can indeed have very different properties depending on which side of the mirror they are found. One of the most dramatic examples is the drug Thalidomide, the molecules of which cannot be superimposed on their mirror images. One form is effective in relieving pain, the other, when present in the body of a pregnant woman, can cause defects in her unborn baby.

KEY POINTS

There are five principal forms of isomerism related to the spatial arrangement of atoms in organic molecules; structural, positional, geometrical, conformational and stereo.

Different isomers may display different physical and/or chemical and/or biological properties.

Enantiomeric molecules are identical in their chemical and physical properties except for the fact that they rotate the plane of polarised light in opposite directions.

Molecules are able to "recognise" other molecules but this recognition may be greatly affected by the handedness of each.

Because of differences in recognition, absolute stereochemistry can have major effects on the biological properties of materials.

When stereochemistry is important in a molecule, it places additional constraints on synthetic pathways.

Many synthetic routes to a given material may exist in economic competition with each other.

Absolute stereochemistry can be determined either by X-ray crystallography or by degradation of a chiral material to simpler chiral derivatives which are identical to those of known configuration.

4.1 TYPES OF ISOMERISM

Isomers are molecules which have the same empirical formulae, *i.e.* they have the same total composition in terms of constituent atoms, but are different in the way the atoms are bonded together or arranged in space. In this chapter we will ignore functional isomers (*e.g.* allylic alcohol versus aldehyde) and tautomers (*e.g.* ketone *versus* enol) and will concentrate on those forms of isomerism relating to the spatial arrangement of atoms. There are five principal forms of such isomerism in organic molecules: structural, positional, geometrical, conformational and stereo. In fact, most of the discussion will concern stereoisomerism.

Figure 4.1 shows some pairs of isomers. Following the usual convention in chemical structure drawing, in this and all other figures in the book, a heavily shaded wedge shaped bond indicates a bond which projects forward out of the plane of the paper whereas a dotted wedge shaped bond represents a bond which projects behind the plane of the paper. A wiggle indicates a bond which could be in either configuration. In those structures where stereochemistry is indicated, a normal straight solid line is used to depict a bond lying in the plane of the paper. If all of the bonds in a structure are drawn in this way, then no indication is being given about stereoisomers.

4.1.1 Structural Isomerism

Structural isomers differ in the way the carbon atoms are bonded to each other. Thus myrcene and α-pinene are structural isomers in that the former is an open chain compound whereas the latter contains a bicyclic ring system.

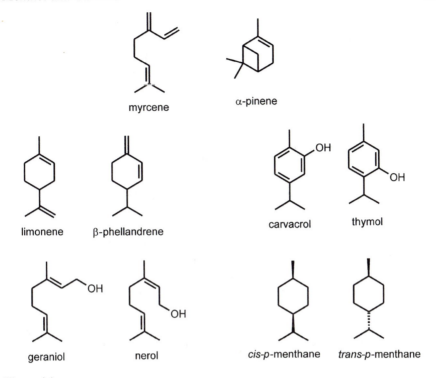

Figure 4.1

4.1.2 Positional Isomerism

Positional isomers are those where the difference lies in the position of a substituent or functional group. Thus, limonene and β-phellandrene are positional isomers because of the different locations of their double bonds. Similarly, carvacrol and thymol differ in the position of the phenolic function in the ring.

4.1.3 Geometrical Isomerism

Geometrical isomers are those in which the distinction lies in the relative placement of substituents across a double bond or ring. The original nomenclature for describing such isomers used the prefix *cis-* to refer to an isomer in which two substituents lie on the same side of a double bond and *trans-* for those in which the substituents are on opposite sides. Therefore, geraniol is considered to be a *trans-*isomer because the main terpenoid skeleton runs across the double bond next to the alcohol function. Nerol, in which both of the chain residues are on the same side of the double bond is, correspondingly, the *cis-*isomer. Exactly the same applies to rings and the two geometrical isomers of *p*-menthane shown in Figure 4.1 as an example of this.

This *cis-*/*trans-*system of nomenclature leads to some difficulties. For instance, in the case of geraniol/nerol, why did we choose the alkenyl residue on carbon 3 rather than the methyl group for determining which isomer is *cis-* and which is *trans-*? The problem would have become even more difficult if one of the substituents had been an atom other than carbon or hydrogen. For example, if one group were an ethyl group and the other a methyl ether, which would have priority when it came to deciding between *cis-* and *trans-*?

In order to surmount these difficulties, an alternative system was developed using the German words *zusammen* and *entgegen* which mean together and opposite, respectively. These are usually shortened to *E-* and *Z-*. In this system, the priority of groups around the double bond is determined by their atomic numbers. For example, let us take the case of geraniol. There are two substituents on the olefinic carbon nearest to the alcohol function. The atoms attached to this carbon atom are a hydrogen atom and another carbon atom. Hydrogen has an atomic number of 1 whereas carbon has a number of 6 and thus takes priority. At the other end of the double bond, the carbon atom is connected to two other carbon atoms and so to determine which has priority, we must go to the atoms attached to each of these and consider their atomic numbers. Thus, the carbon of the methyl group is attached to three hydrogens, each with an atomic number of 1. The other carbon is attached to one carbon (atomic number 6) and two hydrogens (each with atomic number 1). This latter group therefore takes priority. Thus we have defined the higher priority group at each end of the double bond and, since these lie on opposite sides of the double bond, geraniol is designated the *E-*configuration. Similarly, nerol is designated *Z-*. If by moving two atoms away from the double bond does not distinguish between the substituents, then we must proceed along the chain until a distinction does appear. Obviously, the progress along the chain will follow the highest priority group at each step. A detailed account of the rules can be found in the book by Eliel and Wilen (see Bibliography).

4.1.4 Conformational Isomerism

Conformational isomers differ only in the degree of rotation around carbon–carbon bonds. If the substituents at either end of a carbon–carbon bond are in line with each other, they are said to be eclipsed. If they are configured so that each substituent is placed centrally opposite to two of the substituents on the other carbon atom, then they are said to be staggered. The staggered form is the thermodynamically more stable

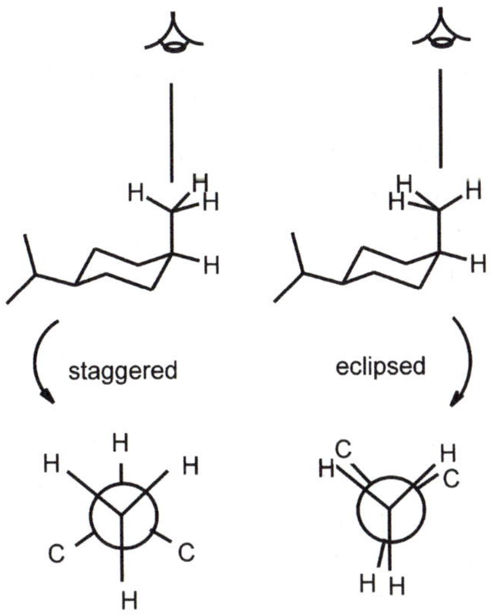

Figure 4.2

of the two. This type of isomerism is illustrated in Figure 4.2, using *cis-p* - menthane as an example. If we look at the methyl group attached to the cyclohexane ring, we can see the two configurations. In the isomer on the left, one of the methyl hydrogens is placed over the ring and is equidistant between the two ring carbons adjacent to it. The other two methyl hydrogens are placed symmetrically either side of the hydrogen on carbon 1 of the ring. This is the staggered configuration. In the isomer on the right, the methyl carbon has been rotated through 60° relative to that of the molecule on the left. This brings the three hydrogen atoms in line with the atoms opposite them on the next carbon atom (*i.e.* the two ring carbons and the hydrogen attached to the ring). This is the eclipsed configuration. Another way of depicting this is through the use of Newman projections and these are shown below the structural diagrams in Figure 4.2. In Newman projections, we view the bond in question from one end. The circle represents the carbon atom at the end nearest to the viewer, represented by an eye in Figure 4.2. The substituents attached to this carbon are shown attached to lines which run into the centre of the circle, depicting the bonds to that carbon atom. The substituents attached to the second carbon are shown as attached by lines which stop at the edge of the circle, depicting bonds whose points of attachment to the second carbon atom are obscured by the nearer, first carbon. The Newman projection in Figure 4.2 therefore represents the view from above the ring, looking down along the bond between the methyl carbon and the ring carbon to which it is attached.

The principles discussed in the paragraph above and depicted in Figure 4.2 refer to open chain compounds where one bond can be rotated independently of all the others. In cyclic compounds, rotation around one bond produces effects all around the ring since the atoms are all connected together and cannot be rotated independently. Usually, it is necessary to rotate at least two bonds simultaneously and the resultant change in shape can have a major impact on the ring and the properties of groups attached to it. Conformational isomers in a cyclohexane ring are illustrated in Figure 4.3 using *cis-p*-menthane as an example. On the left of the figure, we see the ring in what is referred to as the chair conformation, because its shape resembles that of a chair. In this conformation, all of the carbon–carbon bonds of the ring are staggered and thus it represents the lowest energy, hence thermodynamically most favoured configuration of the ring system. Substituents on the chair form of the ring can be in one of two configurations, axial or equatorial. The axial substituents are those which stand either above or below the ring, the bond connecting them to the ring being at right angles to the central plane of the ring. Equatorial substituents lie in the plane of the ring. Inspection of models will reveal that the equatorial position is less hindered. The axial substituents on each side of the ring tend to interfere sterically with each other. In the case of *cis-p*-menthane, one of the substituents is obliged to adopt the axial position if the ring is in the chair form. Since this is a steric effect, the more favourable form is the one in which the smaller of the two substituents, in this case the methyl group, adopts the axial position and the larger, the iso-propyl group in this instance, the equatorial. The other relatively low energy configuration of cyclohexane rings is referred to as the boat configuration, because of the similarity in shape to that of a boat. This is shown on the right of Figure 4.3. In this case, two of the carbon–carbon bonds in the ring are eclipsed, the others being staggered. It is therefore less favourable energetically than the chair form. However, in the case of *cis-p*-menthane, the two substituents can now adopt a configuration known as the bowspit. This resembles the equatorial position in the chair form and thus the adverse

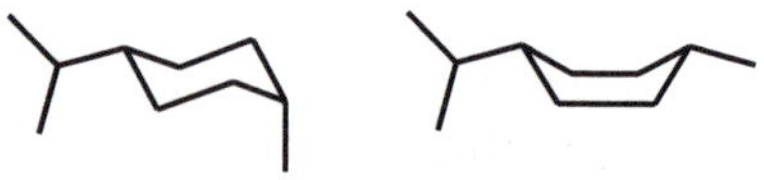

cis-*p*-menthane   cis-*p*-menthane
chair form   boat form

Figure 4.3

trans-annular interactions of an axial group are avoided. The loss in energy on going from the chair to the boat conformation in the ring is therefore compensated for by the gain in energy on reducing *trans*-annular buttressing interactions.

4.1.5 Stereoisomerism

Stereoisomers, also known as enantiomers, differ in that they are not superimposable on their mirror images. Any carbon atom which is bonded to four different substituents will display this property as shown in Figure 4.4. If we imagine the rectangle to be a mirror, we can see that molecule X is indeed a reflection, or mirror image, of molecule Y. If we lift X to the same side of the mirror as Y simultaneously turning it through 180° so that substituents A and B are aligned with the corresponding substituents in Y, we have molecule Z. It is clear that Z cannot be superimposed on Y since in Y, it is substituent C which projects forwards from the plane of the page, whereas in Z, it is D which does so. A few minutes' play with a set of molecular models will soon convince any doubtful reader of the fact that this is so. Such carbon atoms are referred to as asymmetric carbon atoms, chiral centres, centres of chirality or centres of asymmetry. The word chiral is derived from the Greek word for a hand since hands are examples of objects which are not superimposable on their mirror images. In fact, one often hears or reads about the handedness of molecules. A mixture containing equal amounts of opposite enantiomers is called a racemic mixture. The enantiomeric purity of a material is expressed as the enantiomeric excess, ee, which is defined as the excess of the predominant form over the total material present. For example if, in a mixture of 100 molecules, there are 60 with

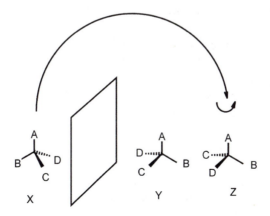

Figure 4.4

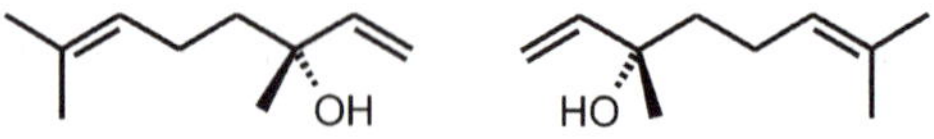

l -linalool   *d*-linalool

R-linalool   *S*-linalool

Figure 4.5

the *R*-configuration, then there must be 40 *S*-isomers. Hence, there are 20 more *R*-isomers than *S*- and so the ee is *R*-/total $= 20/100 = 20\%$.

A typical pair of enantiomers, *d*- and *l*-linalool is shown in Figure 4.5. Like all pairs of enantiomers, these two molecules are identical in their physical and chemical properties. They have the same boiling point and refractive index, they will react in the same way with, for example, acetyl chloride or pyridinium chlorochromate. If subjected to gas liquid chromatography, they will have identical retention times and a mixture of the two isomers will produce a single peak at exactly the same retention time as either pure enantiomer. They will only show different behaviour if they are brought into contact with another asymmetric molecule or if they are exposed to polarised light. One isomer, in this case *d*-linalool, will rotate the plane of the polarised light to the right and *l*-linalool to the left. Because of this ability to rotate the plane of polarised light, the molecule is said to be optically active and stereoisomerism is sometimes referred to as optical isomerism. The prefixes *d*- and *l*- are derived from the Latin words *dexter* (right) and *laevus* (left) in reference to this phenomenon. This nomenclature enables us to label each of the isomers according to this measurable physical property but it does not enable us to tell which way the substituents are actually arranged around the asymmetric carbon. Similarly, knowing the stereochemistry of the molecule will not enable us to predict the direction of rotation of the plane of polarised light since that will also depend on the nature of each of the substituents. The signs + and − are also used to indicate optical activity of materials. The + sign has the same meaning as the letter *d*, *i.e.* something which rotates the plane of polarised light to the right. The − sign has the same meaning as the letter *l*, *i.e.* something which rotates the plane of polarised light to the left.

In order to be able to define the absolute stereochemistry of a molecule, we therefore use another system of nomenclature which was proposed by Cahn, Ingold and Prelog in 1956. This system uses the prefixes *R*- and *S*-, derived from the Latin words rectus (right) and sinister (left) and the same system of priority setting that was discussed above in relation to *E*- and *Z*-isomers. We start by assigning the priorities of the four groups

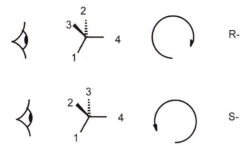

Figure 4.6

attached to the asymmetric carbon atom. Having done this, we then view the molecule looking along the bond between the carbon atom and the lowest priority substituent. We then identify the highest priority substituent and move from it, *via* the second priority to the third. If this movement is clockwise, then the molecule is assigned the *R*-configuration and, if anti-clockwise, the *S*-. This procedure is illustrated in Figure 4.6. Here, the group labelled 1 has the highest priority and 4, the lowest. Both enantiomers are then viewed from behind the carbon with the lowest priority group pointing away from the observer. Following the sequence 1-2-3 on the upper enantiomer involves clockwise movement, and hence it has the R-configuration. Similarly, the movement from 1-2-3 in the lower case is anti-clockwise and hence S-. Applying the rule to the enantiomers of linalool shown in Figure 4.5, we see that *l*-linalool has the *R*-configuration and *d*-linalool, the *S*-. In modern nomenclature, it is usual to name them as (−)-(*R*)-linalool and (+)-(*S*)-linalool, the signs − and + replacing the letters *d* and *l* respectively, to indicate the direction of rotation of plane polarised light.

It is, of course, possible to have more than one centre of asymmetry in a molecule. Each centre can have either the *R*- or *S*-configuration and so with two centres of asymmetry, there will be four possible stereoisomers; *R–R, R–S, S–R* and *S–S*. Such isomers are known as diastereoisomers, or diastereomers. Similarly, if a third asymmetric carbon is introduced, the number of possible isomers doubles again, this time to eight. Simple mathematics reveals that, if there are *n* asymmetric carbon atoms in a molecule, then there are 2^n stereoisomers.

In general, these will exist in pairs since each diastereomer will have another one which is its mirror image. Figure 4.7 illustrates this for a molecule with two contiguous asymmetric centres. The "mirrors" in the figure help to show that the two isomers on the upper line are mirror images of each other and the two on the lower are mirror images of each other. The two isomers on the top line are neither mirror images of, nor

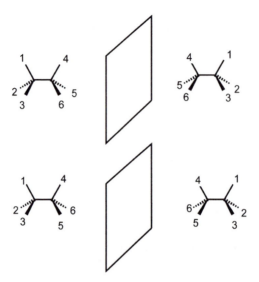

Figure 4.7

superimposable on, either of the isomers on the lower line. In each diastereomeric pair, the enantiomers will possess identical physical and chemical properties and will not be resolved without the help of a chiral agent. However, the two pairs will differ from each other. For example, a compound with two chiral centres will exist as four diastereomers which will resolve into two peaks on a gas chromatogram. Similarly, the two pairs of enantiomers are separable by distillation.

This provides the chemist with a means for separating, or resolving, enantiomers. If a compound exists as a pair of enantiomers and the mixture is reacted with an enantiomerically pure material, the products will be diastereomers and can therefore be separated, for example, by crystallisation or distillation. Reversal of the reaction then produces enantiomerically pure forms of the original material. For example, if we have a terpenoid alcohol as a racemic mixture and wish to obtain the two pure enantiomers, we could esterify it with a single enantiomer of an optically active acid, such as lactic acid. The ester mixture will then contain diastereomers which can be separated. After hydrolysis of the separated esters and recovery of the alcohols, we will have each alcohol in enantiomerically pure form. Such a process is referred to as resolution. Treatment of a mixture of enantiomers with a chiral reagent or catalyst may result in a reaction of one enantiomer only. For example, treating a racemic acetate ester of a terpenoid alcohol with a chiral lipase enzyme might result in the hydrolysis of only one enantiomer. The alcohol thus produced will be optically pure and separable from the unreacted ester. This ester can then be hydrolysed in a simple chemical reaction. The two enantiomers have therefore been resolved. Such a process is known as a kinetic resolution.

Figure 4.8 shows an interesting phenomenon which occurs when each asymmetric carbon in a material with two optical centres contains the same substituents. The two isomers on the top line are enantiomers of each other as before. However, when we construct the other two possible isomers (centre line), we find that each has an internal plane of symmetry. This is shown on the bottom line where a "mirror" is placed across the bond joining the two asymmetric carbon atoms to show that each end of the molecule is a reflection of the opposite one. (In the figure, the central bond has been elongated for clarity of illustration.) If we rotate one of the isomers through 180° as shown in the figure, we find that the molecules are, in fact identical. Such isomers are referred to as *meso*-isomers. Because they are identical and internally compensating, such isomers are not optically active.

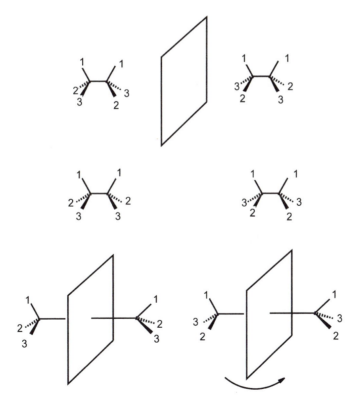

Figure 4.8

For a more detailed review of stereochemistry, Eliel's classical textbook[4.1] is highly recommendable and there is also a useful chapter in Vogel.[4.2]

4.2 MINT COMPONENTS

Having reviewed the principles of stereochemistry, we are now in a position to investigate the chemistry of mint. There are many species and sub-species of mint and their chemistry is dominated by monocyclic mono-terpenoids, mostly alcohols and ketones. Some of the more important ones are shown in Figure 4.9. By far the most important of these are carvone and menthol, both of which will be discussed in more detail later.

The oxygen atom of these materials is found either at the 2- or the 3-position of the *p*-menthane ring. Those with the oxygen atom located on the 2-position are related to carvone and mostly take their names from carvone. Carvone itself takes its name from caraway (*Carum carvi*) the oil of which contains up to 85% of the *d*-isomer of carvone. The same isomer also occurs at levels up to 65% in the oil of dill (*Anethum graveolens*) and is responsible for much of the characteristic flavour of dill pickles. Spearmint oil (*Mentha spicata*) contains up to 75% of *l*-carvone. Spearmint oil also contains about 5% of dihydrocarvone. The alcohol corresponding to carvone is known as carveol and is found in *Buddleia asiatica* and gingergrass (*Cymbopogon martini*) as well as in various mint species. Dihydrocarveol is found in the oils of various mints, caraway and rosemary. The phenol, carvacrol, is a major component of oregano (*Origanum vulgare,* up to 85%), thyme (*Thymus vulgaris*, up to 70%), summer savoury (*Satureja hortensis*, about 50%) and winter savoury (*Satureja montana*, about 50%). Carvacrol is actually a double bond isomer of carvone. If the double bond of the iso-propenyl group of carvone is brought into the ring by acid, base or metal catalysed migration, then the ketone function can enolise to produce carvacrol. Typical aromatic resonance energy is 27 kcal mol^{-1} and so this isomerisation is very favourable thermodynamically. This can present problems when handling or distilling carvone. Unless care is taken to eliminate potential isomerisation catalysts, the purity of the product can suffer. Even more seriously, isomerisation of the bulk of the material could lead to an uncontrollable exothermic reaction with disastrous consequences.

Menthol is 3-hydroxy-*p*-menthane. It occurs widely in mint species and the richest source is cornmint (*Mentha arvensis*, up to 85%). The three unsaturated analogues shown in Figure 4.9 occur in many mint oils. Pulegol also occurs in *Eucalyptus citriodora* and rose oils, isopulegol can account for up to 10% of the oil of *E. citriodora* and piperitol occurs

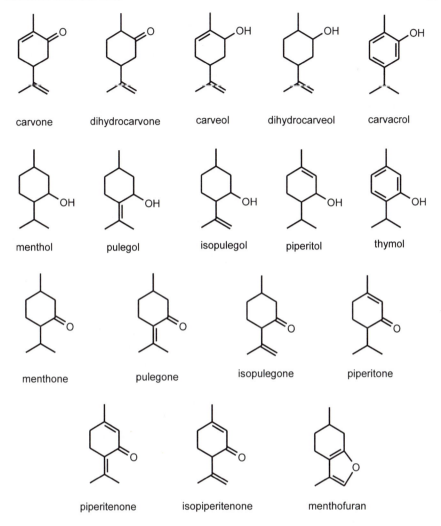

carvone   dihydrocarvone   carveol   dihydrocarveol   carvacrol

menthol   pulegol   isopulegol   piperitol   thymol

menthone   pulegone   isopulegone   piperitone

piperitenone   isopiperitenone   menthofuran

Figure 4.9

in marjoram. The corresponding phenol, thymol, is found in thyme, from which it takes its name, and oregano. It can account for up to 70% of these oils. The balance between thymol and carvacrol in thyme and oregano will depend on the exact sub-species.

The ketones derived from menthol, like the alcohols, all have intense minty odours. All of them occur widely in mints. The levels (typically 25–50%) at which menthone occurs are usually higher than those of the other ketones. Menthone also occurs at low levels in geranium. Pulegone is the major component of Pennyroyal Oil (*Mentha pulegium*) from which it takes its name, occurring at up to 75%. The richest source of isopulegone is Buchu Leaf Oil (*Agathosma crenulata*, 2.5%). Piperitone is found in pennyroyal and lemon balm (*Melissa officinalis*). Piperitenone also occurs in pennyroyal and both it and isopiperitenone are found in pistachio fruits (*Pistacia lentiscus*). Menthofuran, in which the oxygen atom is bonded to both the 3-carbon of the ring and to the 9-carbon, is found in most mint species. We will now look in more detail at the chemistry of the two most important mint compounds, carvone and menthol.

4.3 CARVONE

Carvone contains one asymmetric carbon atom, at the 4-position of the ring *i.e.* the carbon which carries the iso-propenyl group. It therefore exists as two enantiomers as shown in Figure 4.10. As mentioned above, the *d*-isomer is a key component of the flavours of dill and caraway whilst the *l*-isomer is the key component of spearmint. These flavours are of considerable commercial importance and so considerable attention has been given to the synthesis of carvone. Nowadays, most of the dill and caraway flavour tonnage is derived from the natural sources and so the attention of the synthetic chemist has been directed more to *l*-carvone. The synthesis outlined in Figure 4.11 was developed in the nineteenth century and is still the most efficient process in large-scale commercial use today. Quest International is the world's largest producer of carvone.

There are a number of interesting features of this synthesis which are worth discussing. The crucial feature of the synthesis is the use of a chiral starting material, *d*-limonene. This is the major component of orange peel oil and so is in plentiful supply as a by-product of the citrus fruit juice industry. When chirality is important in a large scale synthesis, it is always helpful if nature provides a good, inexpensive chiral feedstock which can be used as a starting point. In the case of the conversion of limonene to carvone, all the chemistry occurs at the top end of the ring and so the chiral centre is safely out of the way at the bottom and will not be affected. However, this is only the case if the regiochemistry of the

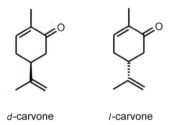

d-carvone   l-carvone

Figure 4.10

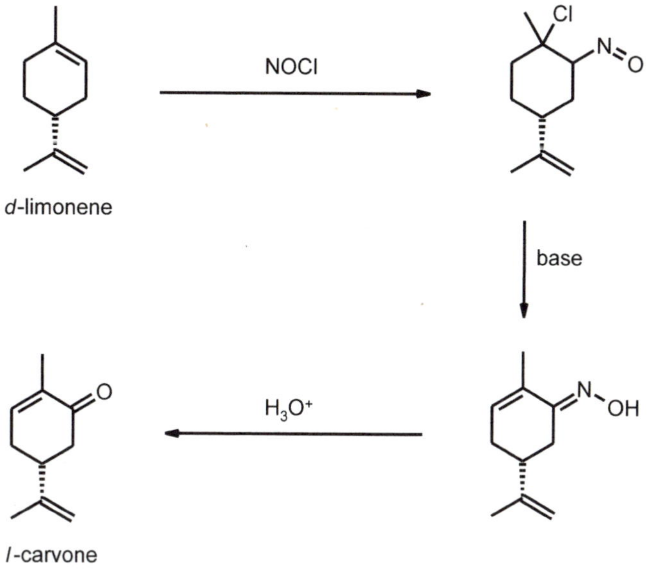

d-limonene

l-carvone

Figure 4.11

reactions is tightly controlled. Any process which was not regioselective, could result in racemisation. For example, the autoxidation of limonene might be seen as a potential way of introducing an oxygen function. The first step in this process would be the removal of an allylic hydrogen atom. Even if this were regioselective, the intermediate radical could pick up oxygen at either side and, as can be seen from Figure 4.12, the chiral purity of the material would be lost. This is, of course, because the chirality of the 4-carbon stems from the asymmetry of the ring. Since the allylic radical is symmetrical, the 4-carbon loses its chirality.

The advantage of using nitrosyl chloride to introduce hetero-atoms into limonene, is that the reaction is highly regioselective and, once the carbon–nitrogen bond has formed, the asymmetry of the ring and hence the chirality of the 4-carbon is assured. Nitrosyl chloride is a toxic, irritant and corrosive gas and so its use on a large scale would require serious safety and handling precautions. It is therefore much more convenient to generate an equivalent *in situ*. This is done by treating a nitrite ester with hydrochloric acid. The nitrosyl cation which is produced is trapped by the double bond of the limonene. It will be noticed that there are two elements of selectivity in this reaction. Firstly, only the more electron rich double bond (the trisubstituted one in the ring as opposed to the disubstituted one in the tail) reacts. Secondly, this double bond reacts only in one direction. The nitrosyl cation adds to the

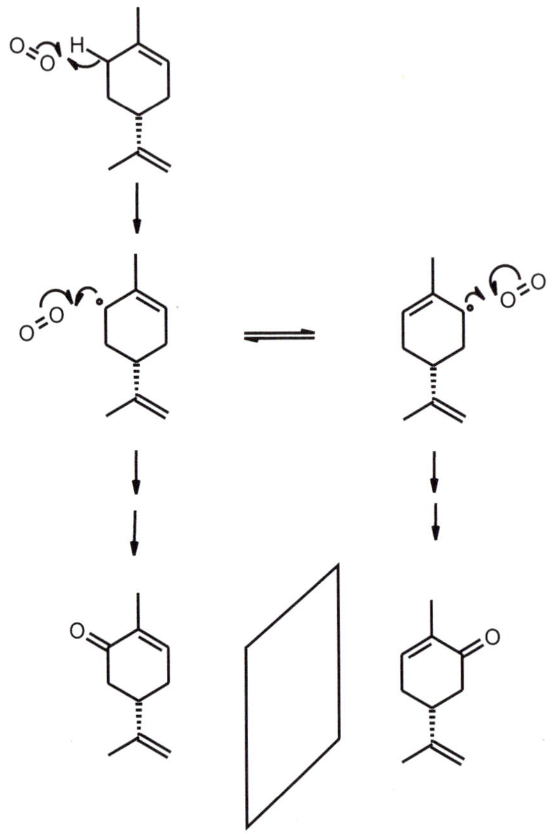

Figure 4.12

less substituted end so that a tertiary, rather than a secondary, carbocation will be produced. This carbocation is then intercepted by a chloride anion from the hydrochloric acid, to give the observed product. This sequence of events is shown in Figure 4.13.

The next interesting feature of this reaction is that the nitrosyl chloride adduct exists in equilibrium with its dimer as shown in Figure 4.14. The monomer is a blue liquid, the dimer is a white solid. Any attempt to separate the two results in re-establishment of the equilibrium. This makes the reaction mixture pretty to look at but difficult to analyse.

The next stage involves the elimination of hydrogen chloride from the adduct and rearrangement of the nitrosyl function to an oxime. The two basic conversions are shown in Figure 4.15. At first sight, this might look straightforward. However, although the mechanism is not fully under-stood, there are a number of factors which argue against the simplistic

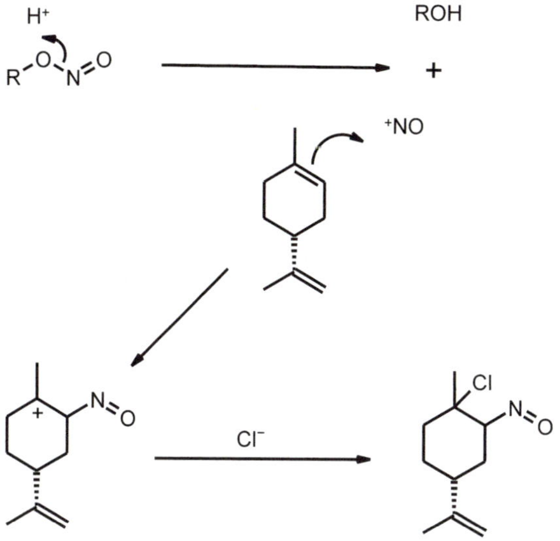

Figure 4.13

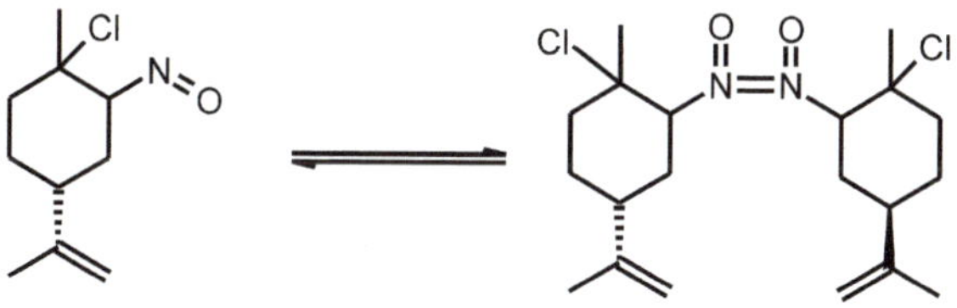

Figure 4.14

scheme of Figure 4.15 and it is fairly certain that these reactions do not occur as shown in the figure.

The base most commonly used for the elimination is urea. This is a very weak base, certainly not strong enough to remove the proton next to chlorine. Moreover, in similar systems, bases remove a proton from the methyl group rather than the methylene group in the ring. This is exactly as one would expect on electronic and steric grounds. For example, as shown in Figure 4.16, treatment of limonene oxide with a strong base,

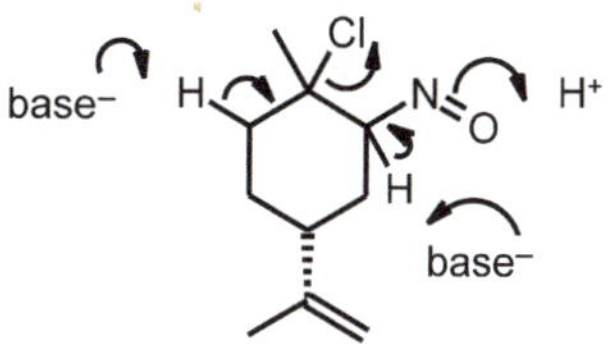

Figure 4.15

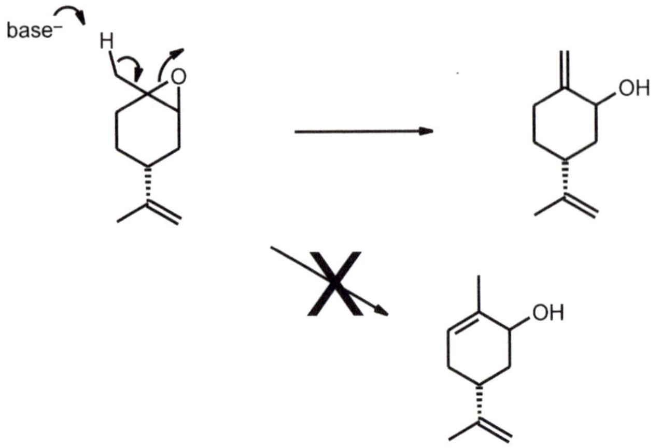

Figure 4.16

such as lithium diethylamide, produces the exocyclic rather than the endocyclic allylic alcohol. The exocyclic olefin is, obviously, the product of removal of a methyl proton. The rearrangement of the limonene-nitrosyl chloride adduct is the only reaction which shows the opposite selectivity.

The most acidic proton in the nitrosyl chloride adduct is the one next to the N=O group. If this were to be removed first, one might expect elimination of the chloride to occur at that point as shown in Figure 4.17. Both the methyl and methylene groups next to the new double bond would now be activated by the N=O group and the *trans-anti*-periplanar rule would guide the elimination in favour of the methylene group. However, there is no sound physical evidence to support this mechanism. Indeed, it is not even known whether the elimination/rearrangement occur from the monomer or the dimer. So, for the present, the mechanism remains a mystery.

The final stage of the synthesis of carvone is the most straightforward. The oxime can be converted to carvone either by hydrolysis using dilute aqueous acid or by acid catalysed *trans*-oximation with another ketone.

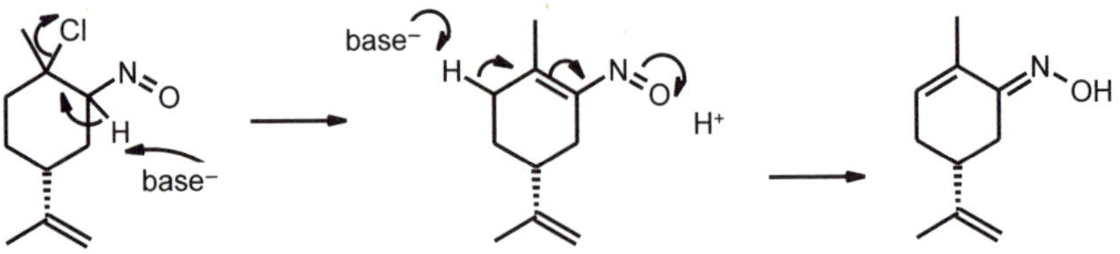

Figure 4.17

4.4 MENTHOL

In stereochemical terms, menthol is more complex than carvone. Menthol has three asymmetric centres, *viz* carbons 1, 3 and 4 of the ring. With three centres, this means that there are 2^3 (*i.e.* 8) individual isomers. Each of the isomers will have a complementary one with which it is enantiomeric. Thus, menthol exists as four pairs of diastereomers. Each diastereomeric pair will have slightly different physical and chemical properties and may therefore be separated from the other three by simple physical methods such as distillation, crystallisation and chromatography. The eight isomers of menthol are tabulated in Figure 4.18. It should be observed that the word menthol has three separate meanings. Firstly, it is used generically to refer to all of the 3-hydroxy-*p*-menthanes. Secondly, it is used to mean specifically the thermodynamically most favourable pair of diastereomers in which all of the

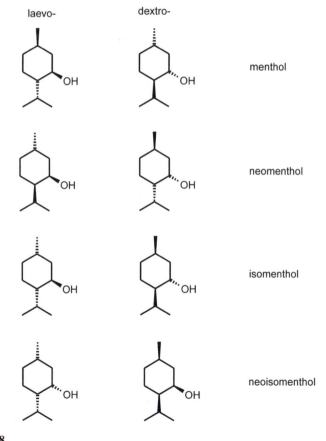

Figure 4.18

substituents on the cyclohexane ring can be equatorial simultaneously. Thirdly, it is often used to refer to the most important single isomer, that is, *l*-menthol. The meaning intended can usually be inferred from the context. The column on the left shows the laevorotatory or *l*-isomers and the column on the right, the dextrorotatory or *d*-isomers. The top row shows the menthol structure, the next row, neomenthol, the third, isomenthol and the bottom row, neoisomenthol.

All eight isomers possess characteristic minty odours. The reason for the importance of *l*-menthol is its ability to interact, not only with odour receptors, but also with the receptors which sense cold. The presence of menthol will induce cold receptors to respond as if they had sensed a drop in temperature. This physiologically induced sensation of cold is used in many products from foodstuffs, such as confectionery, chewing gum, through oral care products such as toothpaste to cosmetic preparations such as shaving products. There is therefore a large market for menthol and, since not all of it can be met from natural mint oils, a demand for synthetic material. Consequently, there has been a lot of work on the synthesis of menthol and all of the synthetic routes considered must take stereochemistry into account as *l*-menthol is always the preferred target.

There are two synthetic routes to menthol which are of major commercial importance and these will be discussed below. These two are the most efficient syntheses of menthol at present and therefore account for most of the synthetic menthol produced today. Three other syntheses will also be discussed. These are interesting in that each starts from a chiral precursor which is readily available in the country of production. Therefore, while they were in operation, these processes demonstrated how local factors, often assisted by import restrictions, could affect the total economic balance.

Figure 4.19 shows the synthesis used by the German company Haarmann and Reimer. This company is named after its two founders. The Reimer concerned is the same person who also gave his name to the Reimer–Tiemann reaction. It is not surprising therefore, that this synthesis starts with a phenolic compound. Alkylation of *m*-cresol with propylene in the presence of an aluminium catalyst, produces thymol. Hydrogenation of thymol gives a mixture of all eight isomers. Since the menthol (as opposed to isomenthol, neomenthol and neoisomenthol) isomers are the thermodynamically most favourable, they predominate in the product. The balance between the pairs of diastereomers in the crude product is 62–64% menthol, 18–20% neomenthol, 10–12% isomenthol and 1–2% neoisomenthol. The advantage of this particular route to menthol lies in the fact that the starting materials are

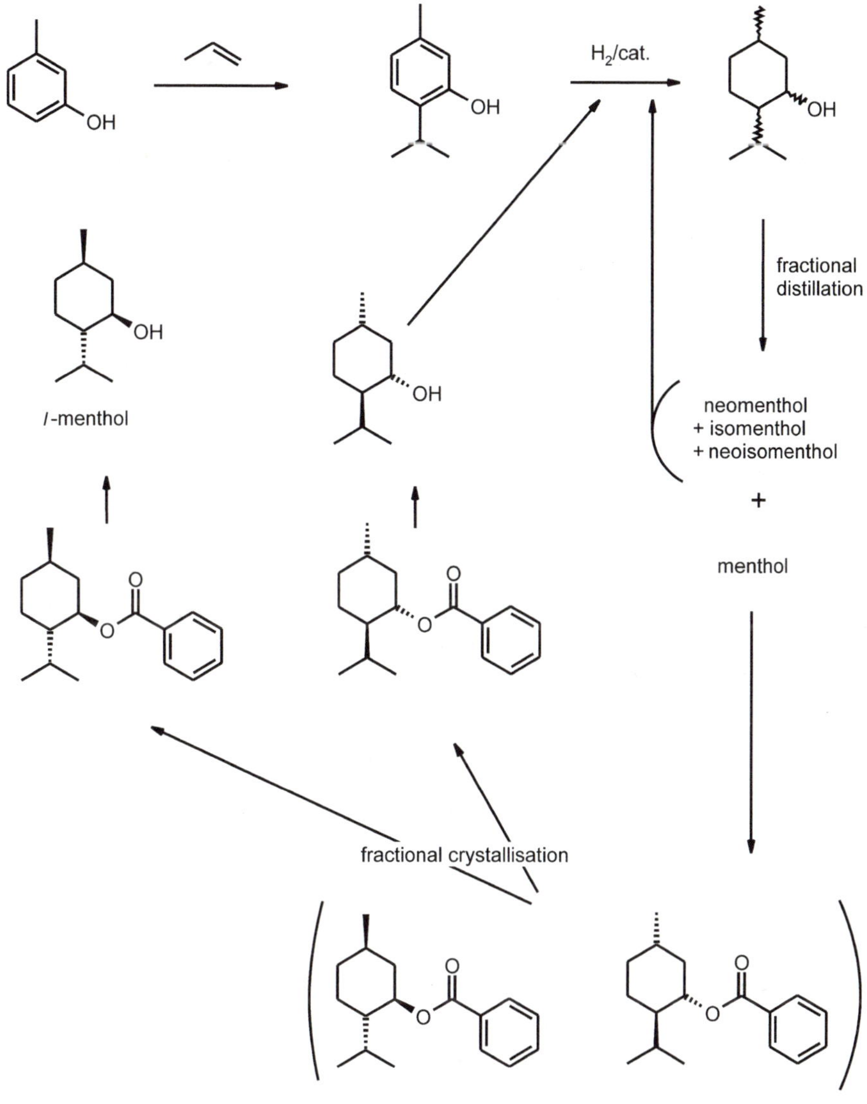

Figure 4.19

inexpensive and readily available and that the basic chemistry is inexpensive. However, this is offset by the fact that it has relatively poor stereoselectivity (68–69% of the bulk is the "wrong" product, *i.e.* not *l*-menthol) and a major task of separation of isomers must now be undertaken.

Unlike enantiomers, diastereoisomers have different physical properties, including boiling points. At atmospheric pressure, *d,l*-menthol boils at 216.5 °C, *d,l*-neomenthol at 212 °C, *d,l*-isomenthol at 218 °C and *d,l*-neoisomenthol at 214.6 °C. The four pairs of diastereomers can, therefore, be separated by fractional distillation through a high efficiency column. The isomenthol, neomenthol and neoisomenthol pairs are then recycled to hydrogenation where they are re-equilibrated under the hydrogenation conditions. The *d,l*-menthol is then converted to its benzoate ester and resolved by fractional crystallisation. Having separated the benzoate esters, they can be hydrolysed separately to produce pure *l*- and *d*-menthol. The unwanted *d*-menthol can be recycled to the hydrogenation reaction where it, like the neomenthol, isomenthol and neoisomenthol isomers, can be re-equilibrated. None of the material is lost in the process but there are costs in terms of the labour, time, plant capacity, and energy consumed in operating all of the separation and recycle processes.

The other menthol synthesis which is of major commercial importance is the one used by the Japanese company Takasago. The synthesis was devised by Professor Noyori of Nagoya University and is part of the work for which he was awarded the Nobel Prize for chemistry in 2002. The basic scheme is shown in Figure 4.20.

The synthesis starts with the pyrolysis of β-pinene to produce myrcene. Subsequent addition of the diethylamide anion to myrcene produces *N,N*-diethylgeranylamine. Only a catalytic amount of the amide is required since the carbanion formed by the addition of amide to myrcene, is protonated by the diethylamine solvent, thus generating another amide anion. The *N,N*-diethylgeranylamine is then isomerised to the *N,N*-diethyl enamine of citronellal. Hydrolysis of the enamine gives citronellal which can be cyclised *via* an ene reaction, catalysed by zinc chloride, to isopulegol. Finally, hydrogenation of the isopulegol produces menthol. The elegance of this synthesis lies in the fact that, in both reactions where new chiral centres are formed, there is tight control of the stereochemistry and an enantiomerically pure product is produced. We will look at both of these reactions in a little more detail below.

The key step of the synthesis is the isomerisation of the allylic amine to the corresponding enamine. This is a simple reaction in which a hydrogen atom is moved from the first carbon to the third of the geranyl skeleton. The ingenious feature of this particular case is that the hydrogen atom is added to that third carbon atom from one face only and therefore results in the formation of a single enantiomer of the enamine. When this is hydrolysed, the citronellal produced, is entirely in the dextrorotatory form. The catalyst which performs this transformation is a complex of the Group VIII metal, rhodium. The catalyst which

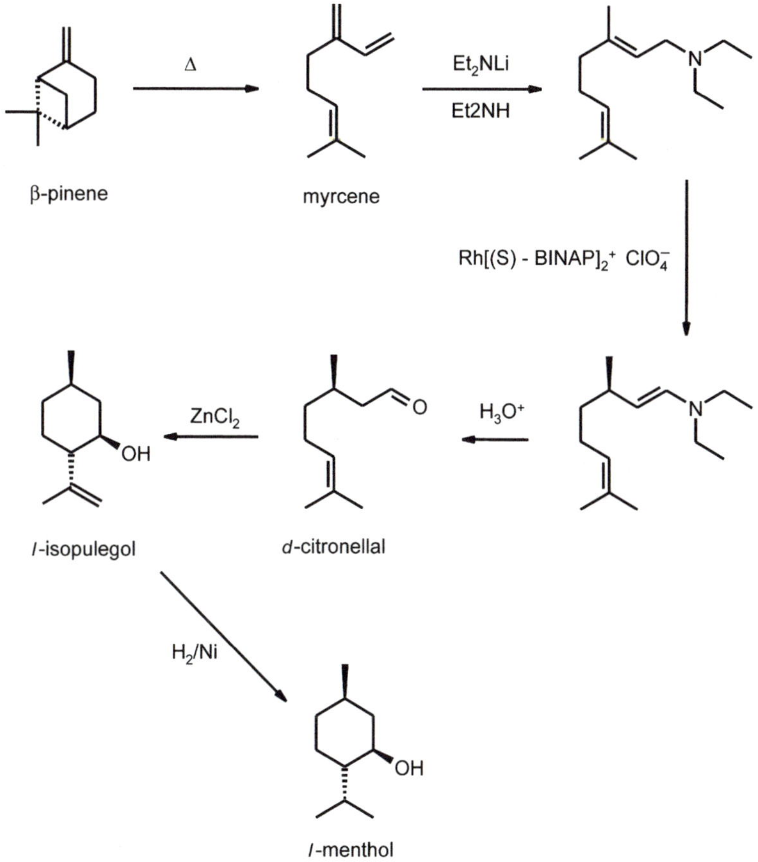

Figure 4.20

imparts chirality to the reaction, is known as **BINAP** or rhodium BINAP. The structure of this catalyst is shown in Figure 4.21. The ligand which is responsible for the chirality of the complex consists of two naphthalene rings joined by their 1-positions. Each of them also carry a phosphorus atom on the 2-position and each phosphorus also carries two benzene rings. The rhodium ion is chelated by the two phosphorus atoms and also to a molecule of cyclooctadiene. The positive charge on the rhodium ion is counterbalanced by a perchlorate anion. In theory, the single bond joining the two naphthalene rings, is free to rotate. However, in practice it is not, since there is insufficient space and flexibility in the molecule to allow the hydrogen atoms on the 8-positions of the two naphthalene rings to pass over either each other or the phosphorus atom attached to the opposite ring. The BINAP molecule therefore exists as two stereoisomeric forms, as shown in Figure 4.21.

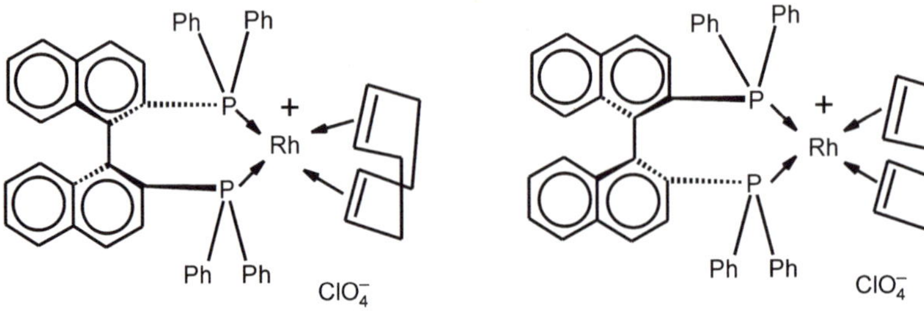

S (-) BINAP Rh COD perchlorate   R (+) BINAP Rh COD perchlorate

Figure 4.21

In the figure, the (*S*)- isomer is shown with the upper phosphorus atom behind the plane of the paper and the lower one in front of it. These are reversed in the (*R*) + form and the two molecules are not superimposable on their mirror images. Construction of molecular models will clearly demonstrate both the restricted rotation around the bond between the two naphthalene rings and also the non-superimposability of the two enantiomers. One form adds the hydrogen atom to one face of the enamine and the other adds it to the opposite one. Thus, using the (*S*) + form of BINAP, only *d*-citronellal is produced.

The other two chiral centres of menthol are formed in the ene reaction when citronellal is cyclised to isopulegol in the presence of zinc chloride. The ene reaction is a 6-centre reaction. As the reaction starts, the six atoms involved come together in a ring and the shape of this ring determines the stereochemistry of the product. This is shown in Figure 4.22. The most stable configuration for the transition state is the one with the methyl, isopropyl and oxygen groups in which all will become equatorial conformations in the newly formed ring. The absolute stereochemistry of the carbon carrying the methyl group, therefore forces an absolute stereochemistry onto the other two centres as they are formed. Fortunately, this stereochemistry is the one which is required for l-menthol.

In the first of the above syntheses, all isomers were formed and the desired one obtained by separation from the others. In the second, we saw chirality being introduced into the molecule through the use of a

Figure 4.22

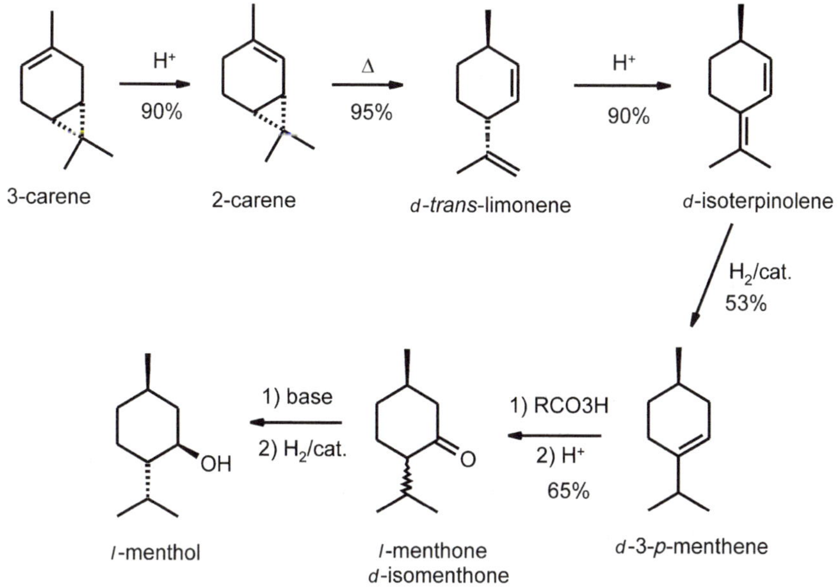

Figure 4.23

chiral catalyst. Another strategy available to the synthetic chemist is to start with a chiral molecule provided by nature. The next three syntheses use starting materials from the homochiral pool.

Indian turpentine is rich in 3-carene and this can be used as a starting material for synthesis of menthol. The overall scheme is shown in Figure 4.23.

Firstly, the 3-carene is isomerised to 2-carene by an acidic catalyst. The driving force for this reaction lies in the conjugation of the double bond with the cyclopropane ring in the product. The strain in the 3-membered ring is relieved through a reaction involving a 6-membered transition state, in which a hydrogen atom is transferred from a methyl group to C-3 of the ring. Obviously, this hydrogen must be delivered from the same side of the cyclohexane ring as the cyclopropane ring. This mechanism is shown in Figure 4.24. The methyl group therefore, ends up on the opposite side of the ring from the isopropenyl group. The absolute stereochemistry of both the methyl group and the isopropenyl group is therefore determined by the cyclopropane ring in the starting natural 3-carene.

Treatment of *d*-trans-limonene with an acid catalyst leads to isomerisation of the terminal bond into the position between the isopropyl residue and the ring. By doing this, the system gains free energy by generating a tetrasubstituted double bond and one which is in conjugation with the ring olefin. Further isomerisation would be

Figure 4.24

possible, but it is easy to stop at this point because the transoid arrangement of the double bonds is energetically favourable. Hydrogenation with a suitable poisoned catalyst, such as the Lindlar catalyst, results in 1,4-hydrogenation of the diene system to give *d*-3-*p*-menthane. Interestingly, chirality has now been lost at both the original optical centres of the molecule but the legacy of that original chirality survives in the carbon carrying the methyl group. Epoxidation and acid catalysed rearrangement of the epoxide produces a mixture of *l*-menthone and *d*-isomenthone. The lack of stereoselectivity in this conversion is not a serious issue as these two compounds can be interchanged via the enolate, as shown in Figure 4.25. Since menthone has both alkyl residues in the equatorial configuration, it is more thermodynamically favoured than isomenthone and thus equilibration with base leads to total conversion to menthone.

The final stage of the synthesis is hydrogenation of the ketone function. The major product of this hydrogenation is the all equatorial *l*-menthol but some of the opposite enantiomers at the 3-position are obtained. The overall yield of the desired isomer from the *l*-menthone/*d*-isomenthone mixture is only about 50%. A little simple arithmetic shows that the overall yield of this synthesis is only about 13% based on the starting 3-carene. This serves to illustrate two important factors in synthesis planning. Firstly, long syntheses with numerous steps will inevitably lead to low overall yields. Even if every yield is 90%, a six stage process will still have a yield of only 50% overall. Secondly, low yields are particularly

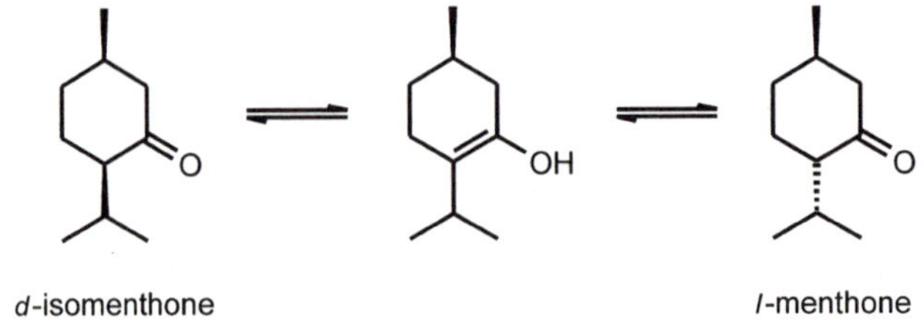

d-isomenthone   *l*-menthone

Figure 4.25

bad in the later stages of a synthesis since most of the material would have been carried through the previous stages only to be lost at the end. This synthesis does not compete with the first two described but it has been operated with commercial success by the Indian company, Camphor and Allied Products. Two factors made this possible; the first being the availability of inexpensive 3-carene in India and the second, and more important, high import tariffs on menthol from other countries.

Pennyroyal (*M. pulegium*) is grown commercially in Southern Europe and North Africa. Its essential oil contains about 75% *d*-pulegone. Hydrogenation of this gives a similar mixture of *l*-menthone and *d*-isomenthone to that produced in the synthesis of *l*-menthol from 3-carene. Using a superior reduction technique, dissolving lithium in ammonia, the Spanish company Destillaciones Bordas were able to produce *l*-menthol in 98% from this mixture. The basic nature of the ammonia used as solvent served to effect the isomerisation *in situ* before the reduction. However, dissolving metal reductions are relatively expensive to carry out and liquid ammonia is a relatively expensive solvent to handle. This synthesis, shown in Figure 4.26, is no longer economic.

The last synthesis of menthol which we will discuss, starts from *l*-piperitone which occurs in various species, especially peppermint. However, it is also a major component of *Eucalyptus dives* which made it an item of commerce in Australia. The Australian company, Keith Harris and Co, used to prepare *l*-menthol from this indigenous feedstock. The overall scheme is shown in Figure 4.27.

Firstly, the *l*-piperitone is reduced, by means of a reagent such as lithium aluminium hydride to a mixture of diastereomeric alcohols, *l-cis*-piperitol (36%) and *d-trans*-piperitol (64%). This mixture is reduced by a Raney nickel catalyst modified by addition of nickel (II) chloride. The hydrogenation is carried out in isopropanol as solvent at 25 °C and a hydrogen pressure of 60 psig. During hydrogenation, the stereochemistry of the isopropyl group controls the stereochemistry of the other centres.

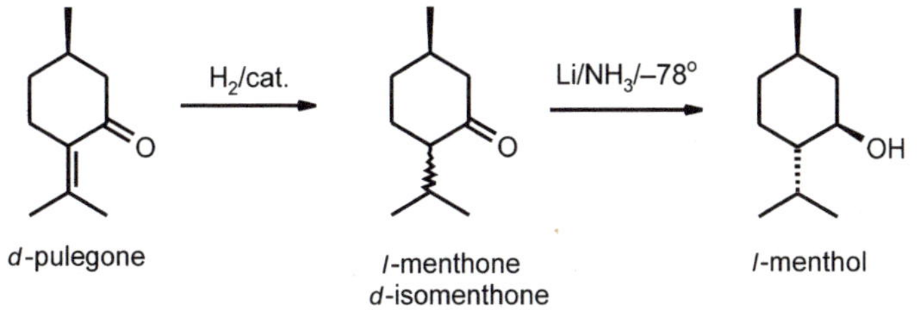

| *d*-pulegone | *l*-menthone *d*-isomenthone | *l*-menthol |

Figure 4.26

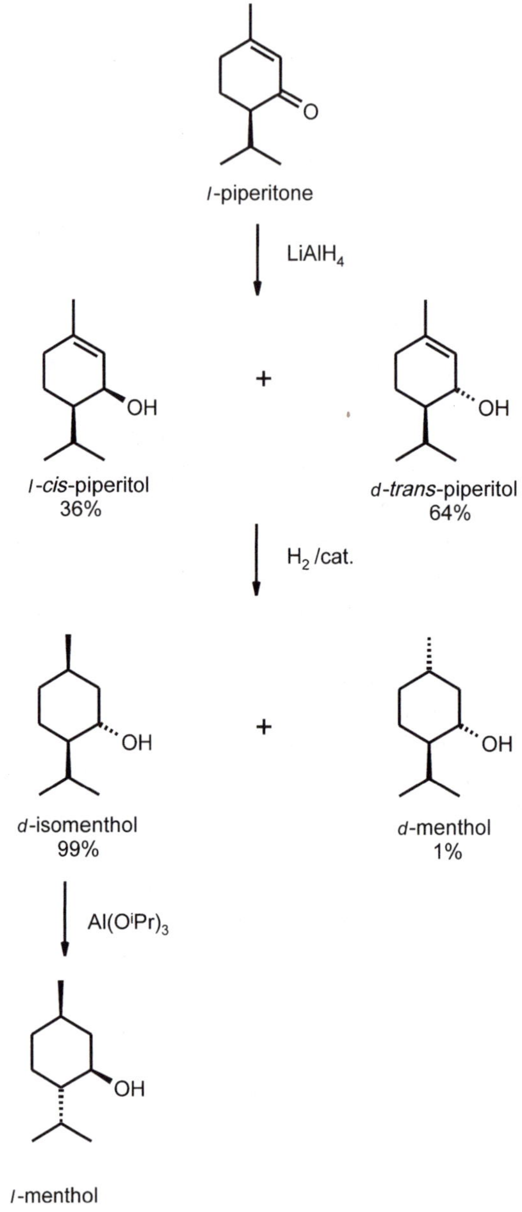

Figure 4.27

The hydrogenation mechanism clearly involves both the double bond and the alcohol function and these must interact with the catalyst in a controlled manner since the major (99%) product is *d*-isomenthol, in which the alcohol is *trans*- and the methyl *cis*- to the isopropyl group. The

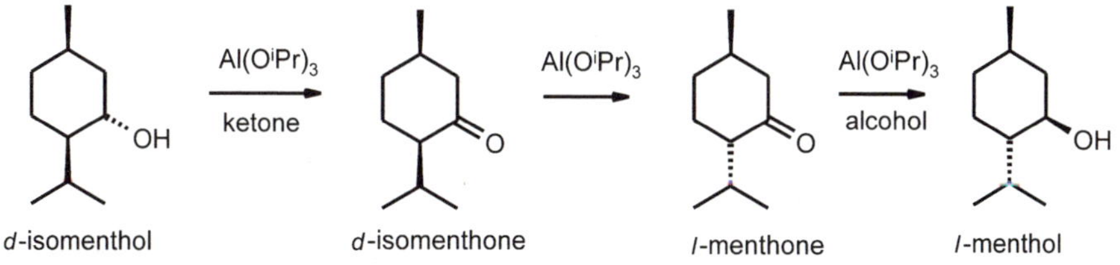

Figure 4.28

major product can be easily separated from the 1% *d*-menthol which constitutes the remainder of the mixture. At first sight, it might seem that this has been a waste of effort since the wrong isomer has been produced in excellent yield. However, the stereochemical ingenuity of this synthesis arises from the fact that the only centre which is not easily isomerised is that carrying the methyl group, and this centre is now fixed in the correct absolute stereochemistry for *l*-menthol. Treatment of *d*-isomenthol with aluminium isopropoxide, gives *l*-menthol in high yield. Again, at first sight this seems very unusual since both C-3 and C-4 of the *p*-menthane ring are saturated sp^3 atoms. The explanation is shown in Figure 4.28.

Aluminium isopropoxide is a Lewis acid and it is also a good catalyst for the Oppenauer oxidation and Meerwein–Ponndorf–Verley reduction reactions. In the presence of a ketone, it will oxidise *d*-isomenthol to *d*-isomenthone (Oppenauer oxidation). The hydrogen atom on C-4 is now enolisable and therefore epimerisation can occur, catalysed by the aluminium isopropoxide acting as a Lewis acid. This will give *l*-menthone. This can now be reduced (Meerwein–Ponndorf–Verley reduction) to *l*-menthol by an alcohol and aluminium isopropoxide. The ketone and alcohol for the redox reactions could be the menthols/menthones themselves or traces of acetone/isopropanol in the aluminium isopropoxide. Obviously, the reactions shown in Figure 4.28 are all reversible. The equilibrium will eventually be driven over completely to *l*-menthol since the latter is the most thermodynamically favoured of all of the isomeric components in the system.

4.5 DETERMINATION OF ABSOLUTE STEREOCHEMISTRY

Nowadays we are in the fortunate position of being able to carry out X-ray crystallography on crystalline derivatives of optically active terpenoids. This enables us to "see" the actual handedness of molecules. In the early days of stereochemistry, this was not possible and absolute stereochemistry could only be determined relative to a standard. Since the initial work on stereochemistry was carried out on sugars and their derivatives it is not surprising that a sugar was selected as the standard to which everything else would be compared. In 1890, Fischer chose *d*-mannose and *d*-glucose as the standards and suggested using the prefixes *d*- and *l*- to relate other structures to these two sugars. However, this leads to various difficulties as the direction of rotation of plane polarised light can vary with changes in substituents as well as changes in the handedness of a molecule. For example, in the case of 2-methylbutanol, the *R*-isomer is dextrorotatory and the *S*-isomer laevorotatory whereas, with the corresponding aldehydes, the *R*-isomer is laevorotatory and the *S*-isomer dextrorotatory. Stereochemical correlations are further complicated by the fact that the same material might be accessible from a precursor in the *d*-series as well as from one in the *l*-series. A partial solution was brought into use in which the prefixes *d*- and *l*- or (+)- and (−)- were used to indicate the direction of rotation of plane polarised light and D- and L- to indicate the relationship to D-glucose. This was still unsatisfactory as glucose and mannose both contain four chiral centres. So, in 1906, Rosanoff suggested that the standard should be changed to the hemi-saccharide, glyceraldehyde (HOCH₂CHOHCHO) which contains only one asymmetric carbon atom. Therefore all of the early work on absolute stereochemistry of terpenoid compounds relied on degradation to glyceraldehyde or to materials whose stereochemistry was known, relative to that of glyceraldehyde. In those days, only the relative stereochemistry was known and not the actual left- or right-handed arrangement in space of the substituents on the chiral centre. Nowadays, we can determine absolute stereochemistry either by using crystallography, by degradation to a compound of known stereochemistry or by NMR using chiral shift reagents. All of the systems of nomenclature can still be found in use today. Only the (+)/(−) and *R*/*S* prefixes are necessary, the others can be regarded as artefacts of outdated systems.

Figure 4.29 shows some examples of how the chiral centres of monoterpenoid molecules can be correlated with each other and with glyceraldehyde.

(+)-Camphor can be prepared from (+)-α-pinene *via* borneol. Acid catalysed hydration of (+)-α-pinene gives (+)-α-terpineol and this can be dehydrated to (+)-limonene. (+)-Limonene can be converted to (−)-carvone as shown in Figure 4.29, and this can be hydrogenated to *trans*-(+)-tetrahydrocarvone. In all of these transformations, C-4 of the cyclohexane ring remains unaffected. Since this carbon atom is the asymmetric centre which imparts chirality to the molecules, we can conclude that they all have the same absolute stereochemistry. Introducing a carbonyl group at C-3 and removing the one at C-2

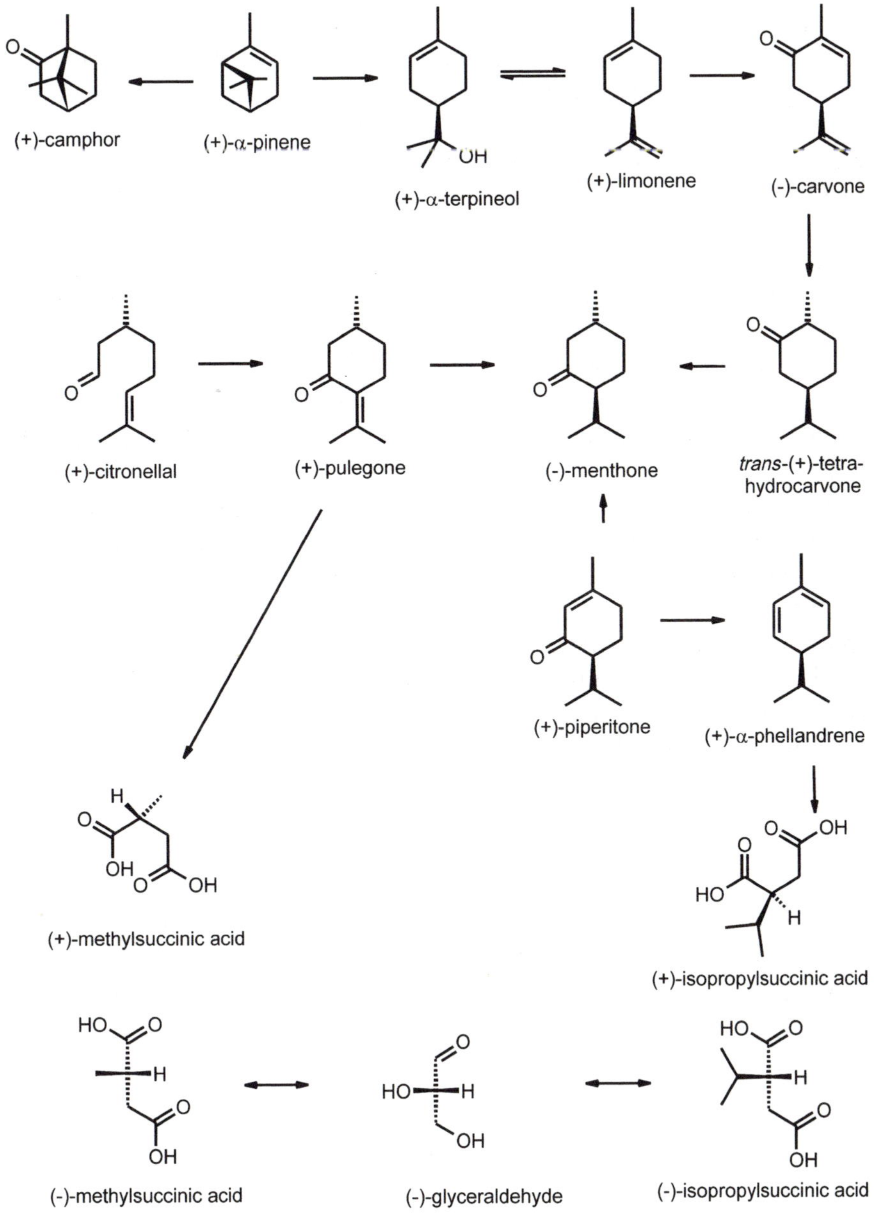

Figure 4.29

converts *trans*-(+)-tetrahydrocarvone into (−)-menthone, which must therefore have the same absolute stereochemistry as those mentioned above. (−)-Menthone can also be prepared from (+)-pulegone which can, in turn be prepared from (+)-citronellal. The absolute stereochemistry at C-1 of (+)-citronellal, (+)-pulegone and (−)-menthone must therefore all be identical. (+)-Piperitone can be hydrogenated to (−)-menthone. Reduction of the ketone of (+)-piperitone and elimination of the resultant alcohol, gives (+)-α-phellandrene. Thus we can now know that the absolute stereochemistry of C-4 in (+)-α-phellandrene is identical with that in the whole series back to (+)-camphor.

Oxidative degradation of (+)-pulegone gives (+)-methylsuccinic acid. This acid is identical to that obtained from (−)-glyceraldehyde except for its effect on polarised light, which it rotates to the same degree but in the opposite direction. Therefore, the absolute stereochemistry of C-1 in (+)-citronellal, (+)-pulegone, (−)-menthone and *trans*-(+)-tetrahydrocarvone are all opposite to that of (−)-glyceraldehyde.

Similarly, oxidative degradation of (+)-α-phellandrene gives (+)-isopropylsuccinic acid. This is identical to that obtained from (−)-glyceraldehyde except for the direction of rotation of polarised light. So the absolute stereochemistry of C-4 in (+)-α-phellandrene is opposite to that of (−)-glyceraldehyde.

Thus we can establish the absolute stereochemistry of the chiral centres in all of the terpenoids shown in Figure 4.29 by correlating them back to the standard reference material, (−)-glyceraldehyde.

REFERENCES

4.1. E.L. Eliel and S.H. Wilen, *Stereochemistry of Carbon Compounds*, John Wiley, New York, 1994.

4.2. I.L. Finar, *Organic Chemistry*, Vol. 2, Longmans, London, 1968.

CHAPTER 5

Bicyclic Monoterpenoids

Investigations into the chemistry of pine
components made invaluble contributions to
our understanding of carbocation chemistry

When is a cation not a cation?

Professor William Motherwell

Occasionally, the bond between a carbon atom and another atom can be broken to leave a positive charge on the carbon atom. The carbocation thus produced can then undergo a reaction, for example with a nucleophile, to give a new molecule. There are also times when old bonds are broken and new ones formed without it being clear at which point, if any, a carbocation exists. In such cases, there is often a synchronous push and pull of electrons across the reaction centre. In examples where the carbocation exists only transiently, if at all, the reaction environment is restricted and the presence of neighbouring groups and other species in the vicinity can have a profound effect on the course of the reaction.

KEY POINTS

There are **4** types of reactions that carbocations undergo (elimination, solvolysis, H-shift and C-shift).

There are **3** driving forces, which affect the course of carbocation chemistry (cation stability, ring strain and steric strain).

There are **2** factors controlling regioselectivity of addition of cations to olefinic bonds (electron density and polarisability).

There is **1** additional rule concerning carbocation reactions (the *trans-anti*-periplanar rule).

5.1 BICYCLIC MONOTERPENOIDS

The commonest bicyclic monoterpenoid skeleta are shown in Figure 5.1. The bicyclic monoterpenoids are all formed from geranyl pyrophosphate

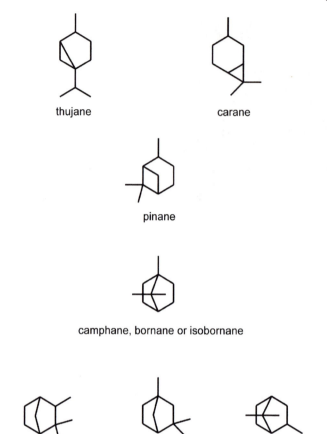

thujane

carane

pinane

camphane, bornane or isobornane

isocamphane   fenchane   isobornylane

Figure 5.1

in nature. The initial cyclisation reaction gives the *p*-menthane skeleton
and then a second ring is formed to give a bicyclic material. In the
case of the caranes and thujanes, the second ring is a three-
membered one, whereas in the pinanes, it is four-membered. In all the
other skeleta shown in Figure 5.1, there is a five-membered ring fused
across the cyclohexane to give the [2.2.1]-bicycloheptyl system, often
referred to as the norbornyl ring system. In the first example, the
camphane skeleton, the outline of the original *p*-menthane ring is still
easily visible, the new bond having been formed between C1 and C8 of
the *p*-menthane. However, in the other three (isocamphane, fenchane
and isobornylane) it is clear that further rearrangement has taken
place as the three methyl groups are no longer in the same relationship to

each other, in terms of bond connectivity. The camphane system is also referred to as the bornane or isobornane system, the difference between the latter two lying in the stereochemistry of functions attached to them.

Some important bicyclic monoterpenoids are shown in Figure 5.2. Camphor occurs very widely in nature. The richest source is the oil distilled from the wood of the camphor tree (*Cinnamomum camphora*). This wood has a long history of use for furniture in China. Linen stored in camphorwood chests is free from attack by insects such as moths owing to the insect repellent properties of camphor. It is also found in a wide range of flowers, such as lavender, and herbs, such as sage and rosemary. Isobornyl acetate occurs in herbs such as sage,

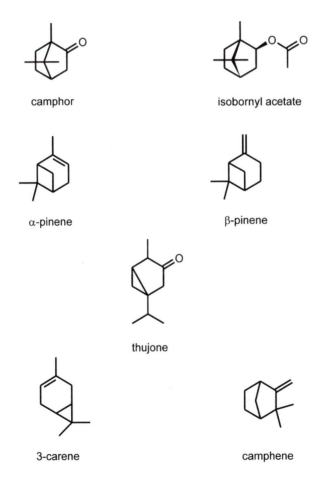

camphor

isobornyl acetate

α-pinene

β-pinene

thujone

3-carene

camphene

Figure 5.2

thyme, basil and oregano and also in flowers, including the endogenous Australian flower boronia (*Boronia megastigma*). The pinenes occur in most members of the pine, fir and spruce families and are readily available as by-products of the wood and paper industries, thus making them important starting materials for the commercial synthesis of other terpenoids, as will be seen in later chapters. The richest sources of thujone include the leaf oil of cedar trees of the thuja family, such as the Western Cedar (*Thuja occidentalis*). This oil contains about 50% thujone. Thujone is also an important component of Dalmatian Sage (*Salvia officinalis*), the oil of which contains 25–30% thujone. Perhaps the most famous thujone rich oils are those of two *Artemisia* species, which are used in the preparation of alcoholic liquors. The liqueurs of the Piedmont region of Northern Italy use the alpine plant *Artemisia genipi* as part of their distinctive cocktail flavour. Even more well known is the use of Wormwood (*Artemisia absinthum*) for the preparation of absinthe, the hallucinogenic liqueur which was very fashionable in artistic circles in nineteenth-century Paris. Wormwood contains 15–20% thujone. 3-Carene occurs in various pines, spruces and firs and is a major component of Indian Turpentine. It also occurs in spices, such as black pepper (*Piper nigrum*); leaves, such as those of the sweet orange (*Citrus aurantia*), and fruits, such as the mango (*Mangifera indica*). Camphene is found in the leaf oils of many trees, the richest source being the eucalypt, *Eucalyptus kirktoniana*. It is also found in ginger (*Zingiber officinalis*) and rosemary (*Rosmarinus officinalis*). Camphene is also important as a synthetic intermediate, for example as in the preparation of camphor from α-pinene as shown in Figure 5.3.

5.2 TWO COMMERCIAL SYNTHESES OF BICYCLIC MONOTERPENOIDS

Figure 5.3 shows the outline of two synthetic schemes which are of commercial importance. As mentioned above, camphor is of use for its insect repellent properties. Turpentine is an inexpensive by-product of the paper industry and so the overall conversion of α-pinene to camphor, through camphene and isoborneol, is a way of adding value to the turpentine derivative whilst increasing the availability of camphor over that from camphorwood. Besides, it is worth pointing out at this juncture that the alcohol function of isoborneol is in the *exo*-orientation on the camphene ring system. Borneol is the isomer with the alcohol function in the endo-configuration. This nomenclature system applies to all compounds containing the fragments bornyl or isobornyl in their

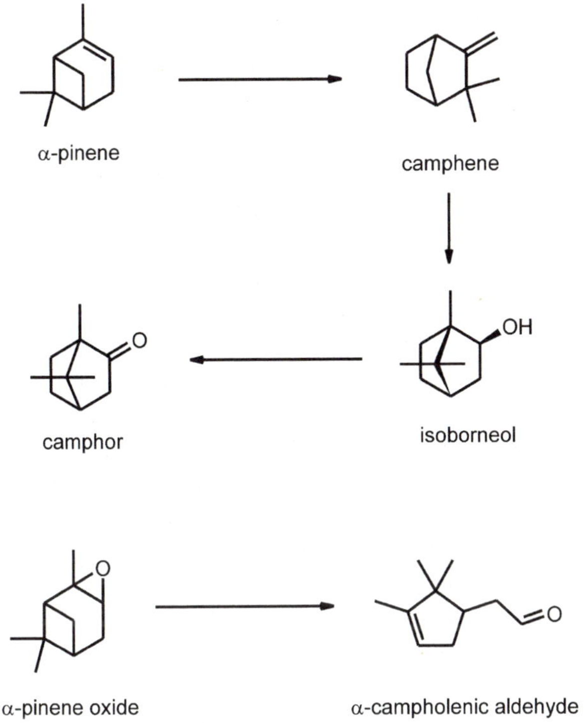

α-pinene

camphene

camphor

isoborneol

α-pinene oxide

α-campholenic aldehyde

Figure 5.3

names. The epoxide of α-pinene serves as a feedstock for the preparation of α-campholenic aldehyde, an important intermediate for the synthesis of synthetic sandalwood ingredients.

All the reactions shown in Figure 5.3 are one-step reactions, yet three out of the four give quite different ring systems in the product compared to those of the starting materials. This may seem rather strange or even incomprehensible at first sight but it is a phenomenon which is quite common in terpenoid chemistry. The explanation lies in the chemistry of carbocations. The major task of this chapter is to explain the principles of carbocation chemistry so that we can understand what is happening in these examples and in the rich chemistry of the terpenoids, some further examples of which will be given in subsequent chapters. The elucidation of carbocation mechanisms was a major intellectual achievement and, before we go on to discuss the fundamentals of carbocation chemistry, let us look at some of the puzzles which confronted the pioneers in terpenoid carbocation chemistry.

5.3 CHEMICAL PUZZLE NUMBER 1

Our first chemical puzzle is shown in Figure 5.4. Addition of hydrogen chloride to α-pinene at low temperature gives pinene hydrochloride. This is a highly regiospecific reaction which gives only one of the two possible directions of addition. The chloride atom invariably ends up attached to the carbon atom which carries the methyl group and not to the one from the other end of the double bond. So, our first question is, "Why?" Pinene hydrochloride is a very unstable molecule and, on standing at room temperature, converts into the isomeric bornyl chloride. Again, this reaction is very selective and only bornyl chloride is formed. There is no isobornyl chloride in the product. So our next question is more difficult and twofold in nature; "Why does it rearrange at all and why is the rearrangement so selective?" Normally, when hydrogen chloride is added to a double bond, the original olefin can be regenerated by treatment of the adduct with a base, the latter inducing elimination of hydrogen chloride. In the case in question, we might expect that we would not recover α-pinene by treatment of bornyl chloride with a base. The rearrangement to bornyl chloride will make recovery of the pinane system more difficult. The product we might expect to obtain would be bornylene. This should, at first sight, be obtainable by elimination of hydrogen chloride from bornyl chloride. However, it is not the product which results in reality. The product of elimination of hydrogen chloride from bornyl chloride is, in fact, camphene. Yet, another rearrangement has taken place and this time it is quite difficult to see what has happened

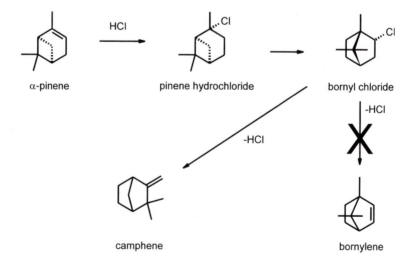

α-pinene   pinene hydrochloride   bornyl chloride

camphene   bornylene

Figure 5.4

in the course of the reaction as so much of the basic structure appears to have changed. So again, we have a twofold question, "Why is bornylene not formed and how did camphene come to be the product?"

5.4 CHEMICAL PUZZLE NUMBER 2

This puzzle is an extension of the first one. We saw above that pinene hydrochloride rearranges to bornyl chloride and that camphene can be formed from it by elimination of hydrogen chloride. If hydrogen chloride is added to camphene, we do not see the reverse reaction to give bornyl chloride, but a simple 1,2-addition to give camphene hydrochloride. As with the addition to α-pinene, the reaction is highly regioselective. Like pinene hydrochloride, camphene hydrochloride is not a stable molecule and it rearranges, but the product this time is pure isobornyl chloride (Figure 5.5). So our second puzzle is, "Why does pinene hydrochloride

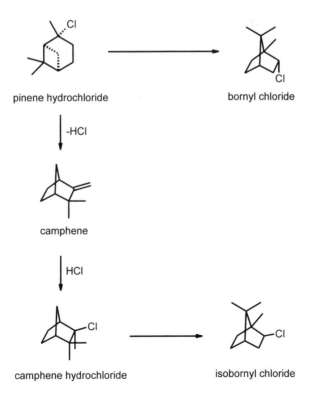

pinene hydrochloride   bornyl chloride

-HCl

camphene

HCl

camphene hydrochloride   isobornyl chloride

Figure 5.5

rearrange to bornyl chloride whilst camphene hydrochloride rearranges to isobornyl chloride?''

5.5   CHEMICAL PUZZLE NUMBER 3

If enantiomerically pure camphene hydrochloride is held at 20 °C, it does not convert to isobornyl chloride but does undergo racemisation (Figure 5.6). Our third puzzle therefore is simply, "How does this happen?"

Unless one is familiar with the basics of carbocation chemistry, these puzzles will be as baffling nowadays as they were to the chemists who were confronted with them and many similar ones in the late nineteenth and early twentieth century. It is now time to look in more detail at carbocation chemistry.

5.6   THE FUNDAMENTALS OF CARBOCATION CHEMISTRY

In studying the chemistry of carbocations, we will consider the reactions which they can undergo, the forces which drive these reactions and the factors which induce selectivity into them. There are four basic types of reactions which carbocations can undergo and there are three forces which drive them. Two factors control the chemo- and regioselectivity of additions to double bonds and the other major factor in determining selectivity is the *trans-anti*-periplanar (TAP) rule. This gives us a simple mnemonic for memorising all of the key features of carbocation chemistry. I will call it the 4-3-2-1 rule. This is summarised in Table 5.1.

We will now look at each of these in turn.

5.6.1   Reaction Type 1 – Elimination

One very simple thing for a carbocation to do, is to drop off an adjacent proton to give an olefin as shown in Figure 5.7. The electrons forming the carbon–hydrogen bond, move in to form a double bond between the

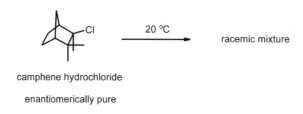

camphene hydrochloride

enantiomerically pure

Figure 5.6

Table 5.1 *The 4-3-2-1 Rule*

4 types of reactions	Elimination
	Solvolysis
	H-shift
	C-shift
3 driving forces	Cation stability
	Ring strain
	Steric strain
2 selectivity factors	Electron density
	Polarisability
1 other factor	*trans-anti*-periplanar rule

Figure 5.7

carbon which is losing the proton and that carrying the positive charge. Obviously, the expression "lose a proton" is somewhat figurative since the energy required to strip off electrons and expose a naked proton is considerable and this is unlikely to occur under the conditions of the type of reaction under discussion. The proton will, of course, be taken up by a base present in the system, perhaps even the solvent may act as the base. I refer to these reactions as eliminations because a proton is eliminated from the molecule. If the cation was formed by heterolysis of a bond between a carbon atom and a group X, then the overall process constitutes an elimination of HX. For example, if a chloride ion were to be lost initially by cleavage of the carbon–chlorine bond, and the resultant carbocation were to eliminate a proton, the overall process would constitute an elimination of hydrogen chloride. Of course, as we have seen, the carbocation which loses the proton is not necessarily the one which was formed by cleavage of the carbon–chlorine bond. The initial carbocation can undergo rearrangement before a proton is lost. The elimination of hydrogen chloride from bornyl chloride to give camphene (Figure 5.4) is an example of this.

5.6.2 Reaction Type 2 – Solvolysis

Another simple reaction for a carbocation is to form a new bond with an anion or nucleophile as shown in Figure 5.8. A pair of electrons from the

Figure 5.8

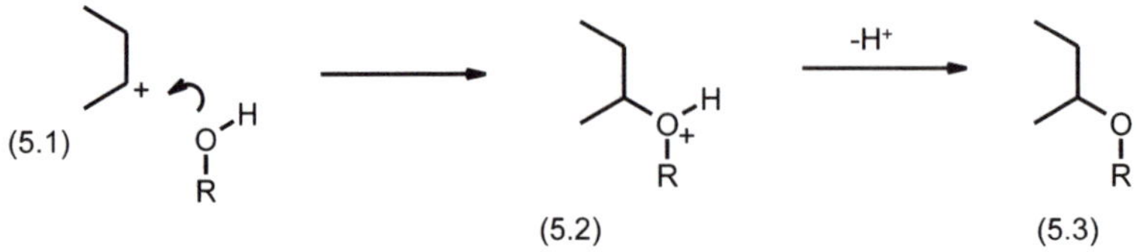

Figure 5.9

nucleophile is used to form a bond with the positively charged carbon atom. If the nucleophile is an anion, then the addition product will be neutral. If the nucleophile is electrically neutral, then a new cation will be formed. Figure 5.9 shows an example of this. A neutral alcohol molecule, possibly from the solvent, adds to the cation (5.1) using one of the lone pairs of electrons on the oxygen atom. This results in a new cation (5.2) in which the positive charge is now on the oxygen atom. Loss of a proton from this species, results in the formation of a neutral product, the corresponding ether (5.3). Strictly speaking, the term solvolysis is restricted to those cases where the nucleophile is a molecule of the solvent. For simplicity, I have used the term to cover nucleophilic substitution and nucleophilic addition reactions also.

5.6.3   Reaction Type 3 – H-shift

It is possible for a hydrogen atom to move from being bonded to one carbon atom to being bonded to an adjacent one which is carrying a positive charge. This is referred to as a 1,2-hydrogen shift and is shown in Figure 5.10. As the hydrogen atom moves from one carbon to another, it is clear that the positive charge must move in the opposite direction to

Figure 5.10

form a new carbocation (5.5). The hydrogen can move to a more remote carbon atom and so other shifts are possible, *e.g* 1,3- or 1,5-shifts, but the commonest is the simple 1,2-shift.

5.6.4 Reaction Type 4 – C-shift

Just as a hydrogen atom can move onto a neighbouring carbon atom bearing a positive charge, so can a carbon atom. A simple 1,2-carbon shift reaction is shown in Figure 5.11. In this example there is a positive charge on C1 of 2,2-dimethylpropane (5.6). One of the methyl groups on C2 moves across onto C1, taking the electrons of its bond with it and consequently, the positive charge is transferred to C2. Carbon shifts have more impact on the structure of the molecule as is clear from this example. The basic skeleton has changed from that of 2,2-dimethylpropane in (5.6) to that of 2-methylbutane in (5.7).

When a carbon shift occurs in a ring system, the change in skeleton can be much more difficult to see. Figure 5.12 shows an example of a simple 1,2-carbon shift which appears much more complex because it transforms a pinane ring system (3,1,1-bicycloheptane) (5.8) into a bornane (2,2,1-bicycloheptane) (5.10). This is particularly confusing when only the usual drawn forms are shown in figures. In this case, a reaction scheme might show only structures (5.8) and (5.10). When examining such mechanisms, it is easiest to initially redraw the new ring system in the same configuration as the original. This has been done in Figure 5.12 in which the new carbocation is drawn first as structure (5.9). This allows us to see that one of the methylene carbons of the four-membered ring has moved along by one position in the six-membered ring. The

(5.6)   (5.7)

Figure 5.11

(5.8)   (5.9)   (5.10)

Figure 5.12

configuration of structure (5.9) is clearly not a favourable one and so we can now redraw it more neatly as (5.10). When redrawing the skeleta, it is important to ensure that all carbon–carbon connectivities are maintained. For example, when comparing structures (5.9) and (5.10), we can see that in each, the newly formed bridgehead atom is connected to the other bridgehead by two 2-carbon bridges and one 1-carbon bridge. We must then check substituents. In this case the only feature is the positive charge and we can see that, in both the drawings, it is adjacent to the new bridgehead and on a two-carbon bridge. In this chapter, for ease of explanation, such examples of redrawn structures will be given different structure numbers. In subsequent chapters, it will be assumed that the reader is familiar enough with the concept that the same number will be used for different representations of the same structure.

Molecular models are very useful when looking at these reactions and I would recommend that every chemist interested in terpenoids should possess a set. Ball and stick models or skeletal models are more useful for following carbocation rearrangements than are space filling ones. There are many types of models available any of which will serve the purpose. The choice is up to the reader's personal preferences and budget. Models are particularly useful for those who have difficulty in visualising structures in three dimensions.

The phenomenon of changing of carbon ring structures by means of carbon shifts, is known as the Wagner–Meerwein rearrangement, after its discoverers.

In Figure 5.12, we saw how a pinane ring can be transformed into a bornane one. Figure 5.13 shows a 1,2-carbon shift in a bornane which produces another bornane. This can appear even more confusing since, if

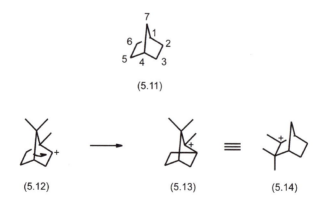

Figure 5.13

both starting material and product are drawn in the conventional way, it seems that all of the substituents have moved around the ring. An example is the rearrangement of the camphene system to that of isocamphane as shown in Figure 5.13.

For clarity in the following discussion, structure (5.11) shows the numbering system of the 2,2,1-bicycloheptyl skeleton. This ring system nomenclature is based on the bridgehead carbon atoms. To name a bicyclic compound, one first identifies the two bridgehead carbon atoms and then counts the number of atoms in each bridge. Thus in structure (5.11) we identify two bridges of two carbon atoms and one of one and give it the name 2,2,1-bicycloheptane as there are seven carbon atoms in total. The numbering of the individual carbon atoms starts on a bridgehead and then proceeds round the largest bridge, continues round the next largest bridge and finally the smallest bridge. To name a tricyclic structure, one first breaks the smallest bridge and names the material as for a bicyclic structure. The location of the fourth bridge is then indicated by a fourth numeral, indicating the number of carbons in it, with two superscript numbers identifying which two atoms of the bicyclic structure are connected by it. For example, tricyclene, (5.87) in Figure 5.45, would be named as 1,7,7-trimethyltricyclo[$2.2.1.0^{2,6}$]heptane.

Normally the reaction shown in Figure 5.13 would be shown simply as structure (5.12) going to (5.14). The first impression is that all three methyl groups have changed their positions in the ring. In reality, none of the methyl groups have moved from the carbons to which they are bonded, it is another bond in the ring which has moved. Initially, C6 is bonded to C1. However, the presence of a positive charge on C2 allows a Wagner–Meerwein rearrangement to occur, with the electrons which form the bond between C1 and C6 moving towards the positive charge and thus simultaneously breaking the bond between C6 and C1 and forming a new one between C6 and C2. The positive charge now resides on C1 and we have the structure (5.13). If we follow bond connectivities around (5.13) and [5.14], we will see that the two are identical (the whole molecule having been rotated through about 120°).

When there are two groups which could undergo a carbon shift and there are no other factors to overcome the effect, there is a general rule that the more heavily substituted carbon atom will move. Thus, for example in the case of carbocation (5.15) in Figure 5.14, it is possible for either the tertiary butyl carbon or one of the two methyl carbons to move across to the carbon carrying the positive charge. The carbon carrying three other carbon atoms is the one which moves, giving the 2,4-dimethylpentane skeleton of (5.16) rather than the alternative

(5.15)   (5.16)

(5.17)   (5.18)

Figure 5.14

2-*t*-butylbutane which would be the product of a methyl shift. The reason for this lies in the greater ability of the more substituted atom to stabilise the transient positive charge through the inductive effect. As the carbon atom moves from one position to another, we can envisage an intermediate in the form of a cyclopropyl ring as in (5.20) with the positive charge spread across all three component atoms. This is shown in Figure 5.15. The greater the ability of the moving group, R, to stabilise the charge, the more likely it will be to move. If there is a benzene ring which can move, then it will do so in preference to any alkyl groups. So, for example, with carbocation (5.17) in Figure 5.14, it is the phenyl group which moves to give cations (5.18). This is because of the ability of the aromatic ring to stabilise the positive charge through delocalisation.

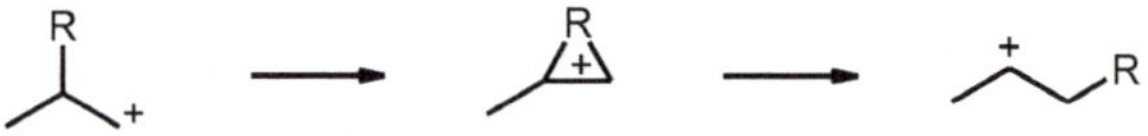

(5.19)   (5.20)   (5.21)

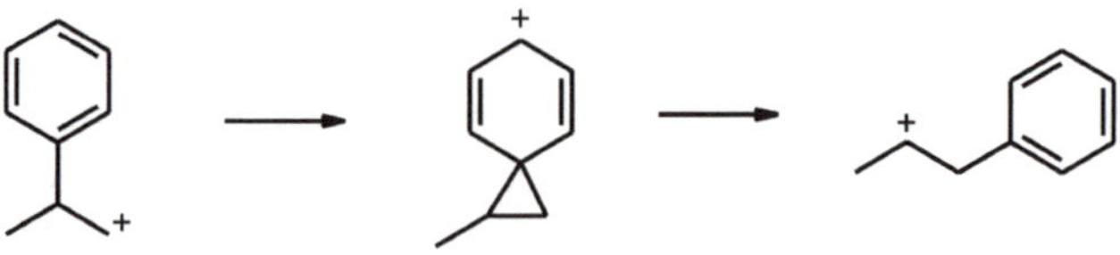

(5.22)   (5.23)   (5.24)

Figure 5.15

For instance, structure (5.23) shows one canonical form with the positive charge located on the opposite end of the benzene ring. Note that the cyclopropane based carbocations (5.20) and (5.23) are both examples of non-classical carbocations, *i.e.* carbocations in which the positive charge is spread across a number of atoms rather than being localised on one.

5.6.5 Driving Force 1 – Cation Stability

The stability of a positive charge on a carbon atom is increased if electrons can be drawn in from neighbouring atoms through the inductive effect. The more electrons there are on the neighbouring atoms, the more charge stabilisation there will be. A hydrogen atom can donate only the two electrons of the bond which holds it to the carbon atom. A carbon atom has more electrons of its own and also can draw in the electrons of the hydrogens to which it is also attached, the effect known as hyperconjugation. Figure 5.16 shows the canonical forms at either extreme of hyperconjugation, the reality being a sharing of the positive charge across the entire system. These factors mean that there is a marked increase in stability on going from a primary to a secondary to a tertiary cation as shown in Figure 5.17.

Let us now look again at the reactions in Figures 5.10 and 5.11. In both cases, the reactions can proceed in either direction. However, there will be a driving force moving them to the right in both instances, since in each case, the products, (5.5) and (5.7), contain tertiary carbocations which are more stable than the primary carbocations of the starting materials (5.4) and (5.6), respectively.

A carbocation will be even more stable if there is a neighbouring oxygen atom which can donate electron density from its lone pairs of

Figure 5.16

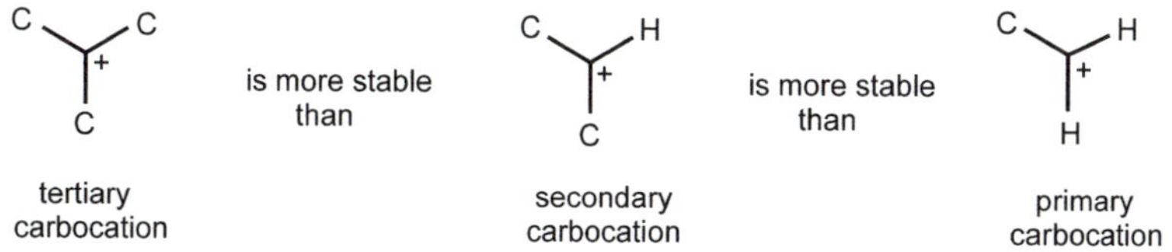

Figure 5.17

Figure 5.18

electrons. The canonical forms of a carbocation stabilised in this way are
shown in Figure 5.18.

As already mentioned, if a charge can be shared over more than one
atom, then the partial charge over each is reduced which lowers the
energy of the system. Thus, in Figure 5.19, the 1,2 H-shift from (5.25) to
(5.26), brings the positive charge into conjugation with the double bond
and thus lowers the overall energy of the molecule.

The carbocation of Figure 5.19 can be written either as one of the two
canonical forms (5.26) and (5.27) or using an electron smear (5.28) as
shown in Figure 5.20. The smear is closer to reality since the molecular
orbitals of the molecule will be distributed across the three carbons of the
allylic cation. However, it will be able to react as if the positive charge
were localised at either end. For each individual reaction, the nature of
the other reactive species, the reaction conditions and the nature of the
product will determine which way round the system will react.

(5.25)   (5.26)

Figure 5.19

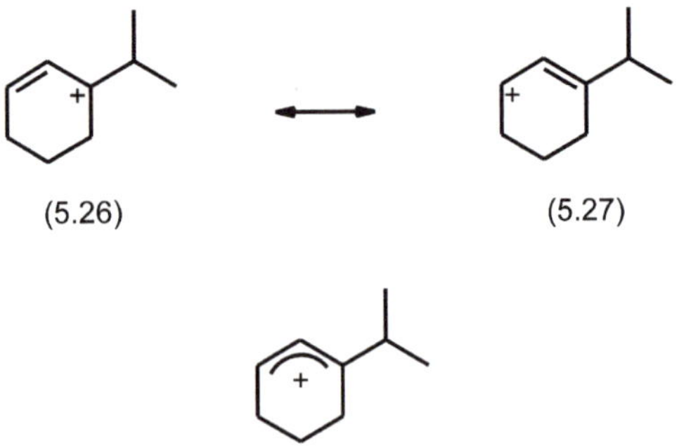

(5.26)   (5.27)

(5.28)

Figure 5.20

5.6.6 Driving Force 2 – Ring Strain

The concept of ring strain was first developed by von Baeyer in 1885. There are two components in ring strain, angle strain and steric strain. The angle between the bonds of a tetrahedral sp^3 carbon atom is $109°28'$ Compare this with the internal angles of regular polygons as shown in Figure 5.21. Constraining a tetrahedral carbon atom into either a cyclopropyl or cyclobutyl ring involves bending the bonding angles considerably. This is known as angle strain. The tetrahedral bond angle is almost right for the regular pentagon of cyclopentane and, with a little puckering of the ring, leads to a stable structure. In the case of the cyclohexane ring, the angle of the regular hexagon is slightly too large, but this is not a problem because the ring can pucker to accommodate the tetrahedral angle. Not only that, but, by doing so, it allows the substitutents on the ring carbon atoms to become staggered and therefore minimise the steric, or spatial, interaction between them. Cyclohexane rings can exist in various conformations but the two most favoured are what are called the chair and boat conformations. These are shown in Figure 5.22. If you make a model of the chair form and look along any of the carbon–carbon bonds in it, you will see that all of the

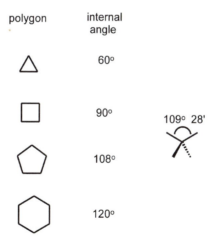

polygon	internal angle
△	60°
▢	90°
⬠	108°
⬡	120°

109° 28'

Figure 5.21

chair   boat

Figure 5.22

other bonds attached to the two carbons, are staggered. The substituents are held either at right angles to or lie in the plane of the ring. The former is referred to as axial and the latter, equatorial. In the boat configuration, the substituents are not so well staggered and, in most cases, this is a less preferred conformation. For ring sizes greater than six, there is no problem with angle strain as the rings can pucker to accommodate the tetrahedral angle. However, by doing so, rings from cycloheptane to cyclododecane are forced to bring substituents on opposite sides of the ring into close proximity across the ring centre. This leads to problems of van der Waals repulsion between them. This is known as steric strain. Larger rings, for example those with 15 or 20 carbons in them, do not have this problem as there is sufficient distance across the ring to allow substituents to point inwards. There is therefore a general principle that five- and six-membered rings are preferred to other sizes. Smaller rings will tend to spring open or enlarge in order to reduce angle strain. Rings between 8 and 12 will tend to break open, reduce in size or form new bonds across the ring in order to reduce the steric crowding.

The reaction shown in Figure 5.12 shows this driving force in action. The starting carbocation, (5.8), contains a four-membered ring and, therefore, there is considerable angle strain in the molecule. By rearranging the skeleton from a 3,1,1-bicycloheptane to a 2,2,1-bicycloheptane, the angle strain is reduced since the product (5.10) contains only five- and six-membered rings.

5.6.7   Driving Force 3 – Steric Strain

Steric problems also exist in open chain structures and between substituents on rings. Bringing atoms close together, when they are not bonded to each other, will lead to adverse van der Waals reactions and hence, steric repulsion. This can happen with substituents on adjacent carbon atoms or with substituents which are brought together through the geometry of the molecule. As far as interactions between substituents on adjacent carbon atoms are concerned, there are two ways of depicting this in illustrations. These are shown in Figure 5.23. The system with the circles around the central atom is known as Newman projection. The idea is that the line of viewing is along the bond between two carbon atoms. The circle represents the profile of the atom closer to the viewer. The three lines radiating from the centre of the circle represent the three other substituents attached to this carbon atom. The other three lines represent the substituents on the other carbon atom, the one hidden from view by the front atom. It is clear from Figure 5.21 that the eclipsed configuration brings these substituents closer together. The larger the

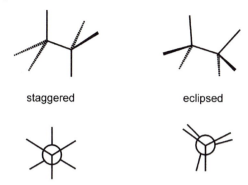

Figure 5.23

Figure 5.24

substituents, the larger will be the repulsion and the greater the energy difference between the configurations.

In Figure 5.24, Newman projections show how, in the staggered configuration, two larger substituents will prefer to be aligned opposite each other, rather than adjacent, since in the latter arrangement, there is still a considerable degree of steric interaction.

5.6.8 Selectivity Factor 1 – Electron Density

Just as different substituents are able to stabilise carbocations by releasing electrons towards them, substituents can release electrons into double bonds or draw them out from it. Electron donating groups such as alkyl groups and ethers will increase the electron density of double bonds to which they are attached. Therefore, in hydrocarbons the more heavily substituted double bonds will be more electron rich. So, if there are two or more double bonds in a molecule, electron deficient reagents such as ozone or peracids will preferentially attack the more/most electron rich olefin.

Limonene (5.29) provides us with an example of this type of selectivity. The endocyclic double bond of limonene is trisubstituted and is therefore richer in electrons than the disubstituted olefin in the isopropenyl group.

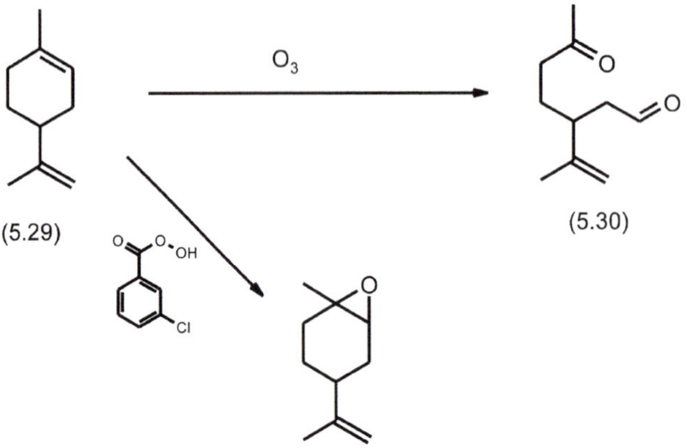

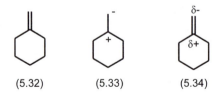

Figure 5.25

Ozone will therefore selectively cleave the ring double bond leaving the other untouched, provided of course, that no more than one molar equivalent of ozone is used. Similarly, one molar equivalent of *m*-chloroperbenzoic acid will selectively give only limonene 1,2-epoxide (5.31). These reactions are shown in Figure 5.25.

5.6.9 Selectivity Factor 2 – Polarisability

If a double bond is more heavily substituted at one end than the other, then the π-electrons will be capable of being polarised so that the electron density moves towards the less substituted end leaving a partial positive charge at the more heavily substituted end. To illustrate the point by taking it to the extreme, Figure 5.26 shows how the exocyclic methylene bond of methylenecyclohexane (5.32), could be polarised to give a carbanion on the primary carbon and a carbocation on the tertiary (5.33). Since the bond is only partially polarised, it is more accurate to

Figure 5.26

write it with partial charges as in structure (5.34). It must, of course, be remembered that polarisability is the capacity to be polarised rather than the actual degree of polarisation in the ground state. Reagents such as Friedel–Crafts reagents, carbon monoxide and formaldehyde, prefer a polarisable olefinic substrate.

Some examples of selectivity through polarisability are shown in Figure 5.27. As with electron density, limonene provides a nice example. The Friedel–Crafts acylation of limonene (5.29) demonstrates two effects of polarisability. Firstly, the acylating species prefers the more polarisable of the two bonds in the molecule and so there is attack only at the isopropenyl group. Secondly, the olefin only polarises in one direction, that which pushes the negative charge to the less substituted end. It is that end, therefore, which adds to the positive centre of the acylating species giving total regioselectivity and hence (5.35) as the sole product.

The second example of selectivity through polarisability is the Prins Reaction of the pinenes. The double bond of β-pinene (5.36) is more easily polarised than that of α-pinene (5.37) and so only the β-isomer undergoes the reaction. As with the Friedel-Crafts reaction of limonene,

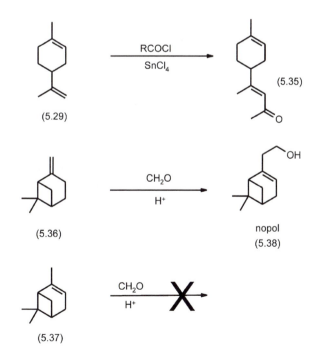

Figure 5.27

the regioselectivity is total because only the more negative end of the olefin adds to the positive carbon atom of the protonated formaldehyde. The product of this Prins reaction is known as nopol (5.38). The reaction is usually carried out in acetic acid and the resulting product acetate is a useful perfumery ingredient.

5.6.10   The *Trans-Anti*-Periplanar Rule

This rule states that all bonds being made or broken in a concerted reaction should preferably be coplanar and aligned in a *trans, anti* geometry relative to each other. The explanation for this lies in the molecular orbitals involved. As illustrative examples, let us look first at the Prins addition of formaldehyde to 1,1-dimethylbuta-1,3-diene (5.39) and then at the elimination of water from an alcohol.

The Prins reaction is shown in two different ways in Figure 5.28. At the top of the figure, we see the traditional form with curly arrows depicting the flow of electrons from the diene system to the protonated formaldehyde to give the intermediate carbocation (5.40). The lower part of the figure shows how, by aligning the p-orbitals of the three double bonds into a TAP configuration, we achieve maximum overlap of them. The electrons are therefore perfectly set up to move from the bonding pattern of the starting materials to that of the product.

In Figure 5.29, we see the elimination of water from a protonated alcohol to give an olefin, water and a proton. Again, the upper part of the figure shows the standard representation. In the lower part of the figure, we can once again see the TAP rule at work. In the starting material, only the orbitals of the breaking single bonds are shown. These are aligned

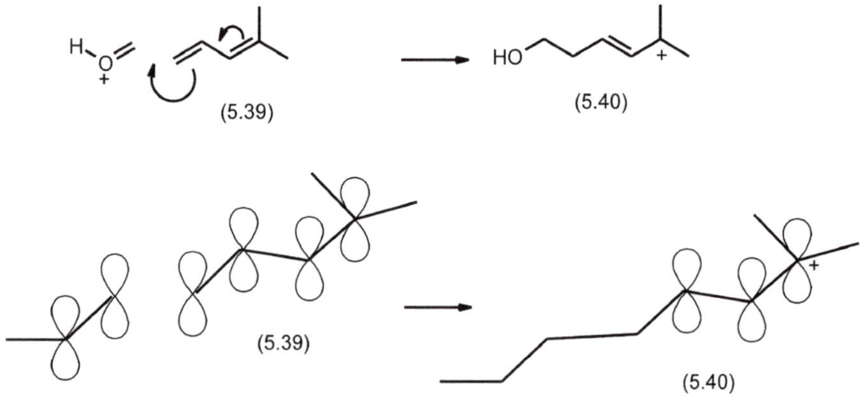

Figure 5.28

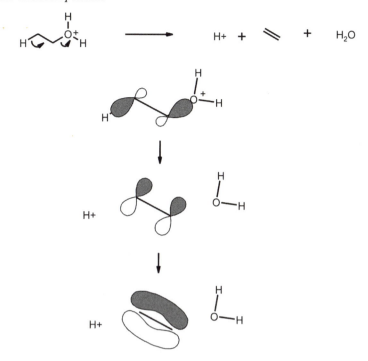

Figure 5.29

TAP which gives the maximum degree of overlap between the smaller lobes at the back of each σ-bond with the larger lobe of the one opposite. This facilitates the formation of the new π-orbital, *via* the p-orbitals, between the two carbon atoms as can be seen in the figure.

Earlier, it was stated that the explanation as to why the staggered conformation is more favourable than the eclipsed, lies in the steric repulsion of neighbouring atoms. There is an alternative explanation which is related to the TAP rule. Figure 5.30 shows the σ-bonds between two geminal carbon atoms and one hydrogen atom attached to each. The bond between the carbon atoms is shown in the staggered conformation and it can be seen how the two σ-bonds lie in the same plane allowing some degree of overlap between the large lobe of each one and the small lobe of the other. It is therefore possible that this stabilisation by overlap is responsible, either totally or in addition to the steric factor, for the preference for the staggered conformation. Evidence to support this theory comes from the fact that, when there is competition between the steric and electronic factors, the electronic one seems to take precedence.

Figure 5.30

5.7 EXPLANATIONS

Having covered the fundamentals of carbocation chemistry, we are now in a position to look again at our three chemical puzzles and find explanations for them.

5.7.1 Explanation for Chemical Puzzle Number 1

The explanation for the first puzzle is shown in Figure 5.31. The addition of hydrogen chloride to α-pinene (5.41) occurs highly regioselectively because of the direction of polarisability of the double bond. If we imagine polarising the double bond completely, *i.e.* putting both the electrons of the double bond onto one of the carbon atoms, we will see that there are two ways in which this could be done. The electrons could, in principle, end up on either the carbon bearing the methyl group or on the carbon bearing the hydrogen atom. In the former case, we would then have a tertiary carbanion and a secondary carbocation. In the latter, we would have a tertiary carbocation and a secondary carbanion. The latter is much more favourable since carbocations prefer to be tertiary and carbanions prefer to be secondary rather than tertiary. The inductive effect pushes electrons in from the other alkyl residues; whilst this will stabilise a carbocation, it will have a destabilising effect on a carbanion. Thus, as the proton approaches the double bond, the latter will polarise in such a way as to direct the proton towards the less substituted end. The intermediate is therefore the relatively stable tertiary carbocation. The reaction could proceed by protonation of the double bond to give a discrete carbocation as described, or it could be that the proton associates with the double bond to give a bridged cation. These two possibilities are shown (for the analogous addition to isobutylene) in Figure 5.32. In the latter case, we can envisage a relatively stable cationic complex existing until the presence of an incoming nucleophile, in this case the chloride anion, disturbs it and causes one of the carbon–hydrogen bonds to break, producing a nascent carbocation which is immediately trapped by the nucleophile. The reality could also be some where between these two extremes. The balance will depend, *inter alia*,

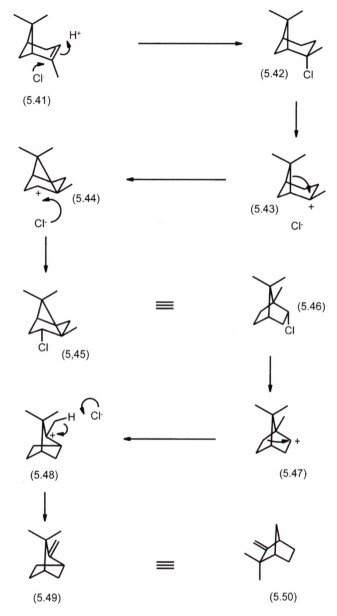

Figure 5.31

upon the reaction conditions. For example, the more polar the reaction medium, the more likely it is that a discrete free carbocation will be formed. A further feature of selectivity is that the relatively bulky choride anion will approach the molecule from the less hindered side. That is, it

Figure 5.32

will add from the side of the CH$_2$ bridge rather than that of the C(Me)$_2$ bridge.

In the case of hydrogen chloride addition to α-pinene (5.41), it is likely that the mechanism has a considerable degree of synchronous character for, if the free cation were to be formed, the molecule would rearrange as happens when pinene hydrochloride is allowed to warm up to room temperature. The reason for the instability of pinene hydrochloride (5.42) lies in the ring strain associated with the cyclobutane ring. This provides a driving force for rearrangement and so, when an opportunity for rearrangement is offered, the molecule jumps at it. The chlorine–carbon bond is polarised and it requires relatively little energy to heterolyse it to give the tertiary carbocation (5.43) and chloride anion. Inspection of the structure of pinene hydrochloride (5.42) reveals that the carbon–chlorine bond is very close to being TAP to the carbon–carbon bond between the adjacent bridgehead carbon atom and that of the dimethylated bridge. Being a quaternary carbon, the latter is also prone to rearrangement. Therefore everything is perfectly set up for heterolysis of the carbon–chlorine bond and migration of the dimethylated bridge to give carbocation (5.44). The chloride ion is already located close to the resultant carbocation and can therefore easily quench it. The resulting 2,2,1-bicycloheptane ring structure is much less strained than the starting 3,1,1-bicycloheptane. The high selectivity for bornyl chloride (5.45) results from the fact that the chloride anion is already on that side of the molecule and adds in an S$_N$2 like fashion relative to the breaking of carbon–carbon bond. As in previous figures, bornyl chloride is drawn in two ways. Structure (5.45) shows its relationship to carbocation (5.44) whereas (5.46) is the more usual representation.

During the reaction, we saw the formation of a secondary carbocation from a tertiary one. This is, of course, all other things being equal, energetically unfavourable. However, in this case the gain in energy from relief of ring strain more than compensates and drives the reaction forward.

For bornylene to be formed from bornyl chloride (5.46), it would require loss of one or other of the hydrogen atoms on the carbon adjacent to that carrying the chlorine. One of these hydrogens is eclipsed relative to the chlorine and the other is skew. Neither can adopt a TAP

Figure 5.33

relationship and so, such an elimination is unfavourable. Heterolysis of the carbon–chlorine bond giving (5.47) allows a second Wagner–Meerwein rearrangement to occur to give the carbocation (5.48), which can eliminate a proton and the product is camphene (5.49), again also redrawn in the more usual way as structure (5.50). This process is shown in Figure 5.31. However, it is also possible that the reaction is initiated by a base, for example a chloride anion, as shown in Figure 5.33. In this figure, it can be seen that the carbon–hydrogen and carbon–carbon double bonds which are broken are TAP, as are the chlorine–hydrogen and carbon–carbon bonds which are formed.

5.7.2 Explanation for Chemical Puzzle Number 2

In our second puzzle, we begin with exactly the same series of reactions from α-pinene hydrochloride (5.46) to camphene (5.50) that we saw in Figure 5.31. They are repeated again in Figure 5.34 to emphasise the difference between the formation of bornyl chloride (5.46) from α-pinene and the formation of isobornyl chloride (5.55) from camphene (5.50).

The addition of hydrogen chloride to camphene (5.50), like that to α-pinene, is highly regioselective. The tertiary chloride is formed and, since the incoming bulky chloride anion will add from the less hindered side, the chlorine atom ends up in the *exo*-configuration giving structure (5.51). (In the 2,2,1-bicycloheptyl skeleton, the substituents on the two-carbon bridge which lie on the side of the one-carbon bridge are referred to as having the *exo*-configuration. Those lying on the side of the other two-carbon bridges, the *endo*-configuration.) This molecule is unstable, not because of ring strain as was pinene hydrochloride, but because there are two serious steric buttressing effects, one between the *exo*-chlorine atom and the *exo*-methyl group on the adjacent carbon and the other between the two *endo*-methyl groups. Heterolysis of the carbon–chlorine bond followed by Wagner–Meerwein rearrangement and then trapping of the newly formed secondary carbocation (5.53) gives isobornyl chloride (5.54) and (5.55). Comparison of camphene hydrochloride (5.51) with isobornyl chloride (5.55) shows that both of the unfavourable

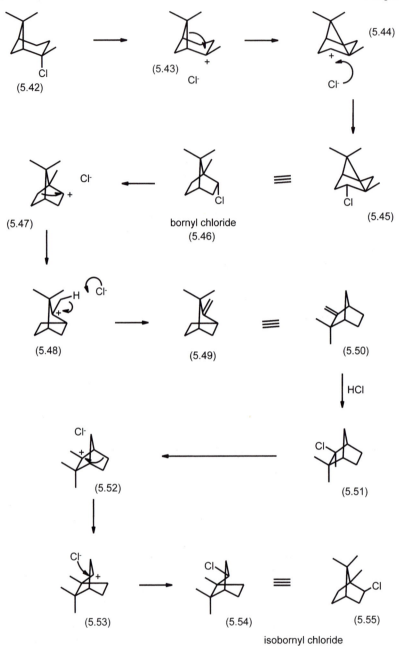

(5.42)
(5.43)
(5.44)
(5.45)
(5.47)
bornyl chloride
(5.46)
(5.48)
(5.49)
(5.50)
(5.51)
(5.52)
(5.53)
(5.54)
(5.55)
isobornyl chloride

Figure 5.34

buttressing interactions have been removed in this rearrangement and this provides the driving force for it.

5.7.3 Explanation for Chemical Puzzle Number 3

As discussed above, camphene hydrochloride suffers from steric strain as a result of two buttressing interactions leading to relatively easy heterolysis of the carbon–chlorine bond. If the material is kept cool enough to prevent the Wagner–Meerwein and subsequent elimination reaction from occurring, what can happen is a simple 1,2-carbon shift reaction. This is shown in Figure 5.35 starting from one enantiomer (5.56). Heterolysis of the carbon–chlorine bond gives carbocation (5.57). The methyl group on the adjacent carbon atom then moves over onto the cationic centre of (5.57) to generate a new carbocation, (5.58). The chlorine now adds back onto the carbon framework to regenerate camphene hydrochloride (5.59). However, it is camphene hydrochloride with a difference. If we rotate the molecule through 180° to give structure (5.51) and then place it next to the original (5.56), we see that it is now the enantiomer of the latter. The overall reaction sequence is, of course,

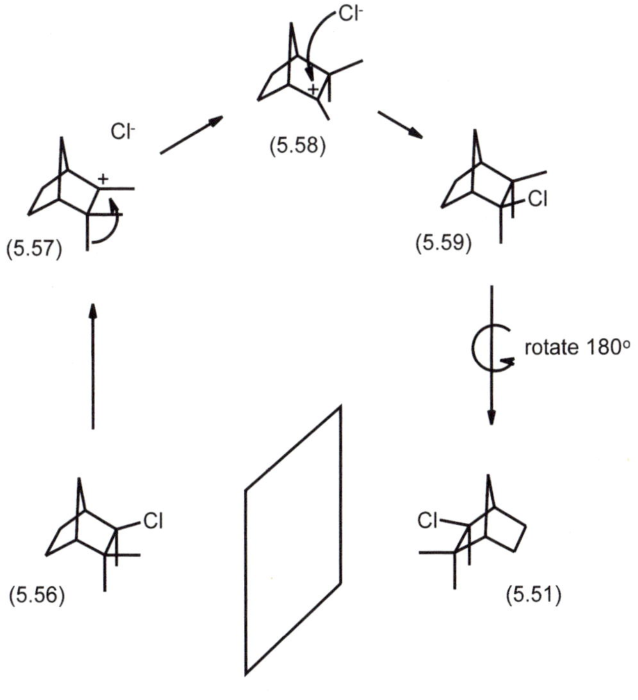

Figure 5.35

reversible and so both forward and reverse reactions will occur and a total racemic mixture will eventually result.

5.8 PREVIOUS CHEMISTRY REVISITED

5.8.1 Carvone

Having seen how the basic principles of carbocation chemistry can explain our three chemical puzzles, it is worthwhile revisiting some of the reactions mentioned in previous chapters to see how the same factors were at work in those cases. Firstly, we can look at the key step in the conversion of limonene to carvone which we encountered in Chapter 4, Figure 4.13. This is shown again in Figure 5.36, but now we can understand why the observed selectivity is found. The nitrosyl cation is very electron deficient and seeks out the more electron rich double bond of limonene (5.29). This is the trisubstituted double bond in the ring. The addition takes place entirely on C2 since the resulting carbocation is tertiary and hence more stable than the secondary carbocation which would be formed if the reaction were to occur in the opposite direction. Therefore only carbocation (5.60) is formed.

5.8.2 Biogenesis of the Acorane Skeleton

Let us now look at some of the biogenetic pathways discussed in Chapter 2. The same factors which control the chemistry of carbocations in the reaction flask also apply in natural systems. Of course, the natural systems are also guided by enzymes which hold the reacting species together and can dramatically affect the outcome of the reaction, by giving preference to what might otherwise be a less favourable alternative. However, for the present, we will consider those which illustrate simple carbocation chemistry in action.

First, let us look at the biosynthesis of the acorane skeleton which is shown as part of Figure 2.10. The part of the scheme leading to acorane

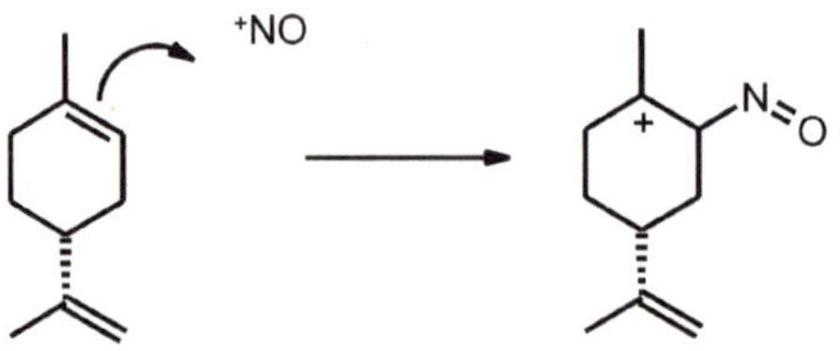

(5.29)   (5.60)

Figure 5.36

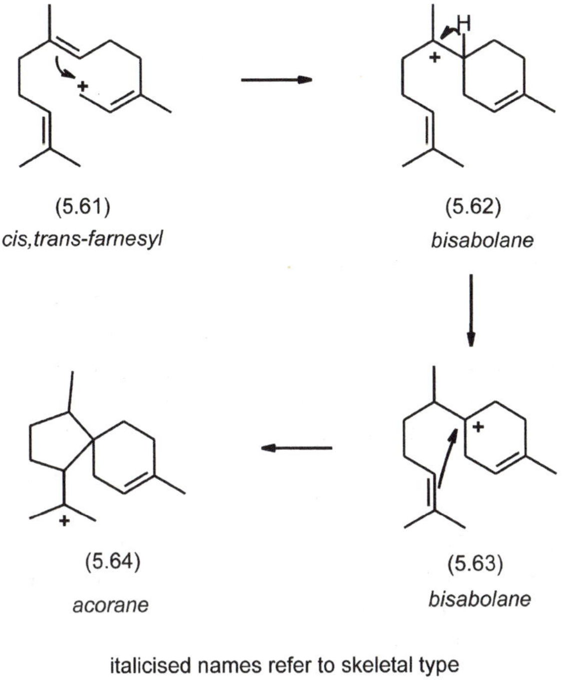

(5.61)

cis,trans-farnesyl

(5.62)

bisabolane

(5.64)

acorane

(5.63)

bisabolane

italicised names refer to skeletal type

Figure 5.37

is shown in Figure 5.37. Heterolysis of the carbon–oxygen bond of *cis,trans*-farnesyl pyrophosphate leads to the corresponding cation, (5.61). This is perfectly placed to cyclise to the bisabolyl carbocation, (5.62). The formation of the six-membered ring is very favourable and the attack at the monosubstituted end of the double bond produces the relatively stable tertiary carbocation, a considerable energy saving over the starting primary carbocation. The double bond at the end of the molecule represents a good source of electrons for further reaction of carbocation (5.62). However, if cyclisation were to occur to the preferred end of the double bond, it would form a strained four-membered ring. It therefore follows a 1,2-hydrogen shift to give the isomeric carbocation (5.63). This is a level process as far as energetics are concerned, but it now means that the cyclisation to the monosubstituted end of the double bond will produce a stable five-membered ring and a stable tertiary carbocation (5.64) having the acorane skeleton.

This is a common feature in carbocation chemistry. Various rearrangements will run, usually in equilibrium with each other, between structures of similar energy, until one structure is reached which represents a minimum in free energy. Eventually all of the material will fall into this free energy "well" and the equilibrium is broken. Occasionally, one of the intermediates might be slightly higher in energy than its precursor but, if the final free energy level is low enough, the energy liberated will be sufficient to drive the entire bulk of the material into the "well".

5.8.3 Biogenesis of the Guaiane Skeleton

It is also possible for the *trans,trans*-farnesyl carbocation (5.65) to cyclise to the 10,11-double bond to give carbocation (5.66) which possess the germacrane skeleton. The 10-membered ring of the germacrane skeleton suffers considerable steric strain and one solution is to undergo a series of hydrogen shifts to produce the isomeric germacrane carbocation, (5.67). This carbocation can now eliminate that steric strain by cyclising across the 10-membered ring to produce two fused rings, one five- and one seven-membered, in the form of the guaiane skeleton of carbocation (5.68). The *trans*-annular cyclisation reaction again proceeded *via* addition of a carbocation to the less substituted end of a double bond, producing a stable tertiary carbocation. This is shown in Figure 5.38 and accounts for the reaction sequence shown as part of Scheme 2.12.

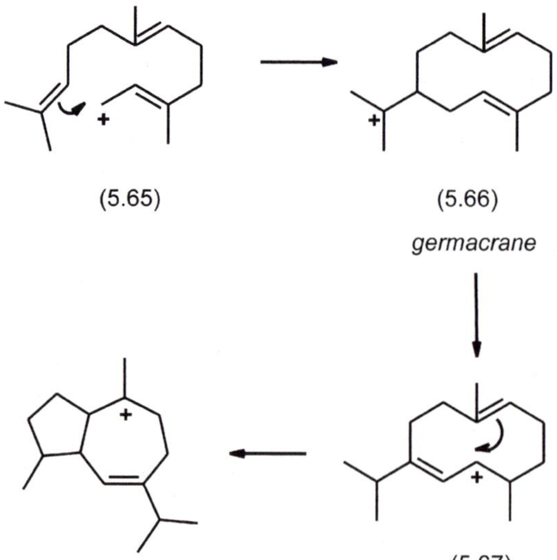

(5.65)

(5.66)
germacrane

(5.68)
guaiane

(5.67)
germacrane

italicised names refer to skeletal type

Figure 5.38

5.8.4 Biogenesis of the Skeleton Caryophyllane and Himachalane Skeleta

In Figure 5.38, we saw the *trans,trans*-farnesyl carbocation cyclise in the preferred way, as far as the bond polarisation is concerned, to the 10,11-double bond. However, this gave a very sterically crowded 10-membered ring. An alternative is to cyclise to the "wrong" end of the double bond as shown in Figure 5.39. Starting from the *cis,trans*-farnesyl carbocation (5.61), this gives a secondary carbocation but a slightly less strained 11-membered ring. Thus the humulane carbocation (5.69) is formed. Two of the possible reactions open to this cation are shown in Figure 5.39. Capture of the carbocation by the electrons of the neighbouring double bond, according to the arrow marked (a), gives the caryophyllane skeleton of (5.70). The caryophyllane skeleton is still very strained as it contains a four-membered and a nine-membered ring. The nine-membered ring is particularly strained because of the presence

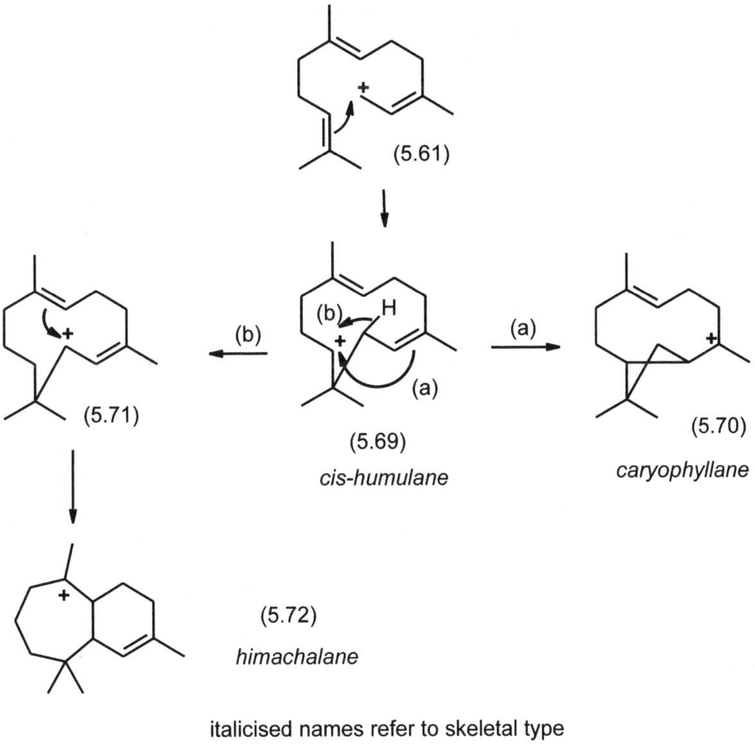

italicised names refer to skeletal type

Figure 5.39

of the *trans*-double bond. The geometry of this bond is not at all easily accommodated in the ring, as molecular models will quickly reveal. This strain is reflected in the chemistry of terpenoids of the caryophyllene family as will be seen in Chapter 8. Another option for carbocation (5.69) is to undergo a 1,3-hydrogen shift as shown with arrow (b). This gives the more stable allylic carbocation (5.71) and allows a *trans*-annular cyclisation to the himachalane skeleton of (5.72). As with guaiane, this fused five,seven-ring system is relatively free of strain.

5.8.5   Biogenesis of the Steroid Skeleton

Figure 5.40 shows the acid catalysed cyclisation of epoxy-squalene (5.73) to the lanosteryl carbocation (5.74). This is one of the loveliest examples of the TAP rule at work. The regiochemistry of each part of the reaction is as would be expected. The protonated epoxide opens so that a tertiary carbocation is formed. Then this cation is trapped by a double bond, the carbocation thus produced is trapped by the next and so on. Each time, the addition occurs at the less hindered end of the double bond to leave a further tertiary carbocation. This is very impressive as a long chain molecule folds up on itself to produce the polycyclic steroid skeleton. However, the greatest beauty of the reaction lies in its total stereospecificity. This is the result of the TAP rule and the fact that all of the double bonds in squalene are *trans*. Inspection of the drawing in Figure 5.40, or even better, of a molecular model of it, shows how all of the bonds involved are aligned in the TAP configuration in this preferred orientation for cyclisation and, by virtue of that fact, the stereochemistry of the steroid ring system is sealed.

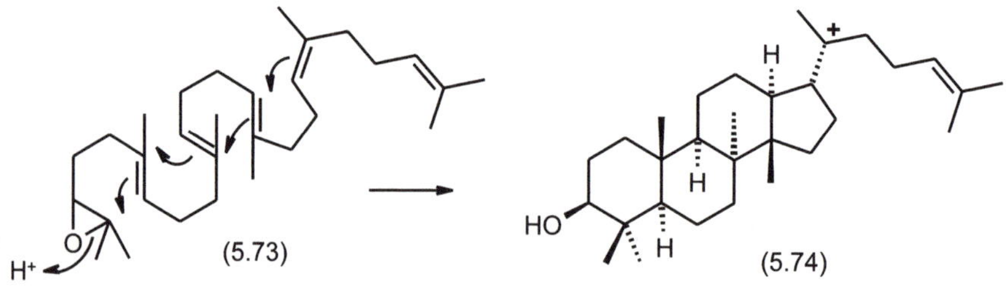

epoxy-squalene

Figure 5.40

The reader would do well to look through all of the biosynthetic schemes of Chapter 2 and rationalise them, in a similar way to the examples above, in terms of the principles described in this chapter.

5.9   AN ANIONIC TRANSANNULAR REACTION

We saw how *trans*-annular reactions are common in carbocation chemistry. The close proximity of atoms in polycyclic systems means that, not only carbocation reactions, but also carbanion and free radical are prone to take place across the ring system. Figure 5.41 shows an anionic reaction in a bicyclic monoterpenoid.

When fenchone (5.75) is treated with allylmagnesium bromide, an intermediate (5.76) is formed which gives a negative Gilman test, showing that the organometallic reagent has been consumed. Addition of water to the product produces the expected allyl fenchyl alcohol (5.77). The intermediate can therefore be assumed to be the magnesium salt of this alcohol. Similar treatment of fenchone with methyl magnesium bromide leads to an intermediate, the existence of which is confirmed by a negative Gilman test indicating that all of the organometallic reagent has been consumed. However, this time, when water is added, it is not the expected alcohol which is liberated from its magnesium salt, but fenchone (5.75). This rather odd result tells us something interesting about the Grignard reaction.

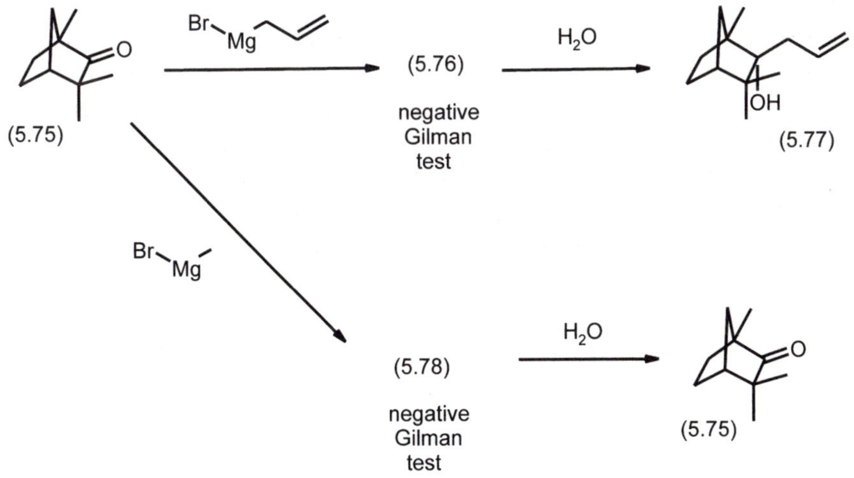

Figure 5.41

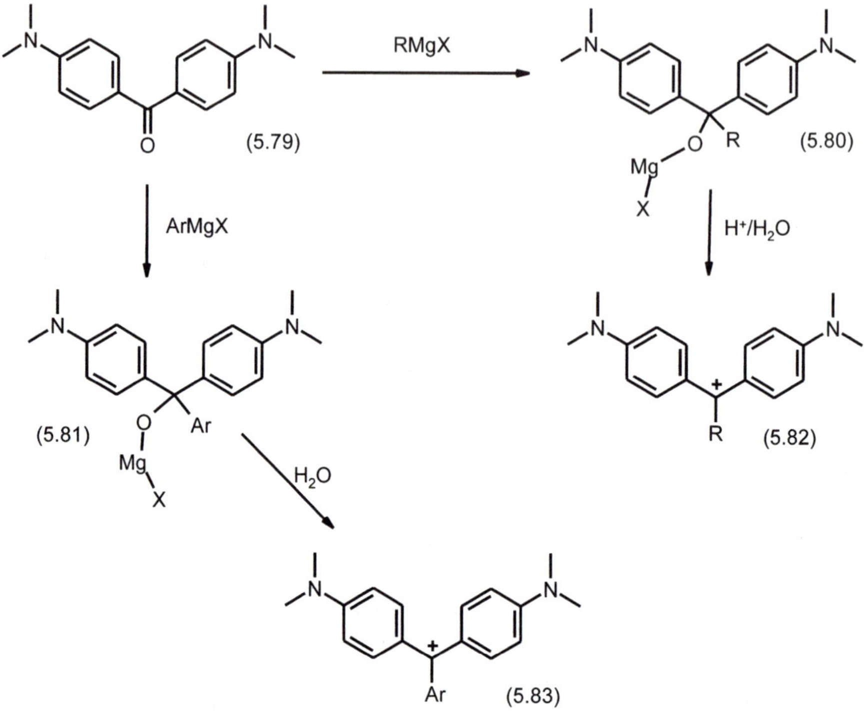

Figure 5.42

Firstly, it might be useful to explain what the Gilman test is. The test is based on the reaction of Michler's ketone (5.79) with Grignard (or other organometallic) reagents as shown in Figure 5.42. Addition of the reagent to the carbonyl group of the ketone gives the magnesium salt of the corresponding alcohol, (5.80) in the case of alkyl Grignards and (5.81) in the case of aryl reagents. With alkyl Grignard reagents, addition of dilute aqueous acid produces the cation (5.82) which has an intense blue colour. The corresponding cation (5.83) which is formed from aryl Grignard reagents, will produce the blue colouration with the addition of just water alone. This occurs because of the increase in stabilisation of the carbocation thanks to the additional aryl ring.

So, in the case of the reaction of fenchone with methyl magnesium bromide, we can be certain that the Grignard reagent has undergone some reaction with the fenchone since there is none left in the system. Furthermore, GC and IR analysis of the reaction medium shows that there is no longer any fenchone present. Normally one would suspect that the Grignard reagent had acted as a base and formed an enolate

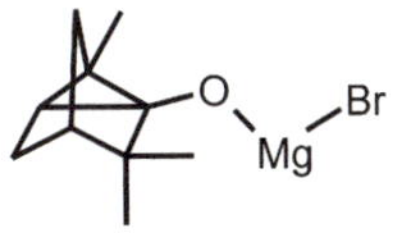

(5.78)

Figure 5.43

by abstraction of a proton α- to the ketone function. However, in this case, there are no protons α- to the ketone. So what could this product be which then releases fenchone on hydrolysis? The answer is that it is the homo-enolate of fenchane, formed by *trans*-annular abstraction of a proton. The structure (5.78) of this material is shown in Figure 5.43.

The explanation for this rather odd behaviour lies in the mechanism of the Grignard reaction as shown in Figure 5.44 and the very compact structure of polycyclic terpenoid structures. The normal Grignard reaction is a termolecular reaction as shown in the figure. It must be remembered that the magnesium atom is bonded, not only to the alkyl group and the halogen, but also to two molecules of ether from the solvent. The six-membered transition state is therefore very crowded and in the case of fenchane, this crowding is too great to be accommodated around such a hindered ketone. The reagent therefore seeks an alternative course. Since fenchane has no enolisable hydrogens α- to the carbonyl group, it is forced to remove one from the far side of the ring with the formation of a *trans*-annular bond. This is possible because of the compact ring structure which places atoms on opposite sides of

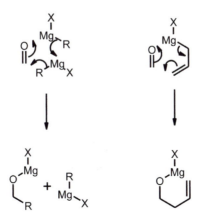

Figure 5.44

the ring in close physical proximity. Allyl Grignard reagents react through a six-membered transition state involving an allylic shift. This is a bimolecular reaction, as shown in the figure, and therefore has a lower steric requirement than that of an alkyl Grignard reagent. This difference is sufficient to allow the reaction with fenchone to proceed in the normal way.

5.10   A NEUTRAL TRANSANNULAR REACTION

A *trans*-annular reaction involving a carbene insertion is shown in Figure 5.45. The starting material is camphor (5.75). Treatment with hydrazine gives the hydrazone (5.84) which can be oxidised to the diazonium salt (5.85) by yellow mercuric oxide. Loss of nitrogen from this gives the carbene (5.86). The electron pair of the carbene then inserts into the carbon–hydrogen bond closest to it, which happens to be on the other side of the ring. The product is therefore, tricyclene, (5.87). The tricyclene molecule suffers from considerable ring strain, as is clearly evident when trying to build a molecular model. It is well known that it will isomerise to camphene, with considerable relief of that ring strain, upon treatment with acid.

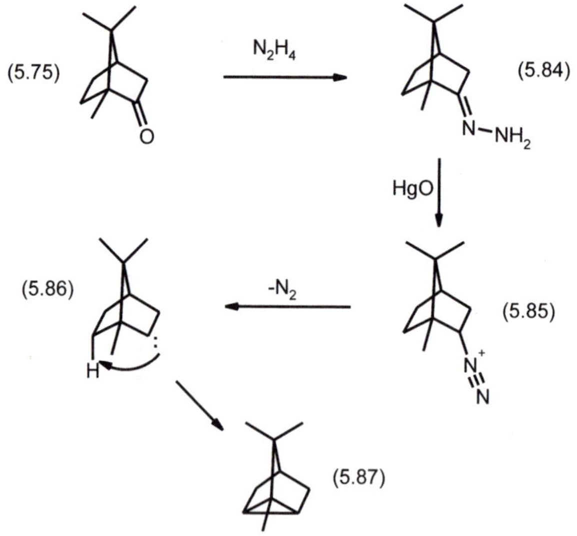

Figure 5.45

© Hulton-Deutsch Collection/CORBIS

CHAPTER 6

Precious Woods

The chemistry of cedar terpenoids is rich
and varied

Quinquireme of Nineveh from distant Ophir
Rowing home to haven in sunny Palestine,
With a cargo of ivory,
And apes, and peacocks,
Sandalwood, cedarwood and sweet white wine.

John Masefield

The odours of sandalwood and cedarwood and their resistance to attack
by fungi and insects have made them prized articles of commerce for both
construction and decorative purposes for thousands of years. Similarly
the oils derived from the woods have a long history of use in perfumery.

KEY POINTS

The same rules of carbocation chemistry apply to sesquiterpenoids
as to monoterpenoids. The fact that the molecules are bigger
means that the reactions can be harder to see and that there may
be more possible pathways for a reaction.

When synthesising a natural product, it is important to ensure that
the synthesis produces the correct isomer.

The course of carbocation reactions is affected by the reaction
conditions and changes to conditions can produce surprising
results.

The stereochemistry of carbanion reactions can be understood
through the use of Newman projections of the transition states.

6.1 SANDALWOOD

In his poem, "Cargoes", Masefield describes a ship of the Assyrian
Empire carrying sandalwood as an article of commerce, imported from

the East Indes. This is not an unlikely scenario. Historical records show that there has been uninterrupted use of Sandalwood Oil in perfumery for at least 4000 years. It is obtained by distillation of the wood of the parasitic tree *Santalum album*. The wood itself is used for furniture and for decoration in addition to its use as a source of the oil. To produce the oil, the wood, taken from trees of about 40 years of age, is sawn and chopped into small pieces and then steam distilled. The largest production in the world is in India, particularly from the state of Karnataka (Mysore). The tree also grows in Sri Lanka and Indonesia. Cultivation is difficult because of the parasitic nature of the tree and the consequent need for a suitable host. Excessive harvesting has endangered the species and control of production is now necessary to prevent extinction.

The major components of the oil are α-santalol (6.1) and β-santalol (6.2). Typically, the oil will contain 45–47% of the α-isomer and 20–30% of the β-isomer. Both isomers contribute to the distinctive woody odour of the oil. The β-isomer is more intense and also contributes to the slightly animalic, urinous character of the oil. It is interesting to note that the odours of sandalwood and urine are closely related. One of the commonest parosmias (where an odorant is perceived differently by different subjects) is that of sandalwood and urine. Similarly, when a secondary alcohol with a sandalwood character is oxidised to the corresponding ketone, there is a high probability that the ketone will smell urinous. There are many other components in sandalwood oil, particularly sesquiterpenoids alcohols, aldehydes and hydrocarbons. Some of these make important contributions to the overall odour effect.

The close structural relationship between the two major components was demonstrated by oxidative degradation of both to the corresponding carboxylic acids, (6.3) and (6.4), respectively, as shown in Figure 6.1. Treatment of the acid (6.3), derived from α-santalol, with hydrogen chloride, gave the acid (6.4), identical to that obtained from β-santalol. This rearrangement is exactly analogous of that from tricyclene (6.5) to camphene (6.6), as shown by inclusion of the structures of these latter two materials in Figure 6.1. (As discussed at the end of Chapter 5, the rearrangement of tricyclene to camphene is well known in monoterpenoid chemistry.) Examination of the two pairs of materials, (6.1) and (6.2) and (6.5) and (6.6), in the figure will reveal that the santalols could be considered to be formed from the monoterpenoids by addition of the five carbon alcoholic chain. It is therefore not surprising that the chemistry of the santalols mirrors that of camphene and tricyclene (apart from the obvious differences which would be expected from the presence of the side chain and its allylic alcohol function).

The way in which the santalols are drawn in Figure 6.1, is used to make clear their relationship to tricyclene and camphene. Often the structures will be drawn in a different way. For example, Figure 6.2 shows two different ways of drawing α-santalol. The first, (6.1), is the way in which it is shown in Figure 6.1. The other, (6.7), is often used to illustrate the close similarity in overall shape between α-santalol and β-santalol (6.8), a feature which is less obvious in the structures, (6.1) and (6.2), used in Figure 6.1. This second way of representing the structures is of particular use when discussing structure/property relationships when patterns of spatial distribution of electron density are being sought. At first sight, structures (6.1) and (6.7) look rather different. Even an experienced eye might not recognise that they are, in fact, different representations of the same molecule. This is an example of the value of molecular models. If a model of (6.1) is constructed, it will be easy to re-orientate it so as to show that it is identical to (6.7) and *vice versa*.

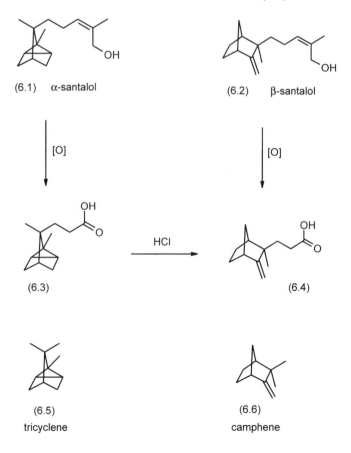

(6.1)   α-santalol

(6.2)   β-santalol

[O]

[O]

(6.3)

HCl

(6.4)

(6.5)
tricyclene

(6.6)
camphene

Figure 6.1

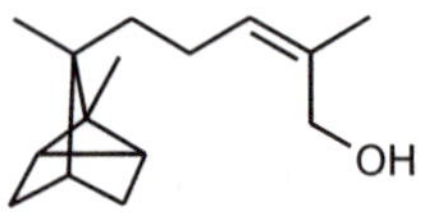

(6.1)

α-santalol

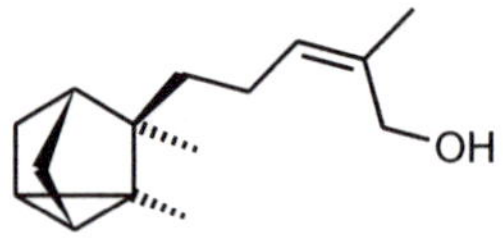

(6.7) α-santalol

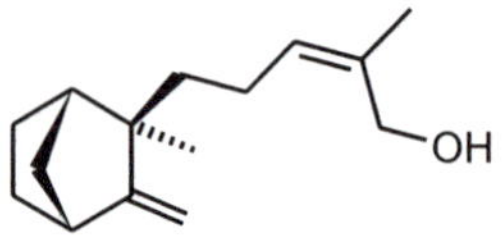

(6.8) β-santalol

Figure 6.2

6.1.1 Synthesis of α-Santalol

The first synthesis of α-santalol was carried out by Guha and Batta-charyya in 1944, and their synthetic scheme is shown in Figure 6.3. Their starting material was camphor (6.9), which is a good choice since it provides the basic framework of the ring system and sufficient functionalisation to enable the rest of the molecule to be built-up around it. The two key changes which are needed, are the addition of the extra five carbons in the side chain and the formation of the third ring. The ketone function of camphor plays a vital role in achieving both these objectives, but the exact nature of its role in each is quite different.

In order to attach five more carbon atoms to one of the geminal methyl groups, it is necessary to functionalise this position. Functionalisation of a saturated hydrocarbon is not easy and requires very aggressive reagents. Guha and Battacharyya decided to sulfonate the methyl group using chlorosulfonic acid. However, there is a problem with this in that the more reactive ketone group and the acidic methylene hydrogens adjacent to it would react faster than the relatively inert hydrogens on the methyl group. The ketone group is needed for the introduction of the third ring, and clearly, if the cyclopropyl ring were to be introduced first, it would not survive the treatment with chlorosulfonic acid. There is also a second problem. They needed to be able to selectively introduce one sulfonic acid function only. Their elegant solution to both these problems was to brominate camphor. The addition of bromine to the enol form of camphor followed by dehydrobromination and re-ketonisation gave α-bromocamphor, (6.10). The presence of the large

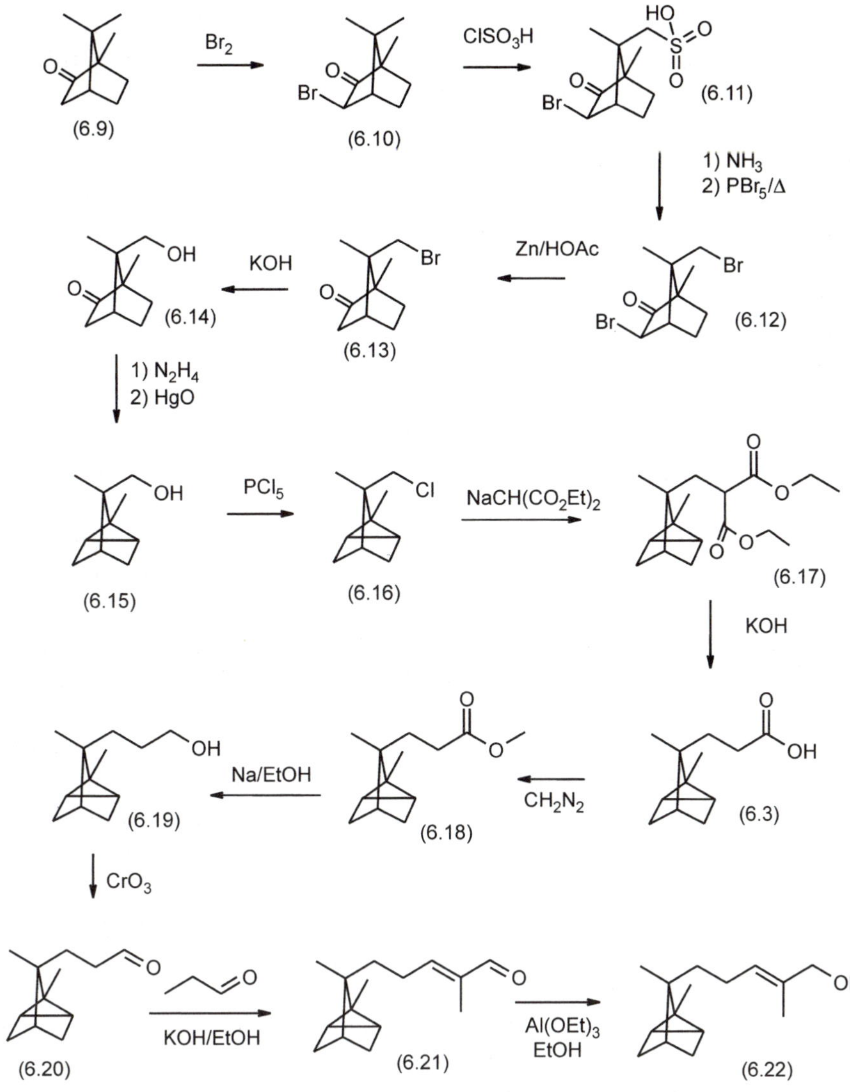

Figure 6.3

bromine atom then serves to protect both the carbon atom α to it and the methyl group on that side of the molecule from attack by the bulky chlorosulfonic acid. Thus sulfonation can be achieved selectively giving the sulfonic acid (6.11). Heating of the ammonium salt of this acid with phosphorus pentabromide then gave the dibromocamphor (6.12). The two bromine atoms display different degrees of reactivity and the one adjacent to the carbonyl group can be reduced selectively by zinc and acetic acid, leaving the other bromine atom unaffected. The resultant 10-bromocamphor (6.13) can then be treated with potassium hydroxide in a nucleophilic substitution reaction to give 10-hydroxycamphor (6.14).

Having served as a protecting group in the first part of the synthesis, the carbonyl group was then used to introduce the third ring. This used the same reaction described in Chapter 5, Figure 5.45. Treatment of the hydrazone of 10-hydroxycamphor with mercuric oxide resulted in generation of the carbene which then inserted into the C2–H bond opposite to give the cyclopropane ring of (6.15).

It was necessary to replace the bromine atom of (6.13) by a hydroxyl group in order to prevent reaction of the halide with the hydrazine used in the cyclisation reaction. However, it was then necessary to re-introduce a halogen in order to provide a leaving group for the next stage. This was achieved by treatment with phosphorus pentachloride to give 10-chlorotricyclene, (6.16). Reaction of this with the sodium salt of diethyl malonate gave the diester (6.17), which could be converted to the acid (6.3) by hydrolysis and decarboxylation. This is, of course, the same acid that can be obtained by oxidation of α-santalol. Esterification of this acid with diazomethane gave the ester (6.18), which could be reduced by dissolving sodium in ethanol to give the alcohol (6.19). Oxidation of the alcohol to the corresponding aldehyde, (6.20), was achieved using chromium trioxide and aldol condensation of this product with propion-aldehyde gave the unsaturated aldehyde (6.21). Meerwein–Ponndorf–Verley reduction using aluminium triethoxide in ethanol gave the alcohol (6.22). However, this alcohol was not identical to the natural α-santalol and had no sandalwood odour.

The answer, of course, is that the natural material is the *Z*-isomer, whereas Guha and Battacharyya had synthesised the *E*-isomer. There was nothing that they could do about this at that time. In 1944, the only method for selective synthesis of *Z*-isomers of olefins was the Lindlar hydrogenation of acetylenes and that is clearly inappropriate in this instance as the target is a trisubstituted olefin. Lindlar hydrogenation is only suitable for the synthesis of 1,2-disubstituted *Z*-olefins.

6.1.1.1 Stereochemistry of the Aldol Condensation. The aldol condensation displays high selectivity towards *E*-olefins as a consequence of the *trans-anti*-periplanar rule. It is worthwhile digressing at this point to look at this mechanism in more detail as, by doing so, we can learn how to achieve the opposite selectivity. A typical aldol condensation is shown in Figure 6.4.

Figure 6.4 shows the reaction between an aldehyde (6.23) as the electrophile and a ketone (6.24) as the nucleophile. The principles to

(6.23)

(6.24)

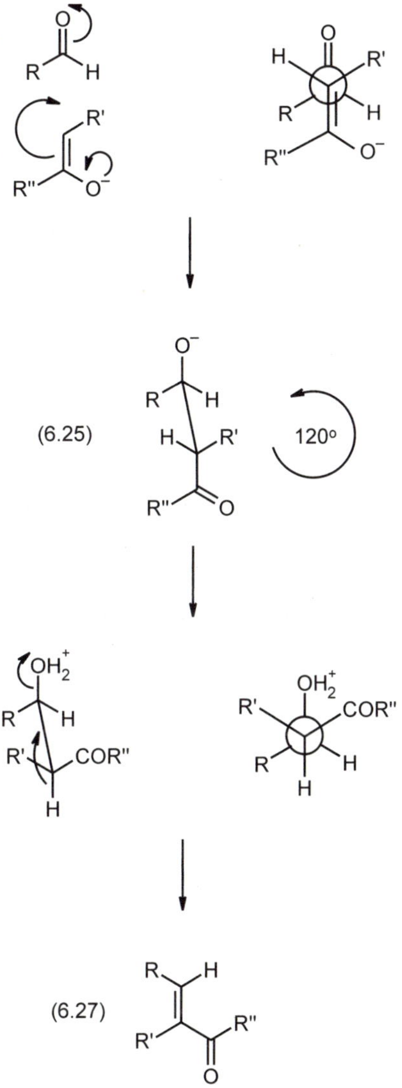

(6.25)

(6.26)

(6.27)

Figure 6.4

be described apply equally to cases where the electrophile is a ketone and those where the nucleophile is an aldehyde. If the electrophile is a ketone, then the smaller alkyl residue would take the place of the hydrogen atom in structure (6.23). The initial step of the sequence is the nucleophilic addition of the enolate anion to the carbonyl bond of the aldehyde. The *trans-anti*-periplanar rule applies and so all four sp^2

carbon atoms will align themselves in that configuration as shown. The Newman projection to the right of the figure shows the view along the bond in the process of being formed. In order to minimise the steric interactions, the alkyl groups R and R′ will align themselves opposite to each other as shown. If the enolate were to adopt the other possible orientation, then the residue R would be squeezed between R′ and the carbonyl function. Thus the aldol reaction product is the protonated form of the intermediate (6.25). If elimination of water is not possible, this aldol product will be the isolable outcome of the reaction with the relative stereochemistry shown. If there remains a proton next to the carbonyl group, then aldol condensation will be possible, to give the unsaturated ketone. The example in Figure 6.4 falls into this category. The alcoholate anion (6.25) is protonated to neutralise it and then protonated again to produce a good leaving group, water. Once again, the *trans-anti*-periplanar rule applies and so the proton to be eliminated will line up *trans-anti*-periplanar to the nascent water molecule as shown in structure (6.26). In order to do this, the molecule must rotate 120° around the newly formed bond as shown in the figure, to give structure (6.26). The Newman projection to the right shows how the alkyl groups R and R′ are now on the same side of the plane of the reacting bonds and that elimination of water will lead to the *E*-isomer of the product as shown.

6.1.1.2 Synthesis of Z-α-Santalol . In 1970, Corey was able to complete Guha and Battacharyya's synthesis to give the natural *Z*-α-santalol, thanks to the reaction discovered in 1953 by Wittig and named after him. Corey's final step in the synthesis is shown in Figure 6.5. This is a variation on the basic Wittig reaction known as the Wittig–Schlosser

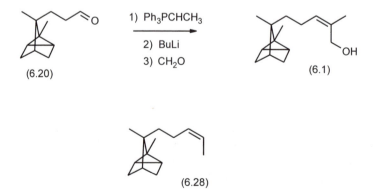

Figure 6.5

reaction. Treatment of aldehyde (6.20) with ethylidenetriphenylphosphorane gave an intermediate which, if allowed to eliminate
triphenylphosphine oxide would produce the olefin (6.28). However,
Corey intercepted the intermediate betaine and treated it with butyl
lithium and then formaldehyde to give, on work-up, Z-α-santalol.

6.1.1.3 The Wittig and Wittig–Schlosser Reactions. In order to understand what is happening here, it would be helpful to first look at the
mechanism of the Wittig reaction. This is shown in Figure 6.6, as an
example, showing the addition of an alkylidenetriphenylphosphorane
(6.29) to an aldehyde (6.23). Obviously, it is possible for the reaction to
occur between more highly substituted reagents, in which case, the
smaller groups on each reagent would occupy the place given to the
hydrogen atoms of structures (6.23) and (6.29).

The initial step of the Wittig reaction is similar to that of the aldol
reaction. The double bonds of the aldehyde (6.23) and the phosphorane

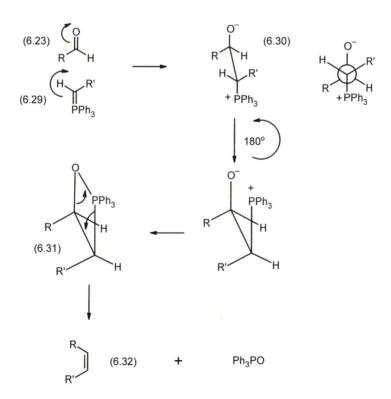

Figure 6.6

(6.29) align themselves in a *trans-anti*-periplanar arrangement with the alkyl groups directed away from each other in order to minimise steric interactions. The new bond forms between them to give the betaine (6.30) which is shown both in stereo structure and as a Newman projection. Betaines are compounds which carry both a positive and a negative charge, but not on adjacent atoms. In this case, the positive charge is on the phosphorus atom and the negative on the oxygen. In order to complete the reaction, it is necessary to eliminate the phosphine oxide. This cannot happen in the same way as for the aldol reaction where there is a flow of electrons across the molecule as shown in structure (6.26) in Figure 6.4. Instead, it occurs through the formation of a four-membered ring containing both the phosphorus and the oxygen atoms. In order to form the ring, the molecule must rotate by 180° around the newly formed bond as shown in Figure 6.6. This brings the positively charged phosphorus atom and the negatively charged oxygen atom close enough to form a bond between them and thus neutralise the electrical charges. A $2+2$ cycloreversion reaction can then proceed to produce the olefin (6.32) and triphenylphosphine oxide. As stated earlier, when the carbon–carbon bond is being formed, the larger two alkyl groups place themselves as far apart as possible in order to reduce steric interactions to a minimum. However, in order to form the four-membered ring of (6.31), they must come together into an eclipsed configuration. This is energetically unfavourable, but the energy penalty of doing so is more than compensated for by the formation of a phosphorus–oxygen bond. Elimination of the phosphine oxide then locks the two alkyl groups into the Z-configuration around the newly formed olefinic bond.

In the Wittig–Schlosser reaction, the betaine (6.30) is prevented from closing to the four-membered ring by keeping it cold (usually −78 °C). A second molar equivalent of butyl lithium is then added, and this removes a proton from the carbon adjacent to the phosphorus atom giving structure (6.33) which now carries one positive charge and two negative ones. This sequence of events is shown in Figure 6.7.

The carbanion centre of (6.33) is planar and so, in principle, can be approached from either side by a suitable electrophile. However, as can be seen from structure (6.33) in Figure 6.7, one side is more hindered as a result of the bulky alkyl group, R. An electrophile will therefore approach from the side of the hydrogen atom as shown. In this case, the electrophile is formaldehyde and the product of addition is the lithium alcoholate (6.34). Rotation by 180° around the carbon–carbon bond allows the formation of the phosphorus–oxygen bond of (6.35) and subsequent elimination of the phosphine oxide to give the allylic alcohol (6.36).

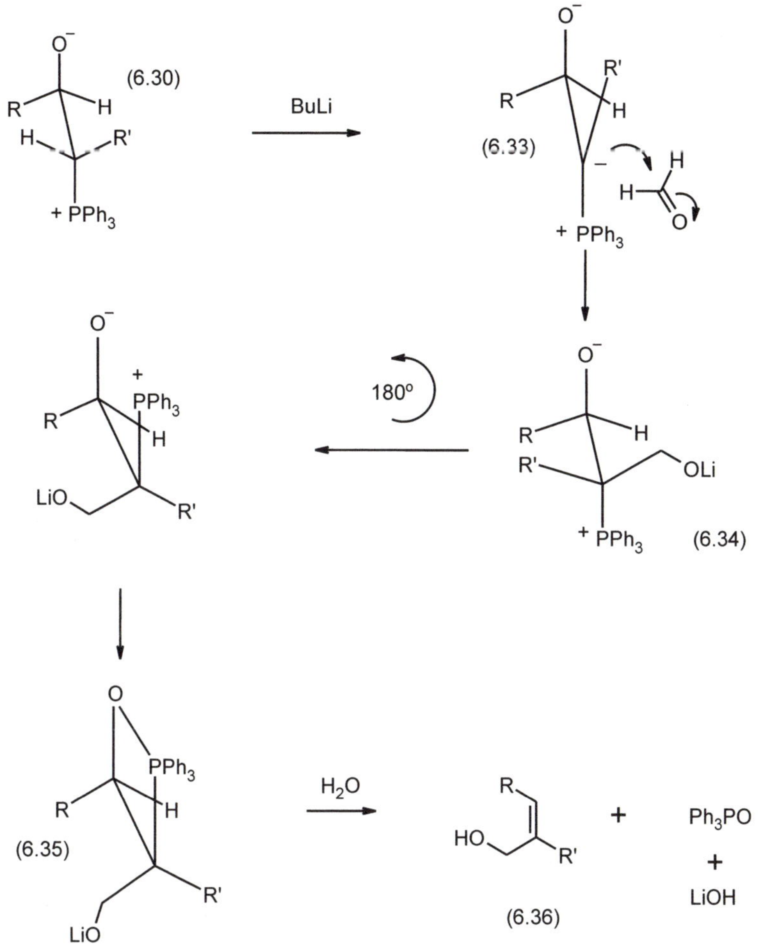

Figure 6.7

As can be seen from Figure 6.7, following the fate of the groups around the reaction centre reveals that the carbon carrying the oxygen atom will end up in the Z-configuration relative to the alkyl group R, at the opposite end of the double bond. Thus, Corey was able to use this reaction to adapt the synthesis of Guha and Battacharyya to produce the desired product with the correct geometry.

6.1.2 Synthesis of β-Santalol

As an example of the synthesis of β-santalol, we will look at that of Erman and Kretschmar, since it also illustrates two interesting points of general relevance. The overall scheme is shown in Figure 6.8.

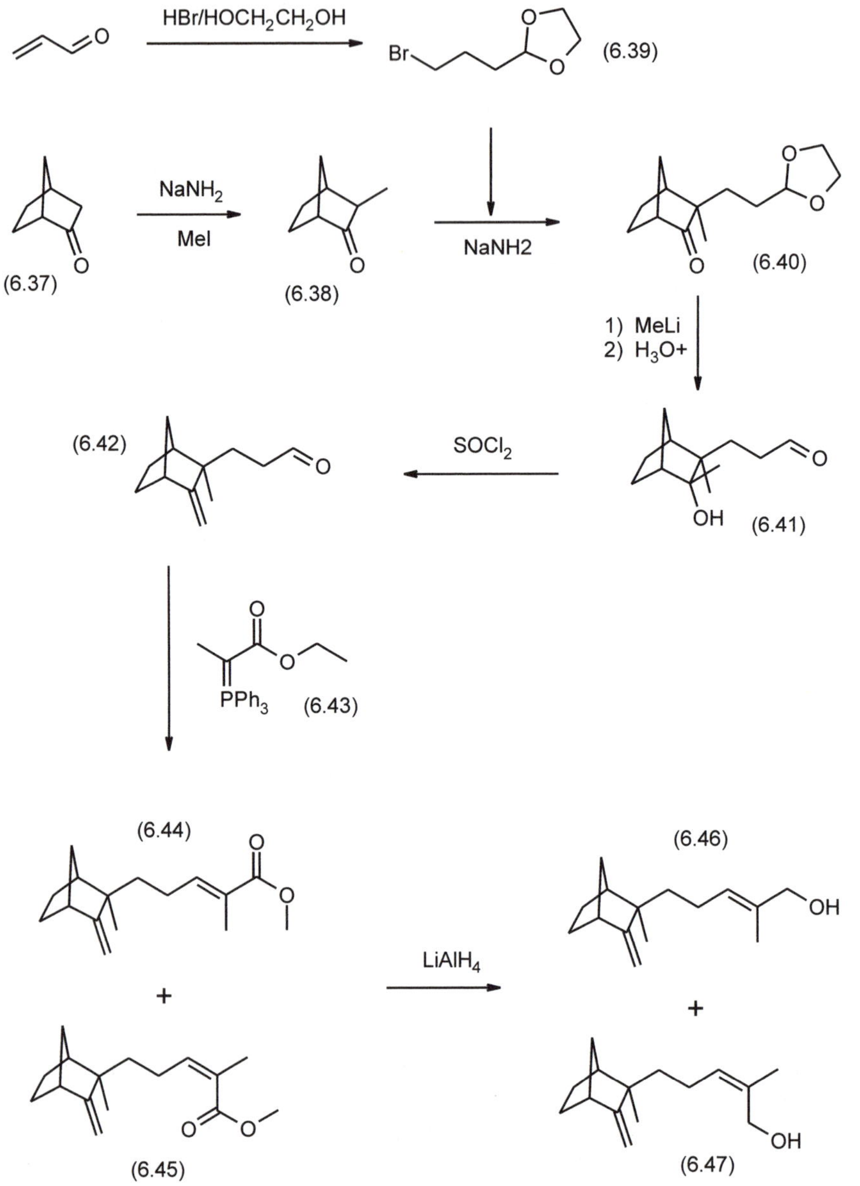

Figure 6.8

The synthesis starts from norbornanone (6.37). This is methylated using iodomethane with sodamide as the base. Alkylation of methyl-norbornanone (6.38), norbornanone (6.37) with Buchi's bromide (6.39) (prepared by addition of hydrogen bromide to acrolein in ethylene glycol), again using sodamide as the base, gives the protected keto-aldehyde (6.40). Only the aldehyde function is protected and so methyl lithium adds totally selectively to the ketone group. The acid in the work-up of this reaction releases the aldehyde function to give the hydroxyaldehyde (6.41). This can be dehydrated using thionyl chloride. Addition of the ylid (6.43) to the resultant unsaturated aldehyde (6.42), gives a mixture of *E*- and *Z*-unsaturated esters, (6.44) and (6.45), respectively. These were reduced by lithium aluminium hydride to give a mixture of *Z*-β-santalol (6.47) and its *E*-isomer (6.46).

6.1.2.1 The Exo-effect. The first interesting feature of this synthetic sequence is what is known as the *exo*-effect. Figure 6.9 shows norbornene (6.48) being approached from either side of the double bond. The face of the double bond next to the one-carbon bridge is known as the *exo*-face and the face next to the two-carbon bridge as the *endo*-face. Whatever be the nature of a reagent which approaches the double bond in order to react with it, it will experience greater resistance on the *endo*-face and so is more likely to add from the *exo*-side.

We see this happen three times in Erman and Kretschmar's β-santalol synthesis. The first is the methylation of the norbornanone enolate (6.49). The iodomethane approaches from the *exo*-face to give the *exo*-isomer (6.38) as shown in Figure 6.10.

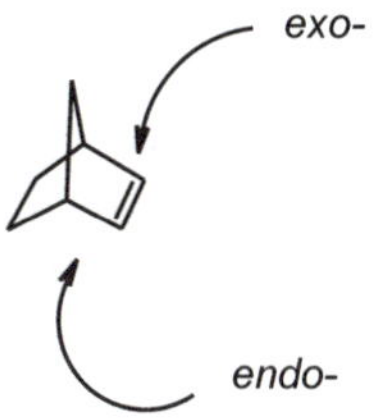

Figure 6.9

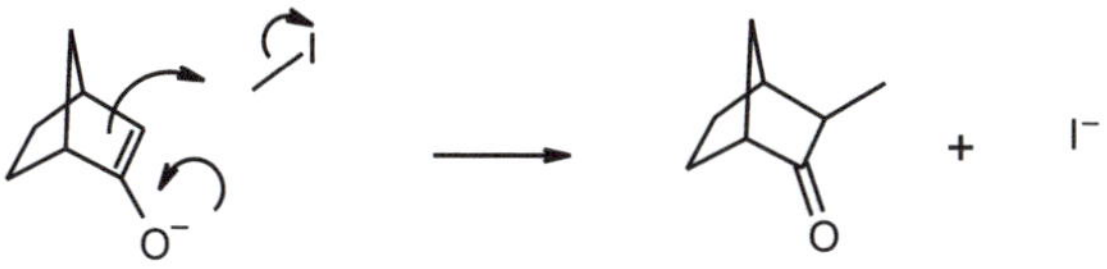

Figure 6.10

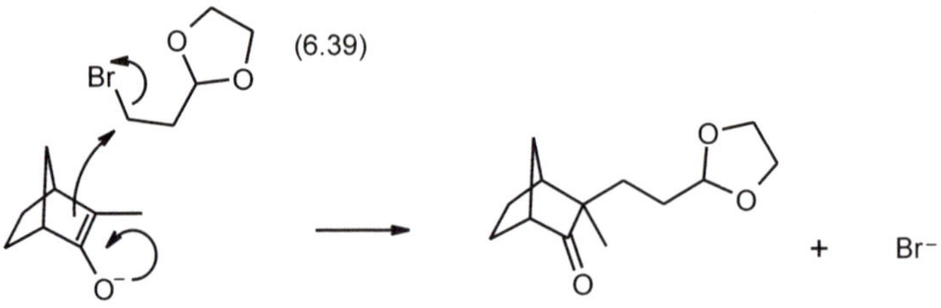

Figure 6.11

When the methylated ketone (6.38) is deprotonated to give the enolate (6.50), we again have a planar olefin-like structure and, once again, the alkylating reagent, this time Buchi's bromide (6.39), adds to the *exo*-face, as shown in Figure 6.11. Consequently, the methyl group is now forced into the *endo*-position giving (6.40) as the product.

When the ketone (6.40) is treated with methyl lithium, the addition again occurs from the *exo*-face. (This time, of course, the electron flow is in the opposite direction from that of the two alkylations.) The initial product, before work-up, is the lithium alkoxide (6.51) (Figure 6.12).

6.1.2.2 The Wittig Reaction with Stabilised Ylids.   The second interesting feature to note is the loss of stereochemical selectivity in the Wittig reaction. This is because the ylid is further stabilised by the ester function. Consider the canonical forms of the two ylids shown in Figure 6.13. The negative charge on the carbon atom in (6.52) is stabilised only by the positive charge on the phosphorus atom. However, in (6.53), the negative charge on carbon is stabilised by both the positive charge on phosphorus and by resonance with the carbonyl bond of the ester function. This makes the anion, and hence the ylid much more energetically favourable and such ylids are referred to as "stabilised ylids".

In a Wittig reaction with a non-stabilised ylid, the carbon–carbon bond formation is essentially irreversible and so the product stereochemistry

Figure 6.12

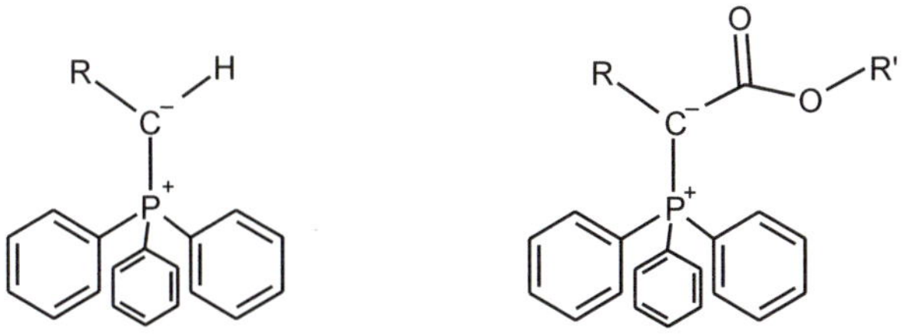

(6.52)　　　　　　　　　(6.53)

Figure 6.13

is determined by the thermodynamics of the transition state formation, as described above. However, when the ylid is stabilised, the adduct can dissociate again to regenerate the ylid and the carbonyl group. Thus, an equilibrium is set up as shown in Figure 6.14, with the aldehyde (6.54) sometimes adding to the anion (6.55) one way round and sometimes the other. If the alkyl groups R and R′ are the bulkiest substituents, then the betaine (6.57) will be the one which is easier to form since the bulky alkyl groups are kept further apart. However, this will result in their having to come close together in order to eliminate the phosphate. The alternative is for the betaine to dissociate again to go back to the starting materials. Elimination is easier from the other betaine (6.56) in which the bulky groups are on opposite sides of the intermediate oxaphosphetane. Since there is an equilibrium between the two betaines, the system will adjust so as to give the most thermodynamically stable product. This is usually the *E*-isomer but, if the energies of the *E*- and *Z*-isomers are close, then a mixture will result. The composition of such a mixture will reflect the difference in free energy of the two isomers.

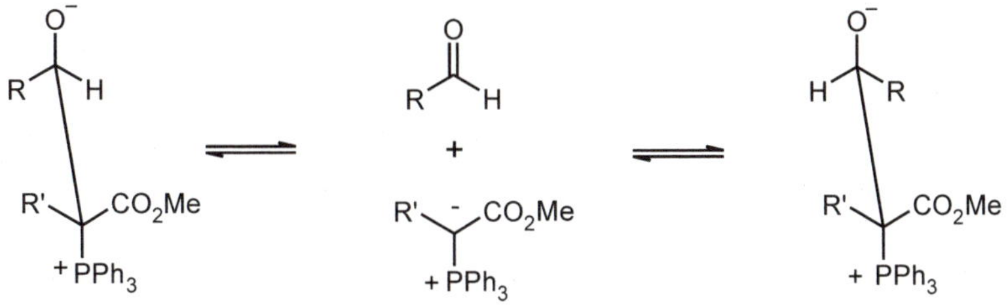

(6.56)　　　　　　　(6.55)　　　　　　　(6.57)

Figure 6.14

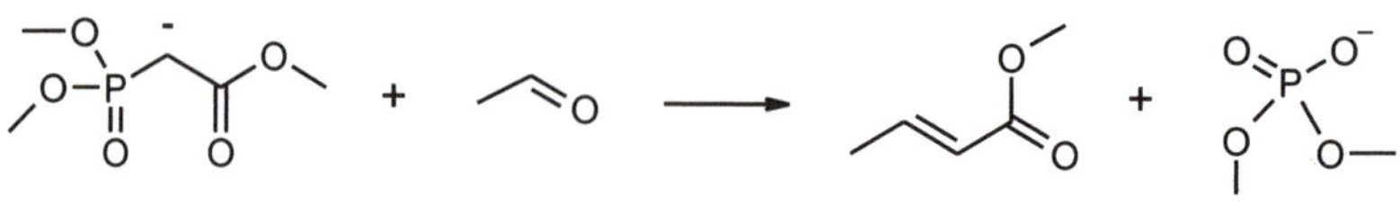

Figure 6.15

6.1.2.3 The Wadsworth–Emmons or Wittig–Horner Reaction. The same loss of *Z*-stereoselectivity occurs in the Wadsworth–Emmons or Wittig–Horner reaction in which the carbanion is derived from a phosphonate carrying an electron withdrawing group (EWG), such as ester, ketone, aldehyde or nitrile, in the α-position. A typical example of such a reagent is the phosphonoacetate (6.58). The advantage of the Wadsworth–Emmons/Wittig–Horner reaction is that the by-product is a water soluble phosphate instead of a rather intractable phosphine oxide. In the example shown in Figure 6.15, the aldehyde (6.59) reacts with the phosphonate anion to give the olefin (6.60) and dimethyl phosphate (6.61).

6.1.3   Sandalwood Substitutes

There are various sandalwood substitutes in the market which can be used to give sandalwood notes in perfumes. It is worth mentioning two classes of these materials because of the chemistry which is used to produce them.

The first group of materials are known as the terpenophenols because they are prepared from camphene (6.6), a terpene, and guaiacol, (6.62) a phenol. The reaction scheme is shown in Figure 6.16. Acid catalysed addition of camphene (6.6) to guaiacol (6.62) gives a mixture of products. The initial carbocation produced by the protonation of camphene undergoes Wagner–Meerwein rearrangement before addition to guaiacol. The three resultant cations: (6.65), (6.66) and (6.67) are shown in the lower part of Figure 6.16. Each of these can add to the guaiacol in any of the four available positions, since each one is either *o* or *p* to an electron releasing group. Since the aromatic ring could add to either face of the cation, this means that there are 24 (*i.e.* 3×4×2) possible products. The resultant mixture (6.63) is then hydrogenated. During the hydrogenation, one of the oxygen atoms of the guaiacol system is lost by hydrogenolysis. This results in the production of a product mixture (6.64) containing no fewer than 128 isomers. This mixture has a strong odour of sandalwood. However, the odour is due to

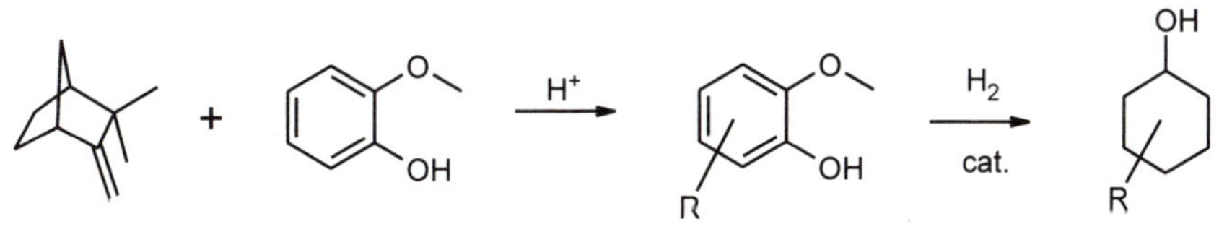

(6.6)   (6.62)   (6.63)   (6.64)

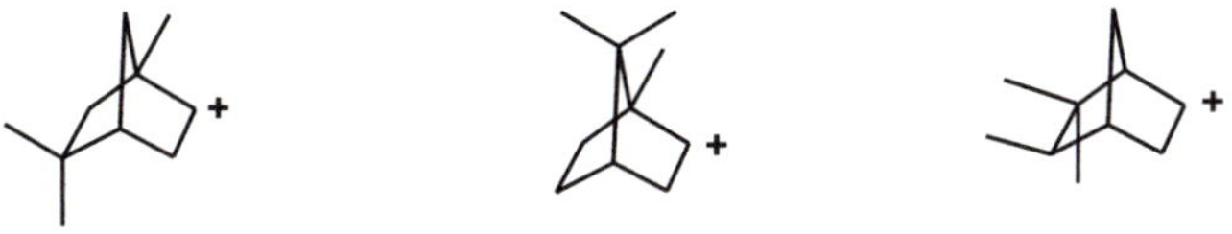

(6.65)   (6.66)   (6.67)

(6.68)   (6.69)

Figure 6.16

only a few of the isomers present. In an impressive feat of practical work, Demole[6.1,6.2] showed that the key odour components were those in which the terpene unit was *exo*-isocamphane, and the hydroxy group and the terpene residue were placed 1,3 and *trans* to each other on the cyclohexane ring. Because of its size, the terpenoid residue takes up an equatorial configuration and the hydroxyl group is therefore forced into the axial configuration because of its 1,3-*trans*-relationship to the former. Thus, the two active isomers are (6.68) and (6.69).

The members of the second family of synthetic sandalwood materials are derived from campholenic aldehyde (6.72). This aldehyde is prepared by treatment of α-pinene oxide (6.71) with a Lewis acid, usually zinc chloride or bromide. α-Pinene oxide is, in turn, prepared from α-pinene (6.70) by treatment with a peracid. The synthesis scheme is shown in Figure 6.17.

Figure 6.18 shows the general scheme for the preparation of sandalwood materials from campholenic aldehyde (6.72) and also three typical members of the family.

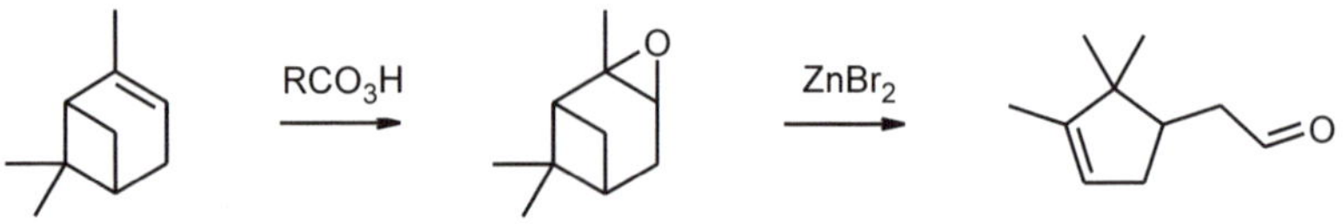

(6.70)   (6.71)   (6.72)

Figure 6.17

Aldol condensation between campholenic aldehyde (6.72) and an aldehyde or a ketone (6.73) produces an unsaturated derivative (6.74). It should be noted that this material contains an *E*-double bond. The reason for this lies in the mechanism of the aldol condensation as described in Figure 6.4 and associated text. The carbonyl group of (6.74) can be reduced to the corresponding alcohol by, for example, a complex metal hydride such as lithium aluminium hydride or sodium borohydride. This produces the sandalwood material (6.75). Three typical examples are shown at the bottom of Figure 6.18. The first, (6.76), is known under various trade names such as Bangalol® (Quest) and is

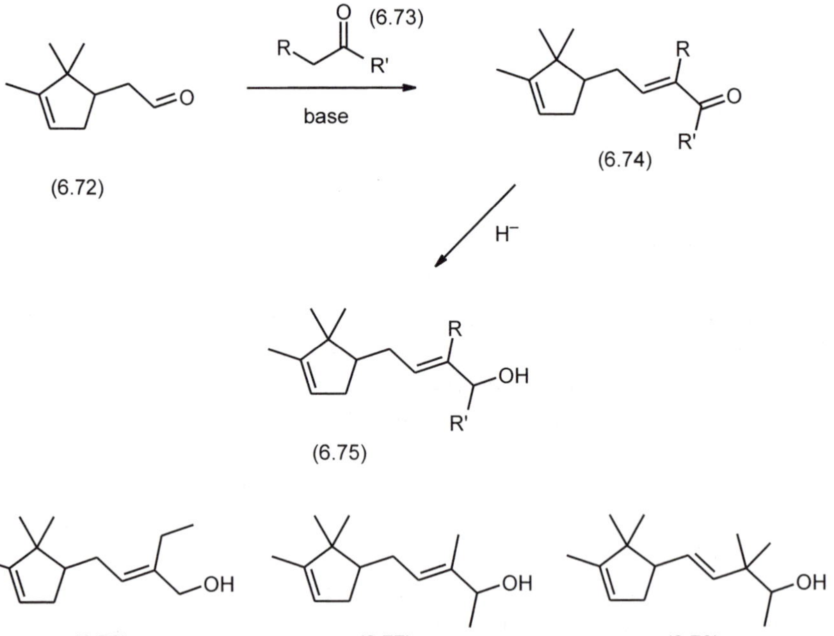

Figure 6.18

formed from campholenic aldehyde (6.72) and butanal. Use of a ketone in the aldol condensation leads to a secondary alcohol as the product. For example, 2-butanone leads to (6.77) which is known under the trade name Sandalore® (Givaudan). The third example, (6.78) Polysantol®. (Firmenich), is interesting because of the gem-dimethyl group. The intermediate ketone (6.79) *en route* to Polysantol® (6.78) is the same as that *en route* to Sandalore® (Givaudan). However, the ketone is methylated under basic conditions before being reduced to Polysantol®. The mechanism for this alkylation is shown in Figure 6.19. The most acidic hydrogen in an α,β-unsaturated ketone (or aldehyde or ester) is the one γ to the carbonyl carbon atom and so this is the one which is removed by a base. The anion produced is, of course, in equilibrium with the one in which the negative charge is located on the α-carbon. The latter is always the one which will react fastest with an alkylating agent and so the new group is not added to the carbon atom from which the proton was removed. In Figure 6.19, we see the starting ketone (6.79)

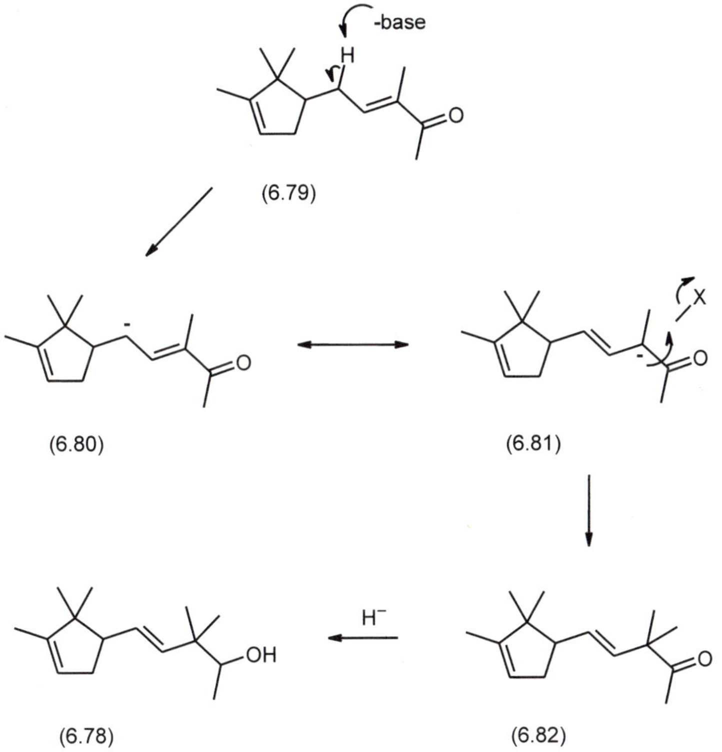

Figure 6.19

being deprotonated to give the anion (6.80) in which the negative charge is located on the γ-carbon atom. However, it is the other canonical form (6.81), which reacts with the methylating agent (*e.g.* methyl iodide or dimethyl sulfate) to give the product (6.82). The double bond has been moved out of conjugation with the carbonyl function and cannot return because the α-carbon atom no longer carries any enolisable hydrogen atoms. Reduction of this ketone gives Polysantol® (6.78).

6.1.4   Alkylation of α,β-Unsaturated Ketones

This type of alkylation reaction is very important in the synthesis of terpenoids as it represents a method for the introduction of gem-dimethyl groups, a common feature of terpenoid structure. An example of this strategy in action is shown in Figure 6.20.

Treatment of methyl cyclohexanonecarboxylate (6.83) with methyl vinyl ketone (6.84) in the presence of a base gives rise to the Robinson

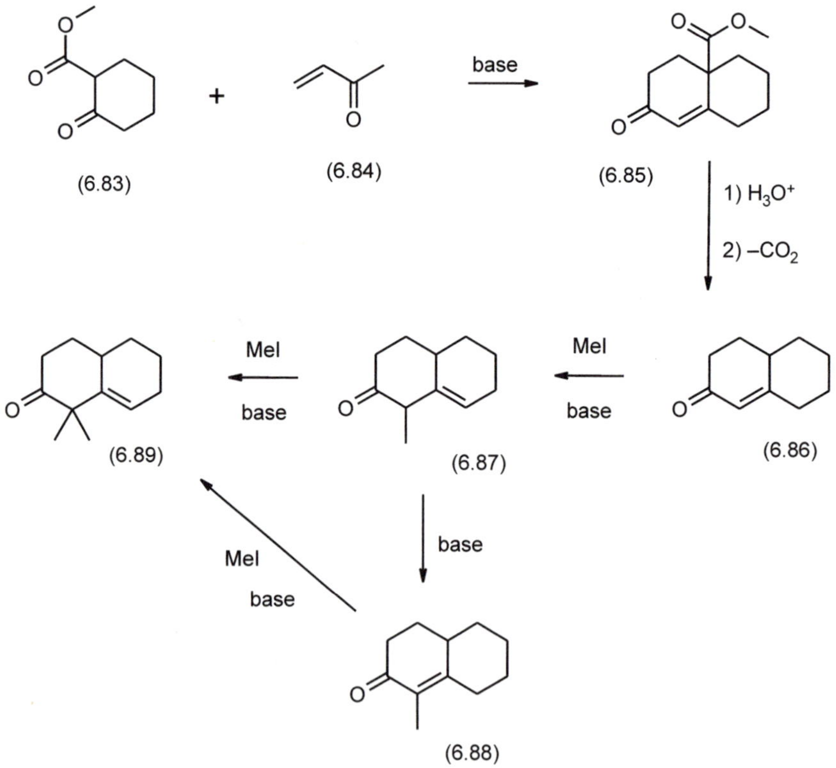

Figure 6.20

annulation reaction. This reaction involves deprotonation of the keto-ester and the Michael addition of the resultant carbanion to the methyl vinyl ketone (6.84). This reaction is followed by an intramolecular aldol condensation to give the bicyclic product (6.85). Hydrolysis and decarboxylation of (6.85) produces the α,β-unsaturated ketone (6.86). Treatment of ketone (6.86) with two molar equivalents of methyl iodide in the presence of a base gives the final product (6.89) containing a gem-dimethyl group. The initial proton abstraction is from the γ-position. Alkylation of this anion gives the α-methyl-β,γ-unsaturated ketone (6.87). This ketone will be deprotonated at the α-position. The resultant anion could be alkylated directly or could re-protonate to give the α-methyl-α,β-unsaturated ketone (6.88). However, deprotonation of (6.88) at the γ-position will reproduce the same anion as that produced by deprotonation at the α-position of (6.87). This anion will alkylate at the α-position and so, whichever sequence of reactions occurs, the product is still ketone (6.89). If the reader turns back to Chapter 2, a quick inspection of the terpenoid structures shown there, will show how frequently a gem-dimethyl group appears and hence how useful this particular synthetic tactic can prove.

6.2 CEDARWOOD

There are many trees which are known as cedars and are harvested for timber and for the production of essential oils. Perhaps the most famous is the Cedar of Lebanon (*Cedrus libani*), which has a long history of use for both purposes. For example, in about 1000 BC, King David used it to build his palace in Jerusalem and his son Solomon used it in the construction of the first temple there. The tree is now a protected species and no good chemical analysis of the wood exists. The commonest cedars nowadays are mostly various *Juniperus* species and belong to the family *Cupressaceae*. Typical examples are Texan Cedar (*Juniperus mexicanus*), Virginian or Red Cedar (*Juniperus virginiana*), Chinese Cedar (*Cupressus funebris*) and East African Cedar (*Juniperus procera*). These species are all related and the oil obtained from their wood usually contains mostly cedrol (6.90), cedrene (6.91) and thujopsene (6.92), the levels of each varying from 15 to 30% depending on the species and environmental factors. Two well-known cedars which have very different essential oils are Atlas Cedar (*Juniperus atlantica*), and Himalayan Cedar (*Cedrus deodara*) both of which contain α- and γ-atlantones (6.93 and 6.94 respectively) and deodarone (6.95). The atlantones are the major components of their oils. These ketones impart a characteristic sweet and warm, almost animalic odour to the oils of Atlas and Himalayan

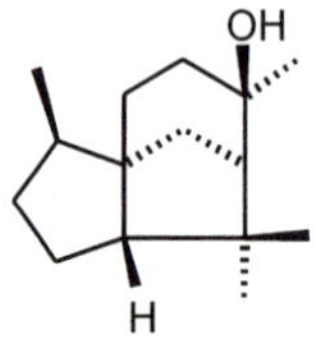

(6.90) cedrol

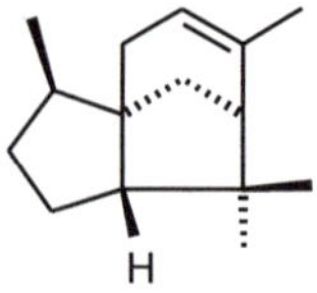

(6.91) cedrene

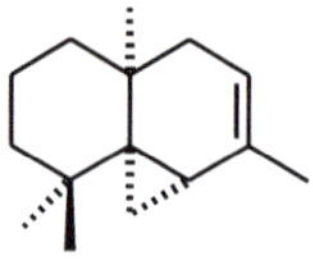

(6.92) thujopsene

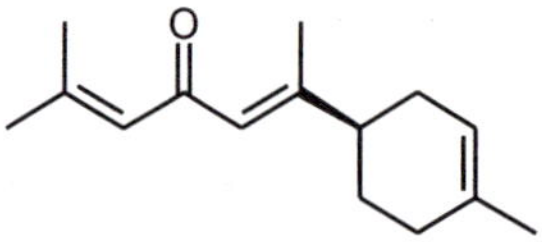

(6.93) α-atlantone

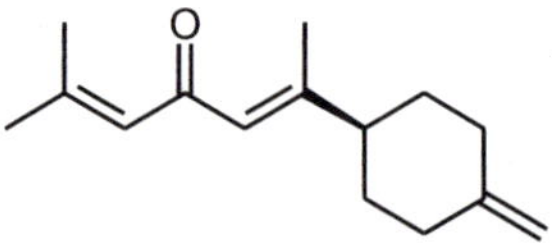

(6.94) γ-atlantone

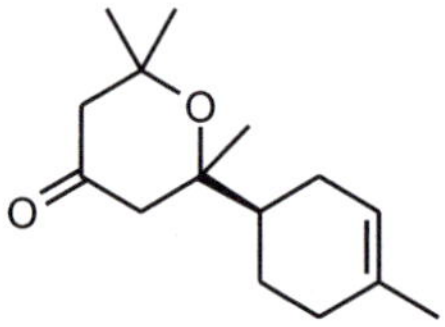

(6.95) deodarone

Figure 6.21

Cedarwood. The structures of all of these major cedarwood sesquiterpenoids are shown in Figure 6.21. Port Orford Cedar (*Chamaecyparis lawsoniana*) is less closely related to the other cedars and its oil contains about 60% limonene. We will not consider it further in this chapter.

A quick glance back at Figure 2.10 will show how the major families of cedarwood sesquiterpenoids are related in terms of their biogenesis, and indeed, how they are related to the sandalwood sesquiterpenoids. Initial 1,6-cyclisation of the *cis,trans*-farnesyl carbocation gives the bisabolane skeleton which is found in the atlantones and deodarones. Starting from this skeleton, different sequences of intramolecular cyclisations and rearrangements lead to the more complex cedrane, thujopsane and santalene skeleta. Following nature's example, let us start our study of the cedarwood sesquiterpenoids with the simpler bisabolane skeleta of α-atlantone and deodarone.

6.2.1   Total Synthesis of α-Atlantone

The scheme used by the English terpene chemist R. C. Cookson,[6.3] for the total synthesis of α-atlantone is shown in Figure 6.22.

The synthesis starts with the addition of senecioyl chloride (6.96) to isoprene (6.97). The acyl cation derived from the acid chloride adds to isoprene with a high degree of regioselectivity, for reasons which will be explained in the next paragraph. The carbocation produced by addition of the senecioyl ion to isoprene is trapped by the chloride anion to give

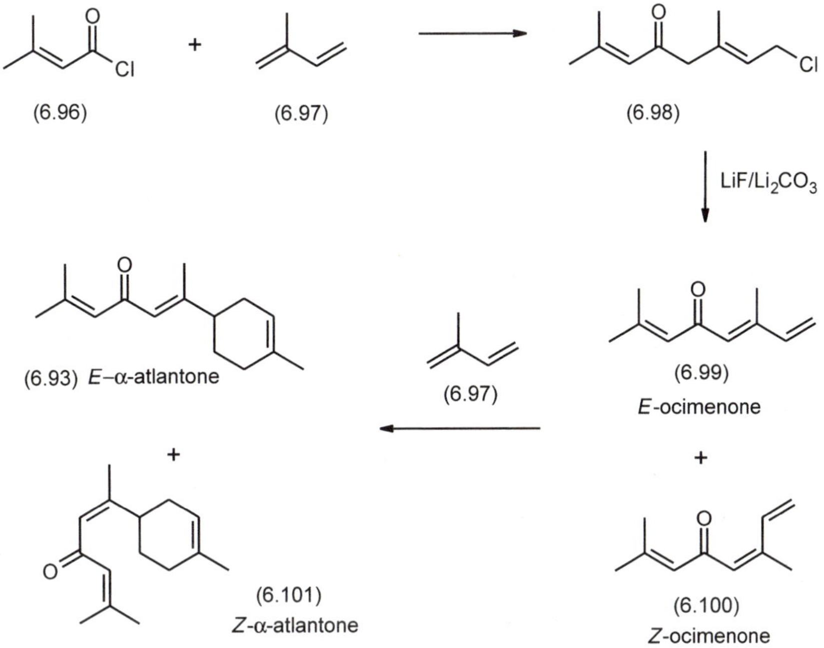

Figure 6.22

the chloroketone (6.98). This chloroketone can be dehydrochlorinated to the ocimenones (6.99) and (6.100) by treatment with a mixture of lithium fluoride and lithium carbonate. The fluoride anion is the base in this reaction and it removes the most acidic hydrogen in the substrate, *i.e.* one of the methylene hydrogens α to the carbonyl function. Allylic displacement of the chloride anion then gives the diene. The hydrogen fluoride produced in this step is neutralised by the lithium carbonate, thus regenerating the fluoride anion. Both geometrical isomers are produced, but the *E*-isomer (6.99) predominates over the *Z*- (6.100) by a factor of 9:1 as a consequence of the *trans-anti*-periplanar rule. These ketones are known as *E*- and *Z*-ocimenone, respectively. They are important odour components of marigold, which is also known in perfumery by its French name, tagete, and the Latin botanical name, tagetes. Diels–Alder reaction of this mixture with a second equivalent of isoprene gives the corresponding mixture of *E*-α-atlantone (6.93) and its *Z*-isomer (6.101). In principle, any of the three double bonds of ocimenone could serve as dienophile in this reaction but, as the Diels–Alder reaction is very susceptible to steric effects, it is the least hindered, terminal bond which takes part. The regiochemistry of the Diels–Alder reaction is, like that of the Friedel–Crafts reaction, highly selective and the same phenomenon accounts for both.

6.2.1.1 Regioselectivity of Isoprene in Friedel–Crafts and Diels–Alder Chemistry. The explanation for the regioselectivity of the Friedel–Crafts reaction of Figure 6.22 is shown in Figure 6.23. The structure of isoprene in that figure shows a partial negative charge on C-1. This results from the polarisability of the 1,2-double bond of isoprene. The partial positive charge on C-2 is stabilised by virtue of its being both tertiary and allylic. This polarisation means that cationic species attack isoprene at C-1. Conversely, electron rich species such as carbanions and free radicals, will attack at C-4.

The same polarisation accounts for the selectivity in the Diels–Alder reaction. Usually, good dienophiles carry an EWG. This group serves to remove electron density from the double bond, and therefore, makes it more attractive to the dienophile. Of course, it is pulling electrons from one direction and this tends to polarise the double bond. For example, if we consider a simple α-methylene ketone (6.102), it can be polarised completely into the canonical form shown as structure (6.103). The more accurate picture of the molecule is therefore a partially charged species, (6.104). When this dienophile aligns with the diene in the transition state of the Diels–Alder reaction, the more negatively charged end of the diene will be attracted to the oppositely charged, positive end of the dienophile as shown in structure (6.105). By joining these two atoms together, the regiochemistry of the Diels–Alder reaction becomes clear. Any diene and dienophile can be treated in the same way. If one considers which of the two possibilities for total charge separation of the diene and dienophile

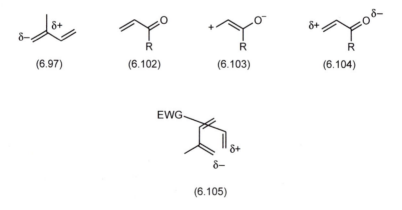

(6.97)   (6.102)   (6.103)   (6.104)

(6.105)

Figure 6.23

would be more favourable, alignment of the opposite charges, *i.e.* negative to positive and positive to negative, on the reagents, will, in most cases, enable the regiochemistry of the reaction to be determined correctly.

Of course, the electronic structure of isoprene is more complex than this simple picture suggests as the two double bonds interact with each other. However, molecular orbital treatment of the entire diene system produces a prediction of reactivity which is similar to that of the simple picture.

A further feature of the stereochemistry of the Diels–Alder reaction is also shown in the transition state (6.105). The EWG prefers to lie across the diene system during the reaction, as shown in Figure 6.23. In the case of isoprene and ocimene, this is of no consequence. However, in the case of cyclic dienes, it explains the preference for the *endo*-configuration of the EWG in the adduct.

6.2.1.2 Synthesis of α-Atlantone from d-Limonene. The total synthesis of α-atlantone from isoprene shown in Figure 6.22, gave positive proof of the structure as far as atom connectivity is concerned. However, what it did not achieve was to demonstrate the absolute stereochemistry of the product as the Diels–Alder reaction was not stereoselective. In order to establish the absolute stereochemistry, Cookson carried out a partial synthesis of α-atlantone from *d*-limonene.[6.4] This is shown in Figure 6.24.

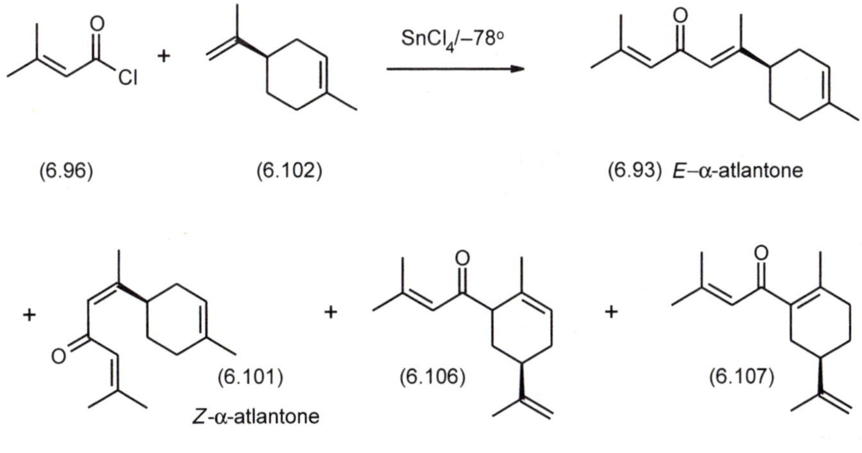

(6.96)   (6.102)   SnCl$_4$/–78°   (6.93) *E*-α-atlantone

(6.101) *Z*-α-atlantone   (6.106)   (6.107)

Figure 6.24

Friedel–Crafts reaction between senecioyl chloride (6.96) and *d*-limonene (6.102) in the presence of stannic chloride at low temperature, gave α-atlantone in one step, along with the isomeric ketones (6.106) and (6.107). The optical rotation of the α-atlantone isomers produced in this way was the same as that of the natural product showing that the absolute configuration of C-4 in the cyclohexane ring of natural α-atlantone is the same as that of *d*-limonene.

The regioselectivity of the Friedel–Crafts reaction in Figure 6.24 is another demonstration of the two factors governing regioselectivity of addition to double bonds (see Chapter 5). The Friedel–Crafts reaction prefers polarisable double bonds and so the 1,1-disubstituted bond in the isopropenyl group reacts faster than the less polarised double bond in the ring even though the latter is more electron rich. This accounts for the relative proportions of products, 60% of (6.93) and (6.101), 20% of (6.106) and 9% of (6.107). The fact that (6.106) predominates over (6.107) is the result of kinetic deprotonation of the intermediate carbocation from the less hindered side. The addition of the acyl cation to either double bond of limonene occurs, as would be expected, at the less substituted end thus giving the more stable tertiary carbocation as the intermediate.

6.2.2   Tagetones, Filifolone and Minor Components of Atlas Cedarwood

Using the same chemistry, Cookson was also able to synthesise two more natural products, *E*- and *Z*-tagetone, (6.108) and (6.109), respectively, by substituting isovaleroyl chloride (6.110) for senecioyl chloride (6.96) in the Friedel–Crafts reaction with isoprene (6.97). Like the ocimenones, the tagetones are important odour components of marigold flowers. Similarly, the use of acetyl chloride gave the ketone (6.111) which could be reacted with isoprene to give a mixture of the cyclic ketones (6.112) and (6.113). The former had been identified by him as a component of Atlas Cedarwood Oil. There exists the possibility that this ketone is an artefact of hydrolysis of atlantone during the steam distillation. However, Cookson's experimental work suggested that the ketone (6.112) is stable enough to survive distillation and must therefore have been formed in the wood. One other natural terpenoid which was easily accessible through this work was filifolone, (6.114). This bicyclic ketone, which occurs in various species of *Artemisia*, was easily prepared by treatment of the ocimenones with aluminium chloride. Figure 6.25 shows all these synthetic routes.

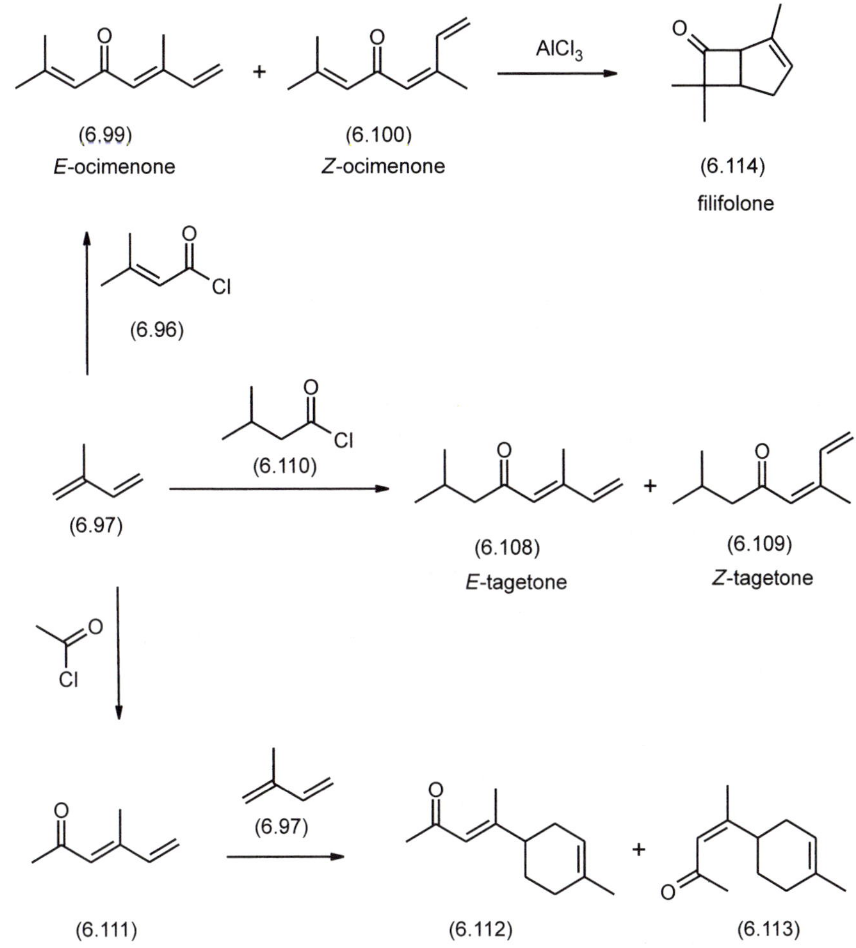

(6.99)
E-ocimenone

(6.100)
Z-ocimenone

AlCl₃

(6.114)
filifolone

(6.96)

(6.110)

(6.97)

(6.108)
E-tagetone

(6.109)
Z-tagetone

(6.111)

(6.97)

(6.112)

(6.113)

Figure 6.25

6.2.3 α-Atlantone and Deodarone

The relationship between the atlantones and deodarone is obvious. One can imagine the Michael addition of water to the two activated double bonds of α-atlantone (6.93) followed by dehydration of the resultant diol to produce the ether link of deodarone (6.95). Confirmation of the structure of deodarone relied on such interconversions. The first to show the relationship between the two ketones was the great Indian natural products chemist, Sukh Dev.[6.5] He carried out a sequence similar to this hypothetical scheme. The hydroxide ion is a very poor nucleophile in Michael reactions and so Sukh Dev used the hydroperoxide anion

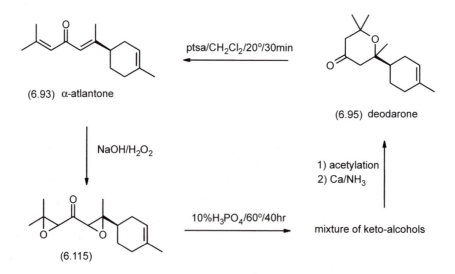

Figure 6.26

instead. This adds readily to α,β-unsaturated ketones but the resultant carbanion then attacks the peroxide bond, displacing hydroxide and producing an olefin, as shown in Figure 6.26.

When this reaction is repeated on both activated double bonds of α-atlantone (6.93), the result is the di-epoxyketone (6.115). This is, of course, at too high an oxidation level for deodarone and a reduction step will be necessary. The next step of Sukh Dev's synthesis was to open the epoxide to the diol with dilute phosphoric acid. This hydration then produces the two tertiary alcohols from which the ether linkage can be made by elimination of water. Acetylation of the secondary alcohols and reduction of the resultant acetate esters with calcium in ammonia, then gives deodarone (6.95). However, the reaction sequence is far from clean and deodarone was only one of many compounds produced. Cookson therefore decided to prove the relationship by eliminating the ether linkage of deodarone to give atlantone. This he achieved easily by treating deodarone with *p*-toluenesulfonic acid in dichloromethane at room temperature for half an hour. Both schemes are shown in Figure 6.27.

ptsa/CH$_2$Cl$_2$/20°/30min

(6.93) α-atlantone

(6.95) deodarone

NaOH/H$_2$O$_2$

1) acetylation
2) Ca/NH$_3$

(6.115)

10%H$_3$PO$_4$/60°/40hr

mixture of keto-alcohols

Figure 6.27

It is interesting to compare the ease of conversion of deodarone to atlantone with the reverse conversion. The explanation lies in the greater stability of atlantone. In this molecule, there is considerable resonance stabilisation because of the delocalisation possible in the $\alpha,\beta,\alpha',\beta'$-unsaturated ketone. Furthermore, there is some steric strain in deodarone which is relieved when the tetrahydropyran ring is opened.

6.2.4   Cedrol and Cedrene

The ring system of cedrol is much more complex than that of the atlantones and this adds some degree of interest to its chemistry. Some of the reactions of cedrol and cedrene proceed in a straightforward manner, giving the products which would be expected by analogy with a simple acyclic system. Figure 6.28 shows a selection of such reactions.

Since it contains a tertiary alcohol function, cedrol (6.90) is easily dehydrated in the presence of acids. This dehydration proceeds in a simple manner to give cedrene (6.91), which is thermodynamically preferred of the two possible olefinic products. This suggests that the dehydration is under thermodynamic, rather than kinetic, control. This is explained in Figure 6.29, which shows the equilibrium that is set up during the reaction. Protonation of the alcohol and elimination of water gives the carbocation (6.120). This can eliminate a proton from either the methyl group or methylene group adjacent to the positively charged carbon atom. The former will be easier since the methyl group is more accessible. However, the olefin produced, (6.121), contains a highly polarised double bond which is easily re-protonated to give the carbocation (6.120) again. Elimination of a proton on the other side of the carbocation gives cedrene. If re-protonation were slow, the first product to be formed, *i.e.* the exocyclic olefin (6.121), would predominate in the reaction mixture. In such a case, it is said that the reaction would be under kinetic control. However, in this case, the equilibrium is established and, through successive protonation and proton elimination steps, a mixture of products results which reflects the thermodynamic stability of the two products. In this example, this means a high preference for α-cedrene (6.91). It is not entirely clear how much cedrene exists in the wood relative to the amount of cedrol. It is quite possible that some, or even most, of the cedrene in the oil results from dehydration during extraction.

Treatment of either cedrol or cedrene with methanol in the presence of acid, results in the formation of cedryl methyl ether (6.116), which possesses a strong woody-amber odour. The intermediate from either starting material is the carbocation (6.120). Treatment of cedrene with

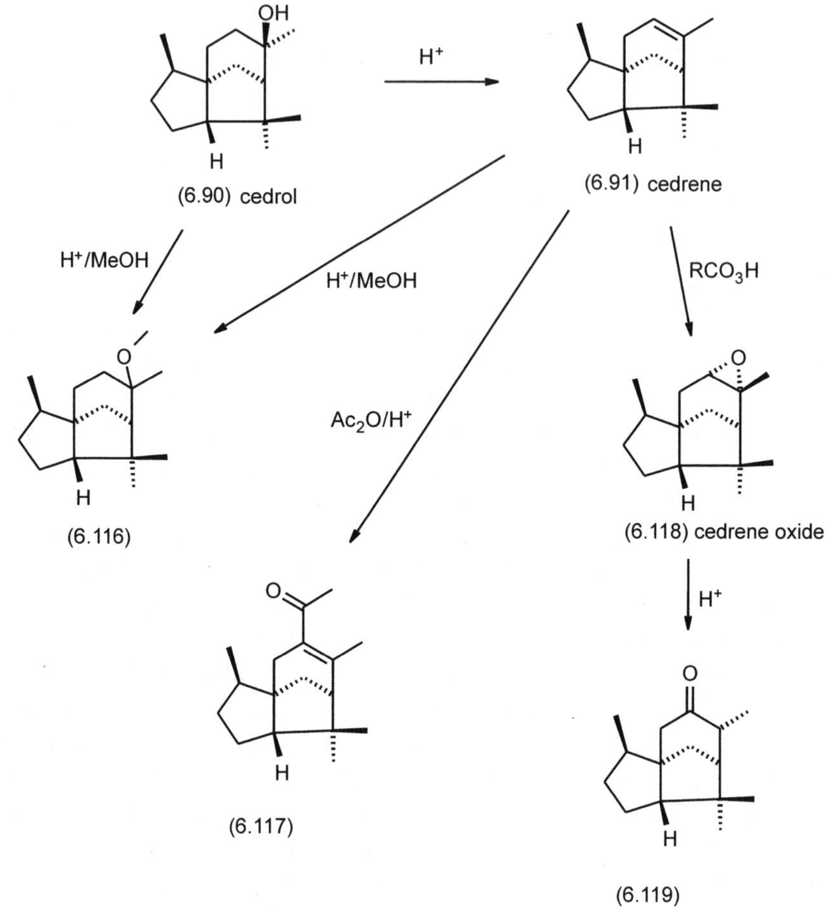

Figure 6.28

acetic anhydride in the presence of a Bronsted acid, results in a straight-forward Friedel–Crafts reaction to give acetylcedrene (6.117). Cedrol, or a mixture of cedrol and cedrene, can also be used in this reaction since the acid anhydride is a desiccant and will convert cedrol to cedrene *in situ* and the Friedel–Crafts reaction can then ensue. There are numerous perfumery ingredients based on acetylcedrene. Sometimes they are simply called acetylcedrene but, more often, they are given tradenames such as Lixetone® or Vertofix®. There will be more discussion of the actual composition of these products later. Treatment of cedrene with a peracid produces the epoxide known as cedrene oxide (6.118). This material has an odour which is somewhat reminiscent of patchouli. The peracid approaches cedrene from the less hindered side to give the

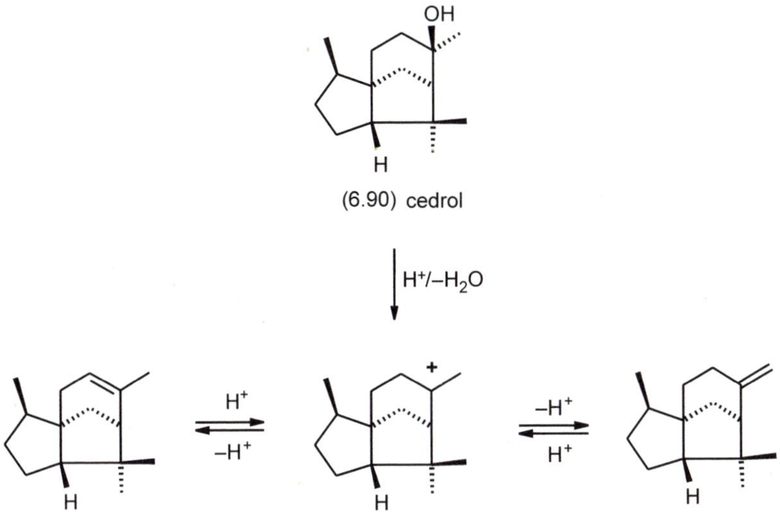

Figure 6.29

product stereochemistry shown in structure (6.118), another example of the *exo*-effect. Cedrene oxide can be rearranged, by an acid catalyst, to the ketone (6.119) which has a vetiver like odour. The methyl group of the ketone adopts the stereochemistry shown in structure (6.119). The position at which this methyl group is attached to the ring, is epimerisable, being adjacent to the carbonyl function. Since epimerisation would be possible under the reaction conditions, and since the hydrogen atom was originally located on the other side of the ring, this must imply that the given stereochemistry is thermodynamically more favourable. Indeed, models show that an *endo*-methyl group in this position, for example, as in the epoxide (6.118), buttresses against the methyl group on the opposite bridge.

6.2.4.1 Friedel–Crafts Acylation of Cedrene. Let us now look in a little more detail at the Friedel–Crafts acylation of cedrene. This reaction can be carried out using acetic acid or acetic anhydride with an acid catalyst such as formic acid, phosphoric acid or sulfuric acid. The species which attacks the double bond of cedrene is the protonated acid or anhydride molecule. When sulfuric acid is used as a catalyst with acetic anhydride as the reagent, it is thought that the sulfuric acid first sulfonates the acetic anhydride to give sulfoacetic anhydride (6.122) as the acylating species. This results in a faster reaction as the sulfoacetate anion (6.126) is a better leaving group than acetate because of the electron withdrawing effect of the S=O double bonds, which helps to stabilise the negative charge. The mechanism is shown in Figure 6.30, using the sulfuric acid/ acetic anhydride system. Substitution of the sulfoacetic residue in structures (6.122), (6.123) and (6.124) by H or CH_3CO will represent the mechanism of the acid catalysed additions of acetic acid or acetic anhydride, respectively.

Protonation of the carbonyl group of sulfoacetic anhydride (6.122) serves to generate a strong partial positive charge on the acid carbon atom and this adds to the double bond of cedrene. The addition, as usual, takes place on the less substituted end of the double bond thus generating the more stable tertiary carbocation (6.123). The elimination of a proton will regenerate the olefinic link. Elimination of the sulfonoacetate anion (6.126) from (6.124) leads to the protonated ketone (6.125) and removal of a proton from this gives acetylcedrene (6.117). The two proton removal stages could, of course, occur in the reverse order and this will lead to the same overall conversion. The elimination of sulfoacetic acid in this scheme is shown as proceeding without the assistance of a base. Obviously, a base (presumably the hydrogen sulfate anion, since the reaction medium is sulfuric acid) could remove the hydroxyl proton from (6.124) to give an anion which would eliminate sulfoacetate and give acetylcedrene. The overall effect would be the same as the sequence shown in Figure 6.30 and possibly both mechanisms operate.

Cedrene can also be acylated by acetyl chloride in the presence of a Lewis acid. This is shown in Figure 6.31, using aluminium chloride as the Lewis acid. The aluminium chloride complexes with the acetyl chloride, activating the acid carbon atom in the same way that protonation did in the previous figure. The double bond of cedrene then adds, as before, to this positively charged carbon atom to give the intermediate (6.127). This species can now eliminate a proton and a chloride anion to give the aluminium chloride complex of acetyl cedrene (6.128). This aluminium complex is stronger than that with the acid chloride, and so, the metal is not freed to induce a second reaction by complexing with another molecule of acetyl chloride. This is an important point in preparative chemistry. Lewis acids, such as aluminium chloride, do not behave as catalysts for the Friedel–Crafts reaction, rather they behave as stoichiometric reagents. It is necessary to use at least one molar equivalent of the Lewis acid and it cannot be recovered. In order to free the product from the complex, it is necessary to add water, or an equivalent reagent, to destroy the complex by providing something which bonds even more

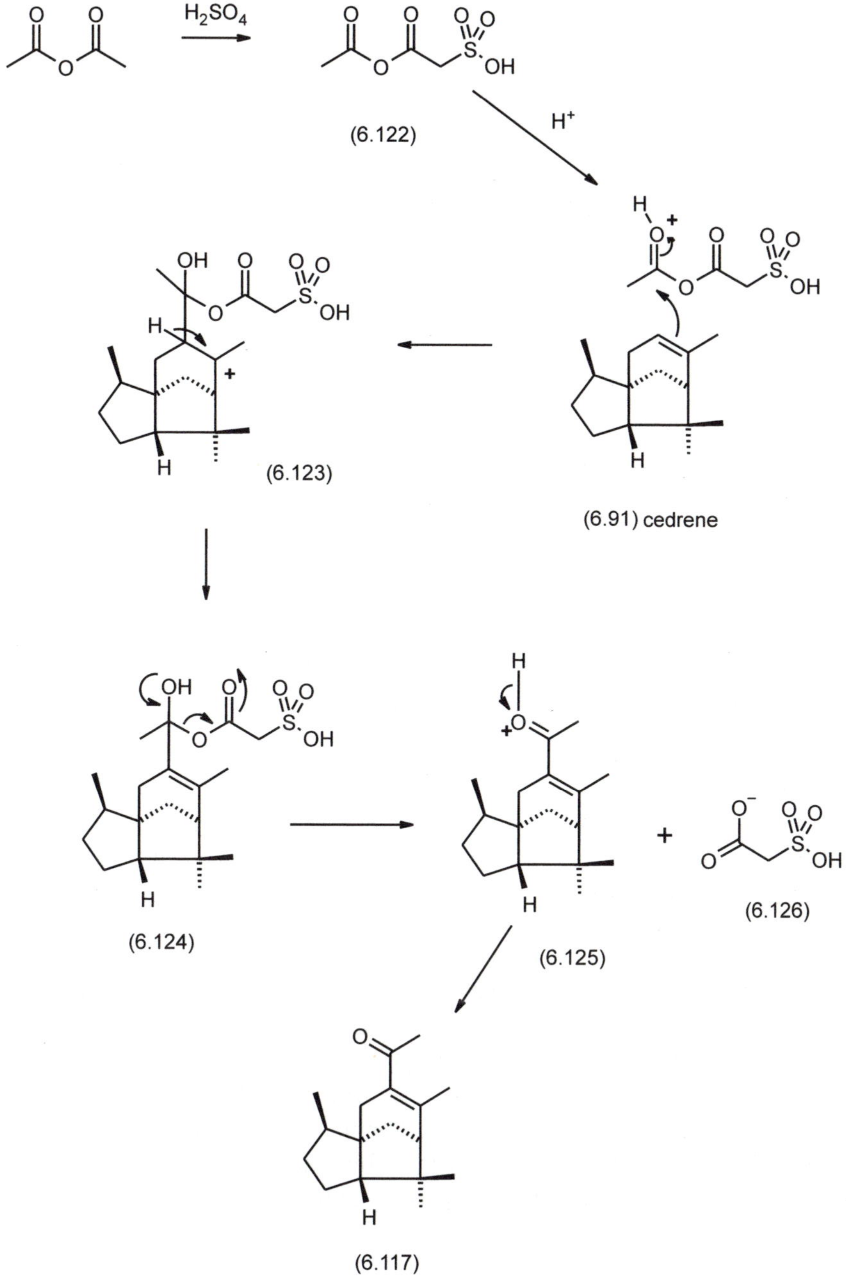

Figure 6.30

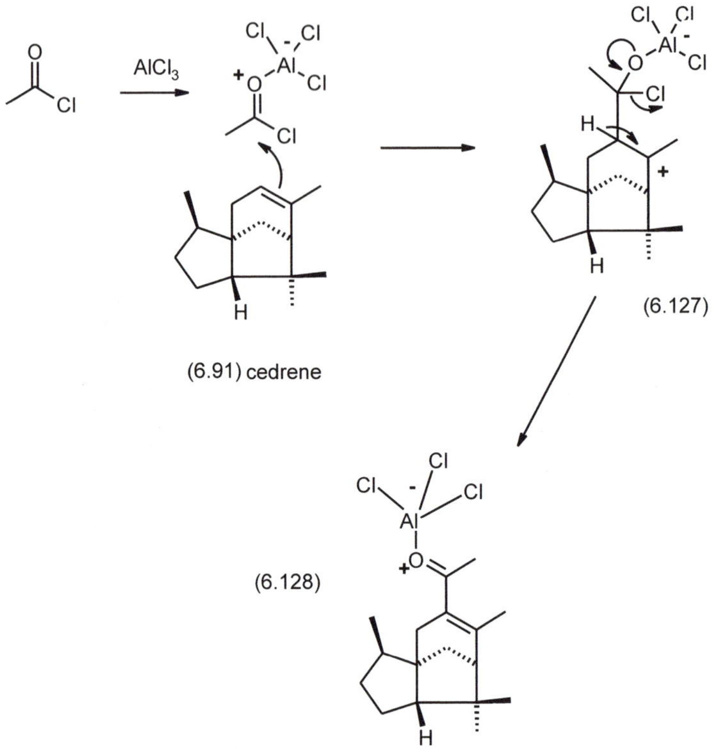

Figure 6.31

strongly to the metal than does the ketone. This completely deactivates the Lewis acid as far as Friedel–Crafts chemistry is concerned.

Both the above methods of preparing acetylcedrene suffer from serious effluent problems on a commercial scale. The use of sulfuric acid as a solvent/reagent results in large quantities of dilute sulfuric acid after the reaction is quenched with water. This effluent also contains sulfoacetic acid and many minor organic by-products which arise from other reactions of cedrene with sulfuric acid and the other reagents present. The use of aluminium chloride or other metallic Lewis acids results in metal salts as effluents after quenching the reaction. Since acetylcedrene is a valuable product, it is not surprising that the fragrance industry has invested a great deal of research into alternatives for these reaction systems.

6.2.4.2 Anomalous Behaviour of Cedrene under Friedel–Crafts Conditions. Sesquiterpene chemistry is always full of surprises. For example, during some investigations into alternatives for the existing acetylcedrene processes, an attempt to use titanic chloride in place of aluminium chloride, resulted in a hitherto unknown compound.[6.6] The puzzle was even more intriguing initially as the unknown product appeared, by virtue of a strong infrared absorption at 1675 cm^{-1}, to be an α,β-unsaturated ketone, yet there were no signals in the NMR spectrum which would corroborate this. The explanation for this apparent inconsistency is that vinyl ethers have intense infrared absorptions in the same region of the spectrum as do α,β-unsaturated ketones. The unknown was therefore a vinyl ether, structure (6.129) to be precise. The mechanism of its formation is shown in Figure 6.32.

For simplicity, the structures in Figure 6.32 do not show the titanium complexes in full, as these would make the crucial changes in the carbon skeleton more difficult to see. Also for the same reason, the cedrane skeleton is drawn in a different way from that used in previous figures. (Any reader who does not see that the drawings of (6.91) in Figures 6.31 and 6.32 represent the same molecule, should build a molecular model to convince him/herself that this is the case.) The first stage involves the addition of an acyl cation equivalent to cedrene to give the

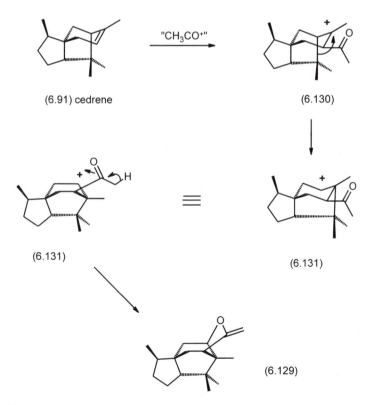

Figure 6.32

carbocation (6.130). When this addition occurred with either sulfuric acid or aluminium chloride as reagent, the proton adjacent to the ketone group eliminated to give the unsaturated ketone function of acetylcedrene. However, with titanic chloride as reagent, the molecule undergoes a Wagner–Meerwein rearrangement, through a 1,2-carbon shift, to give a new carbocation, (6.131). This transformation results in the lowering of ring strain through formation of the 2,2,2-bicyclooctane skeleton. An experiment with molecular models will demonstrate this nicely. In all skeletal rearrangements in complex systems, it is helpful to first move the bonds using the original way of drawing the molecule, even if this leads to rather odd bond angles and distances. Having got the correct atom connectivity for the product, it is then easier to tidy it up into a neater and/or more representative shape. This procedure is adopted in Figure 6.32 where the structure is drawn firstly in a manner resembling the starting material (6.130) (allowing the bond movement to be clearly seen) and then in a way which shows a more realistic representation of the shape of the molecule. It is recommended that students use this technique when tackling the problems at the end of the book. In the second representation of structure (6.131) it is easy to see that the oxygen atom of the ketone function is very close in space to the newly formed carbocation so its lone pair of electrons can easily add onto the carbocation and concomitant elimination of a proton from the methyl group then furnishes the vinyl ether (6.129). Another interesting point to note is the fact that, at one point, the reaction proceeds from a tertiary to a secondary cation. This is, by itself, an energetically un-favourable move. However, in this case, this loss in energy is balanced by a gain in terms of strain energy and the fact that the oxygen atom is there ready to stabilise the positive charge. In fact, the presence of the oxygen atom could encourage the Wagner–Meerwein rearrangement to take place and the overall rearrangement from (6.130) to (6.129) could be essentially synchronous. The explanation for the anomalous behaviour in the presence of titanic chloride must lie in a similar phenomenon, *i.e.* the degree to which bonds are polarised and charges distributed onto individual atoms. (When is a cation not a cation? – see Chapter 5).

6.2.5   Thujopsene

Acetylated Cedarwood Oil is a valuable perfumery ingredient. It is prepared by acetylating the entire oil and the major component is therefore usually acetylcedrene. In fact, the mixture is often referred to as acetylcedrene. However, Cedarwood Oil contains many components and the acetylation reaction is not entirely selective, so the product is actually quite a complex mixture. In a brilliant set of papers published contiguously in the *Journal of Organic Chemistry*, a group of chemists working for the Swiss perfumery company, Givaudan,[6.7,6.8] together with a team from the University of California,[6.9] described how they analysed this mixture, characterised many of its components, identified those making the greatest contribution to the odour and elucidated the reaction pathways by which they are formed. Their conclusions were confirmed by a variety of methods including independent synthesis of the products, isotopic labelling and X-ray diffraction. The original papers are well worth studying. What follows here is a synopsis of some of the key points.

As far as the odour of acetylated Cedarwood Oil is concerned, the most important component of the mixture is not derived from cedrene (6.91) at all, but from the next most significant hydrocarbon in the natural oil, thujopsene (6.92). In fact thujopsene accounts for almost as much of the oil as does cedrene, both being present at about 30%. Thujopsene also occurs in Hibawood Oil (*Thujopsis dolobrata*), of which it is the major component. It is therefore not surprising that acetylated Hibawood Oil has a very intense smell of cedar.

The product of acetylation of thujopsene is the ketone (6.132). At first sight, it is difficult to see how this can be the case since the basic structures are rather different. The answer lies in the series of rearrangements which thujopsene undergoes before finally reacting with the acylating species. This reaction sequence is shown in Figure 6.33.

Initially the double bond of thujopsene (6.92) is protonated, at the less substituted end, to generate the tertiary carbocation (6.133). The positive charge in this molecule is located right next to a cyclopropyl group and so the ring strain of the three-membered ring can be relieved by opening it to a seven-membered ring and thus carbocation (6.134). Six-membered rings are intrinsically more stable than seven and so a 1,2-carbon shift occurs to give the spirane (6.135) which contains only six-membered rings. This loses a proton to give, primarily the α-isomer (6.136). In the presence of acid, the α- and γ-isomers, (6.136) and (6.137), respectively, are in equilibrium through the intermediate carbocation (6.135). Protonation of the other double bond of the γ-isomer (6.137) allows a ring closure reaction to occur through the addition of electrons of the *exo*-methylene double bond to the carbocation, generating the tricyclic material (6.139). Elimination of a proton from this carbocation gives the tricyclic olefin (6.140). The direction of elimination is controlled by ring strain. Elimination in the opposite direction would place the double bond in the bicyclooctane ring system and this would increase the

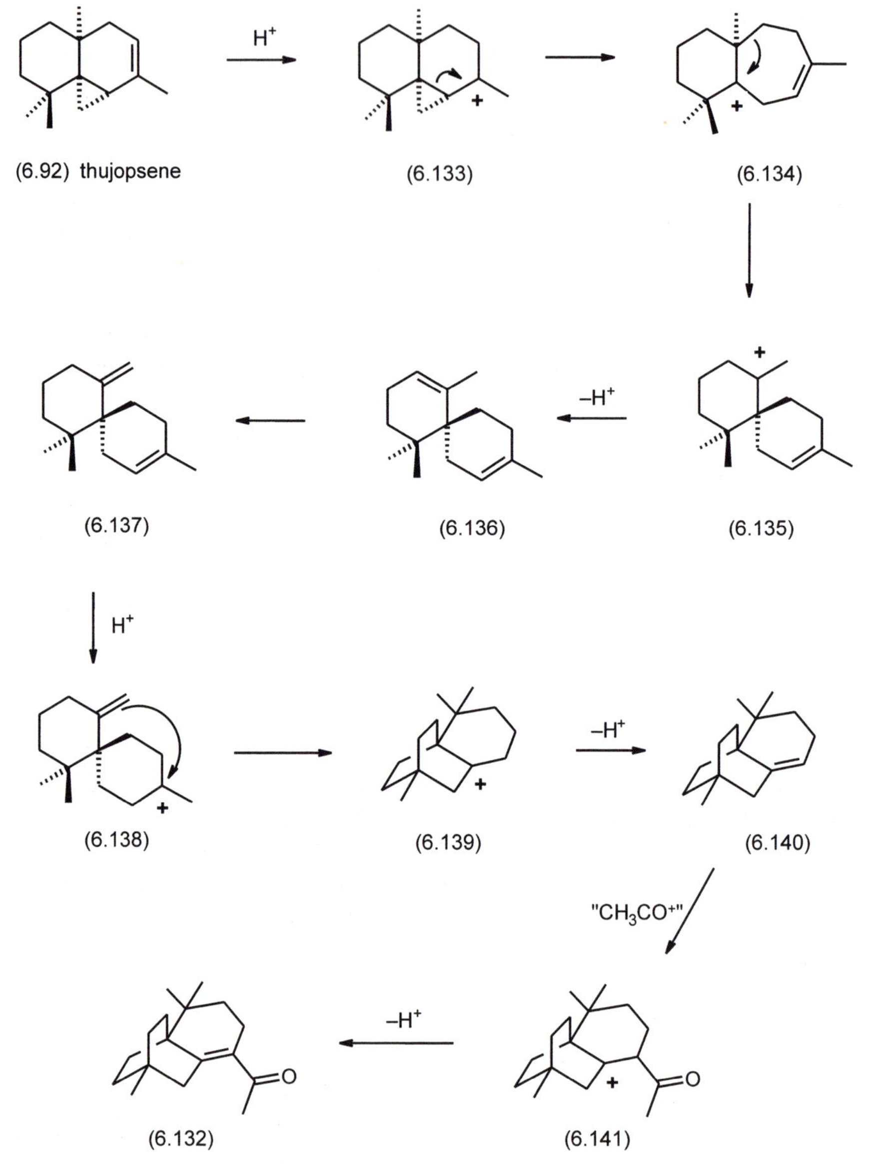

(6.92) thujopsene   (6.133)   (6.134)

(6.137)   (6.136)   (6.135)

(6.138)   (6.139)   (6.140)

"CH$_3$CO$^+$"

(6.132)   (6.141)

Figure 6.33

strain in the molecule. Once again, the use of molecular models will demonstrate this well. This last olefin, (6.140), is the one which undergoes Friedel–Crafts acylation with the acyl anion equivalent to give (6.141) and elimination of the proton next to the carbonyl group gives the final product, (6.132).

6.2.6 Another Anomaly, Vilsmeier–Haack–Arnold Formylation of δ-Selinene

δ-Selinene is another sesquiterpene found in woods. Its behaviour in the Vilsmeier–Haack–Arnold reaction is, like that of the acylation of cedrene using titanic chloride as a catalyst, another interesting and surprising piece of sesquiterpenoid carbocation chemistry.

The Vilsmeier–Haack–Arnold reaction involves the addition of an imminium ion to an olefin as shown in Figure 6.34. The imminium ion (6.143) is formed from dimethylformamide (6.142) by the action of phosphorus oxychloride. Phosphorus trichloride, phosphorus penta-chloride or thionyl chloride can also be used in place of phosphorus oxychloride. This cation adds to 1,1-disubstituted double bonds and to dienes. In the figure, it is shown adding to a 1,1-disubstituted olefin (6.144) to give a carbocation (6.145). Loss of a proton from the position α to the cationic centre and elimination of chloride leads to the product imminium salt (6.146) which, on hydrolysis, gives an α,β-unsaturated aldehyde (6.147).

When subjected to Vilsmeier–Haack–Arnold conditions, δ-selinene (6.148) was found to give a saturated aldehyde[6.10] (Figure 6.35). This was

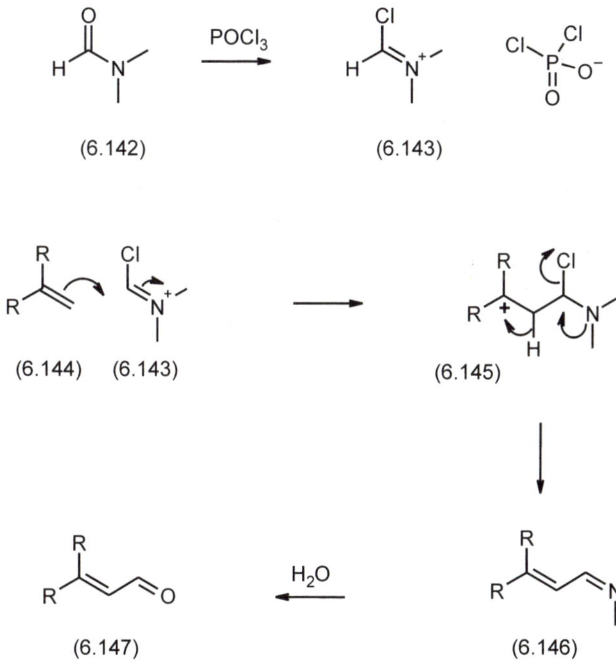

Figure 6.34

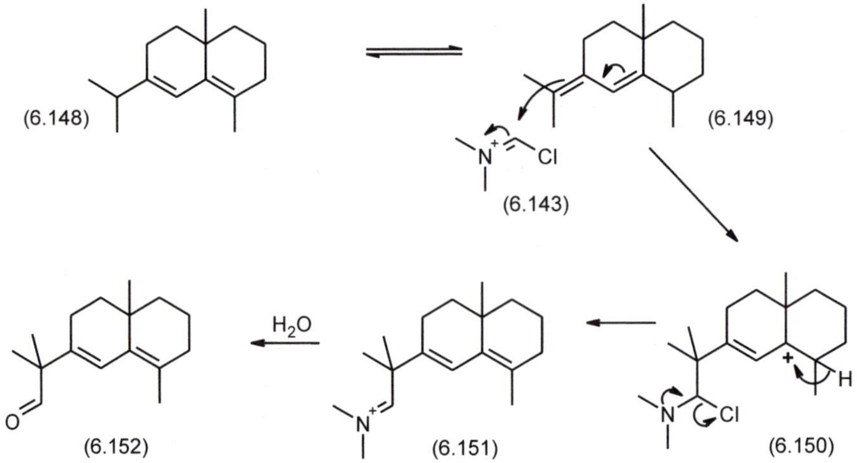

Figure 6.35

very surprising as the products of all previously reported Vilsmeier–Haack–Arnold formylations of olefins and dienes had been conjugated aldehydes. It would seem that δ-selinene itself is inert to the Vilsmeier–Haack–Arnold reaction, presumably because of the steric hindrance and the low polarisability of this particular diene system. The Vilsmeier–Haack–Arnold reaction medium is very acidic and so isomerisation of the double bonds is quite likely. Therefore it was postulated that δ-selinene must have isomerised to diene (6.149). One end of the diene system is now out of the decalin ring system and therefore more sterically accessible. It reacts with the imminium ion (6.143) to give the carbocation (6.150). This carbocation can eliminate chlorine in the usual way but the normal α-elimination of a proton is not possible as the carbon atom bearing the formyl group is quaternary in this case. It is therefore the δ hydrogen which is lost and the diene function is thus restored to the original position it occupied in δ-selinene. The imminium ion produced is therefore (6.151) which, on hydrolysis, yields the observed saturated aldehyde (6.152).

REFERENCES

6.1.   E. Demole, *Helv. Chim. Acta*, 1964, **47**, 319.

6.2.   E. Demole, *Helv. Chim. Acta*, 1969, **52**, 2065.

6.3.   D.R. Adams, S.P. Bhatnagar, R.C. Cookson and R.M. Tudenham, *J. Chem. Soc. Perkin Trans. I*, 1975, 1741.

6.4.   D.R. Adams, S.P. Bhatnagar and R.C. Cookson, *J. Chem. Soc. Perkin Trans. I*, 1975, 1502.

6.5.   R. Shankaranarayanan, S. Krishnappa, S.C. Bisarya and Sukh Dev, *Tetrahedron Lett.*, 1973, 427.

6.6.   B.A. McAndrew, S.E. Meakins, C.S. Sell and C. Brown, *J. Chem. Soc. Perkin Trans. I*, 1983, 1373.

6.7.   H.U. Daeniker, A.R. Hochstetler, K. Kaiser and G.C. Kitchens, *J. Org. Chem.*, 1972, **37**, 1.

6.8.   H.U. Daeniker, A.R. Hochstetler, K. Kaiser and G.C. Kitchens, *J. Org. Chem.*, 1972, **37**, 6.

6.9.   W.G. Dauben, L.E. Friedrich, P. Oberhaensli and E.I. Aoyagi, *J. Org. Chem.*, 1972, **37**, 9.

6.10.   C.S. Sell and B.R. Hart, *Chem. Ind.*, 1979, 59.

CHAPTER 7

Other Woody Odorants

Humulene, obtained from hops, is one of
many sesquiterpenoids with woody odour

The woods are lovely, dark and deep

Robert Frost

Frost was not thinking of perfume when he wrote this line in his poem
"Stopping by woods on a snowy evening". However, his words do apply
very well to the role of woody notes in a fragrance. The woods sit down
under the heart of a composition and give it richness and depth. In this
chapter, the term "wood" is used in the perfumistic rather than the
botanical sense since it covers not only wood (pine) but also roots
(vetiver), leaves (patchouli) and flowers (cloves and hops).

KEY POINTS

As in Chapter 6, we will see that the same rules of chemistry apply to
sesquiterpenoids as with monoterpenoids. The fact that the molecules are bigger means that the reactions can be harder to see and
that there may be possible pathways for a greater variety of
reactions.

Confirmation of structure through synthesis is very important in
natural products chemistry since structural determination through
degradation and spectroscopy can sometimes give wrong answers.

Complex rearrangements such as the santonin rearrangement are
best followed through using molecular models.

Some basic rules of planning a synthesis are: choose reactions with
high conversion; choose reactions with high selectivity; aim for
the fewest number of steps possible; try to run as many steps in
parallel as possible; and avoid low yielding or unreliable steps late
in the synthesis.

The use of elegance in design can enable complex structures to be synthesised in a very efficient way, as demonstrated by Gilbert Stork's synthesis of β-vetivone.

Physical proximity of atoms in molecules can radically affect the course of reactions.

7.1 VETIVER

Vetiver (*Vetiveria zizanoides*) is a tropical grass with large leaves springing from a dense root system. Extraction of the roots yield an oil with a rich, warm, woody and nutty smell. Vetiver oil is highly prized in perfumery and much attention has been given to the chemistry of its components. It contains over 150 sesquiterpenoids in all. The three most significant ones are two ketones, α-vetivone and β-vetivone, and an alcohol, khusimol. Together, these three materials constitute about 30–35% of the oil.

7.1.1. Initial Structural Determination of the Vetivones

Initially, it was thought that α- and β-vetivone were the most important components of vetiver oil in terms of odour. The former was considered to be the most important of all, possessing a pleasant, strong, warm odour which is characteristic of vetiver oil. The latter was seen as less important as its odour is relatively weak and has greenish aspects and a styrax like note. The initial work on determination of the structures of the vetivones was carried out in the 1940s by two teams of chemists, one team comprising Naves and Perrottet[7.1] and the other Pfau and Plattner.[7.2,7.3] Both teams came to the same conclusion based on their findings concerning the structure of the two ketones. Their findings from degradation and a little spectroscopy are summarised below.

Parameter	α-vetivone	β-vetivone
Molecular formula	$C_{15}H_{22}O$	$C_{15}H_{22}O$
	Two rings	Two rings
	one ketone C=O double bond	one ketone C=O double bond
	two C=C double bonds (one conjugated to the ketone)	two C=C double bonds (one conjugated to the ketone)
On ozonolysis	Produces acetone	Produces acetone
On dehydrogenation	Produces vetivazulene and eudalinol	Produces vetivazulene, eudalinol and vetivalene

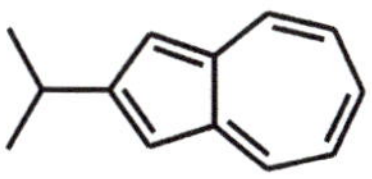

(7.1) vetivazulene 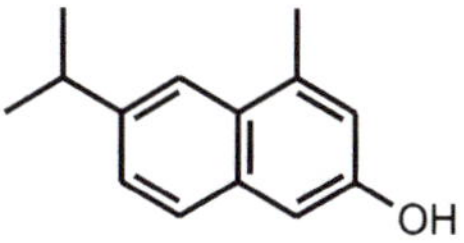   (7.2) eudalinol

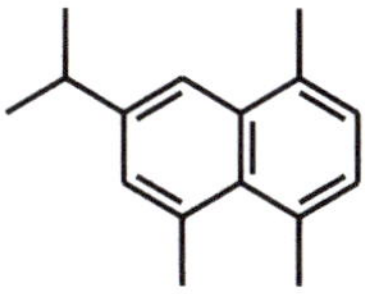

(7.3) vetivalene 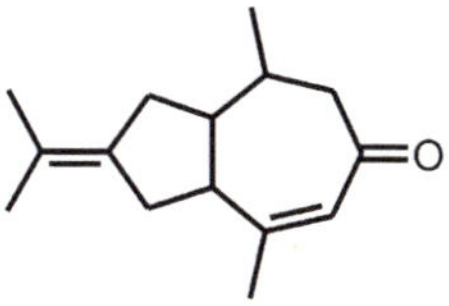   (7.4)

Figure 7.1

The structures of vetivazulene (7.1), eudalinol (7.2) and vetivalene (7.3) are shown in Figure 7.1. Based on these findings, both teams concluded that α- and β-vetivone were different stereoisomers of the hydroazulenone shown as structure (7.4) in Figure 7.1. Examination of this structure reveals how it does appear to fit the facts determined by Naves and Perrottet, and Pfau and Plattner. Its formula is $C_{15}H_{22}O$, contains two rings, one ketone group, two double bonds one of which is conjugated to the ketone. The isopropylidene group would produce acetone on ozonolysis and the observed dehydrogenation products would be consistent with those expected from (7.4). Neither group confirmed their structural determination by synthesis and their findings were accepted as correct for over 20 years.

7.1.2   The Initial Structural Determination Disproved

In the 1960s, Marshall and his team of chemists at Northwestern University in Illinois decided to test the accepted structure of the vetivones. Since the proposed structure (7.4) represented a serious synthetic challenge (largely because of the seven-membered ring), they decided to synthesise a more accessible material, but one which could be obtained unambiguously by degradation of the natural product. If the structure of Naves and Perrottet, and Pfau and Plattner were correct, then it could be converted to the much simpler bicyclic ketone (7.8) as shown in Figure 7.2. The four reactions in this scheme are all highly selective and the pathway illustrated is the only one possible.

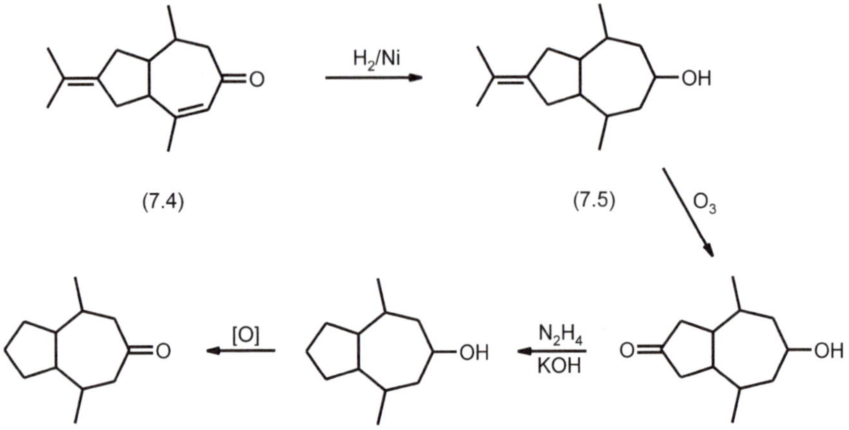

(7.4)   (7.5) O_3

(7.8)   [O]   (7.7) $\xleftarrow[\text{KOH}]{N_2H_4}$ (7.6)

Figure 7.2

In 1940, Pfau and Plattner had found that hydrogenation of β-vetivone using a nickel catalyst, reduced the unsaturated ketone function to the corresponding saturated alcohol (7.5), and Marshall and co-workers took advantage of this fact as a starting point for the simplification of the structure. Ozonolysis of ketone (7.5) produced acetone and ketone (7.6). Wolff–Kishner reduction of the ketone function of (7.6) gave the saturated alcohol (7.7) and oxidation of this gave the symmetrical saturated ketone (7.8). This represents a much easier synthetic target than structure (7.4), and Marshall, Andersen and Johnson synthesised it as shown in Figure 7.3.[7.4]

The starting point for the synthesis is the hydrindanone (7.10). This ketone is readily available, for example, by Robinson annulation of cyclopentanone with methyl vinyl ketone. Treatment of (7.10) with excess methyl iodide in the presence of a base (potassium *t*-butoxide) gave the geminally dimethylated ketone (7.11). (The reader will recall an explanation of this synthetic tactic given in Chapter 6, Figure 6.20 and associated text.) Hydrogenation of (7.11) over a palladium catalyst gave the saturated ketone (7.12). Palladium has a much greater affinity for carbon–carbon double bonds than for carbon–oxygen double bonds and so the reduction proceeds without affecting the ketone function. There is now only one active position in the molecule as far as bromination is concerned and so, treatment of ketone (7.12) with bromine in acetic acid gave the α-bromoketone (7.13). Dehydrobromination in the presence of base gave the α,β-unsaturated ketone (7.14). Reaction of (7.14) with methyl Grignard reagent in the presence of copper (I) gave 1,4-addition

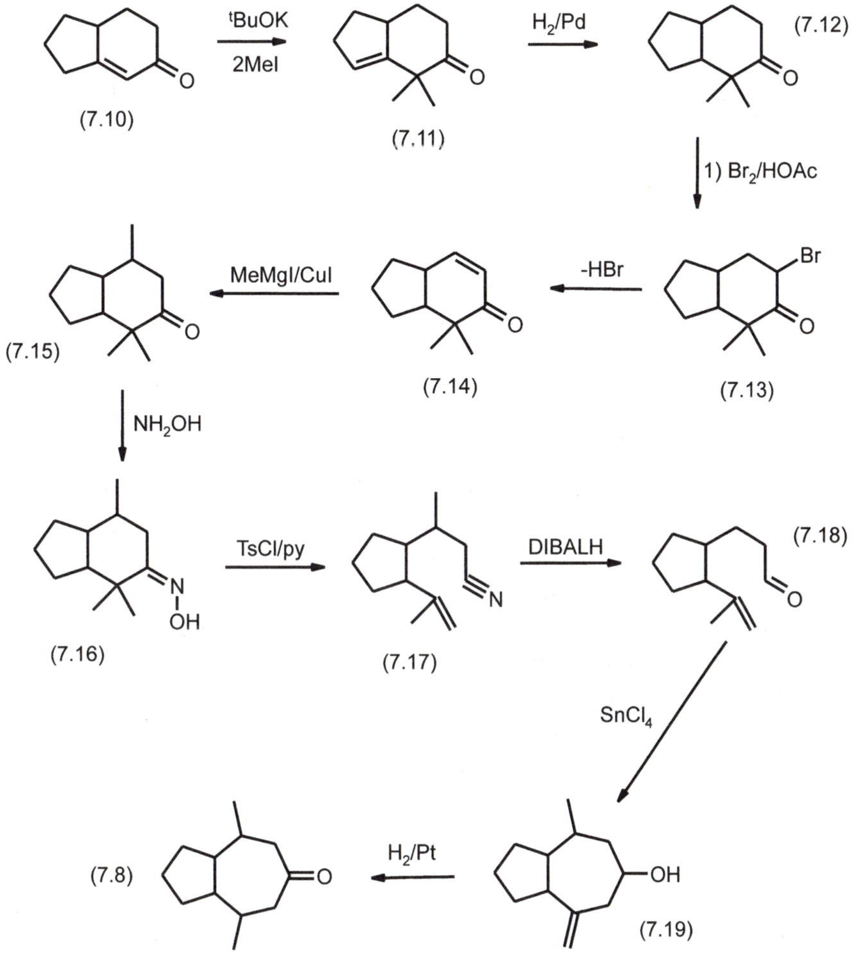

Figure 7.3

as expected, to produce the α,α,β'-trimethylperhydroindnone (7.15). Reaction with hydroxylamine gave the oxime (7.16) and treatment of this with *p*-toluenesulfonyl chloride in pyridine gave the nitrile (7.17). The latter reaction involves the formation of the oxime tosylate (7.20) and subsequent base catalysed elimination, the mechanism of which is shown in Figure 7.4. Reduction of the nitrile function to the aldehyde can be achieved selectively using di-isobutylaluminium hydride (DIBALH) and so the unsaturated aldehyde (7.18) was produced. Ene reaction of this aldehyde gave the unsaturated alcohol (7.19); the mechanism of this is shown in Figure 7.5. Strictly speaking, this reaction should proceed simply by heating the starting material as it is an electrocyclic reaction.

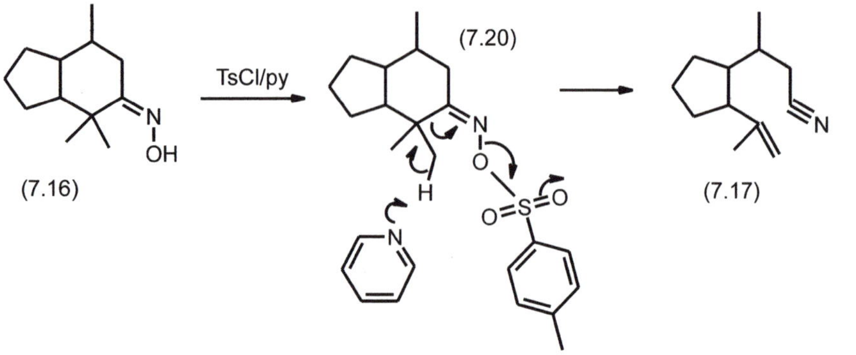

Figure 7.4

However, Marshall *et al.* used stannic chloride as a catalyst in order to keep the conditions as mild as possible to prevent unwanted side reactions such as elimination of the alcohol function from the product. The role of the stannic chloride is to polarise the carbonyl group of the aldehyde and lower the activation energy of the nucleophilic attack by the double bond. Treatment of the resultant unsaturated alcohol with platinum in the presence of hydrogen resulted in hydrogenation of the carbon–carbon double bond, but dehydrogenation of the carbon–oxygen bond to give the saturated ketone (7.19). Unlike palladium, platinum has quite an affinity for oxygenated functions and will dehydrogenate alcohols.

Now comes the crucial point of this work of Marshall and his co-workers. Ketone (7.8) has been synthesised by an unambiguous route, there is no possibility that they have produced anything other than the target material. If the structure (7.4) of β-vetivone according to Naves and Perrottet, and Pfau and Plattner is correct, then the sequence of degradation reactions shown in Figure 7.2, which is also totally unambiguous, should give an identical ketone to that synthesised. When Marshall *et al.* compared the two materials, they found them to be

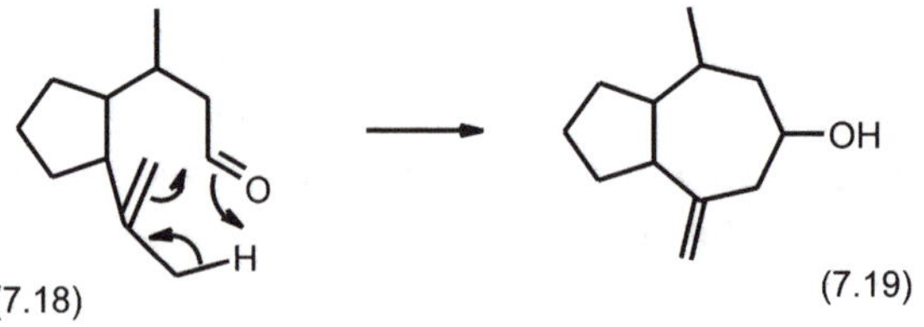

Figure 7.5

different. Therefore, the accepted structure of β-vetivone was shown, beyond doubt, to be incorrect.

7.1.3 The Correct Structure of β-Vetivone Established

Having shown that the accepted structure of β-vetivone was incorrect, Marshall looked again at the evidence. In particular, he re-investigated the infrared spectrum of the natural material and suggested that the α,β-unsaturated ketone system is present in a six-membered rather than a seven-membered ring. This then led him to propose that β-vetivone contained a spirane structure, (7.21) to be precise (Figure 7.6).

In order to test this structural assignment, Marshall *et al.*[7.5] used exactly the same strategy that they had used to disprove the previously accepted structure, *i.e.*, unambiguous degradation of the natural product to a material which is easily and unambiguously synthesised. Their degradation scheme is shown in Figure 7.6. The hydrogenation of β-vetivone over a nickel catalyst selectively reduced the α,β-unsaturated ketone function to the corresponding saturated alcohol (7.22). This was then oxidised to the ketone (7.23) using Jones' reagent and subsequent Wolff–Kishner reduction of this ketone gave the hydrocarbon (7.24). Ozonolysis of hydrocarbon (7.24) gave the saturated ketone (7.25). The synthesis of this last ketone was achieved as shown in Figure 7.7.

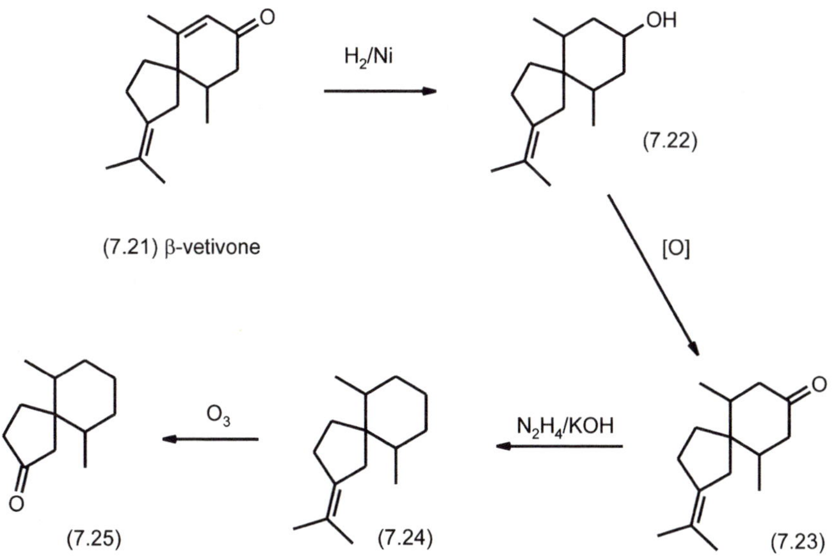

Figure 7.6

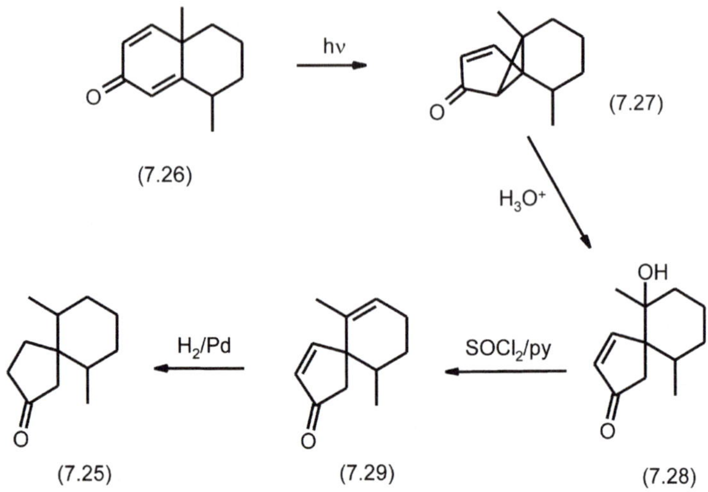

Figure 7.7

For simplicity and to make the key structural changes clearer, the structures in Figure 7.7 have been drawn with no attempt to represent stereochemistry, which will be discussed later. The starting material was the dienone (7.26). Upon irradiation with ultraviolet light, this ketone rearranged to the tricyclic ketone (7.27) in a rearrangement known as the santonin rearrangement. This name originates from the sesquiterpenoid santonin (7.30) which rearranges to lumisantonin (7.31), as shown in Figure 7.9, when similarly irradiated. Acid catalysed ring opening of the tricyclic ketone (7.27) gives the spirocyclic keto-alcohol (7.28). Dehydration of (7.28) using thionyl chloride in pyridine, gives the olefin (7.29) and hydrogenation of the latter gives the target ketone (7.25). This time, the synthetic ketone was identical to that obtained by degradation of natural β-vetivone, thus providing strong support for the structure proposed by Marshall *et al.* In order to confirm their structural assignment, they then carried out a total synthesis of β-vetivone. Before we go on to look at this synthesis, it will be useful to look in more detail at the synthesis of the dienone (7.26) and the chemistry of the santonin rearrangement.

7.1.4   Bloom's Synthesis of the Decalindienone

The dienone (7.26) is the key for both the synthesis of (7.25) as above and in Marshall's total synthesis of β-vetivone and its stereochemistry is important in the subsequent steps of the syntheses. Hence, it is worthwhile looking in more detail at its synthesis. Bloom prepared this dienone in four steps from 2-methylcyclohexanone (7.32) in 1958, as shown in Figure 7.8.[7.6]

The first stage in Bloom's synthesis is the formylation of (7.32) using ethyl formate in the presence of a base. This has the effect of adding a carboxaldehyde function to the 6-position of the cyclohexane ring. The initial addition could occur at either the 2- or the 6-position, but only the latter can allow for formation of the enol form of the product aldehyde. It is the enol form which is shown as (7.33) since this is the tautomer which exists predominantly. The enol form is stabilised both by overlap of the π-electrons of the ketone and olefin double bonds, and by hydrogen bonding of the enolic hydrogen to the carbonyl oxygen atom. Alkylation of the enolate anion of (7.33) gives (7.34) in which the methyl groups are *trans*- to each other across the cyclohexane ring. The reason for this can be seen in the transition state (7.37). The methyl group already on the ring will prefer to adopt the equatorial conformation and

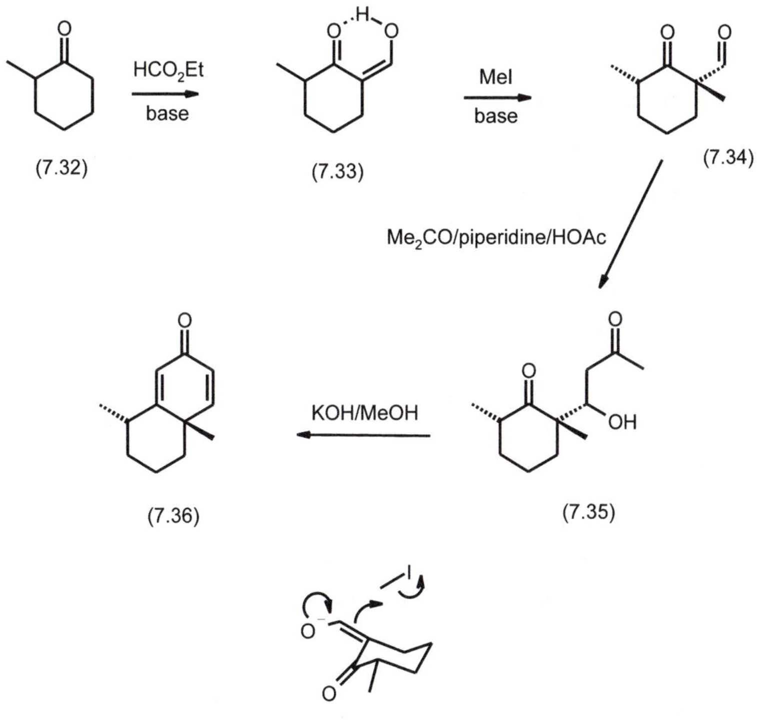

Figure 7.8

this directs the incoming alkylating agent to the opposite side. In all the subsequent structures, the carbon carrying the first methyl group is epimerisable, but the *trans*-configuration is always the preferred one and so this stereochemistry is retained. Two aldol reactions follow subsequently. The aim is to add one molecule of acetone across the two carbonyl groups of (7.34). If random aldol condensations were allowed, a large number of different combinations are possible and a complex product mixture would result. Using different reaction conditions, Bloom was able to achieve fine control over the process. Initially, he used piperidine/acetic acid as the catalyst system. This combination is buffered to a pH close to neutral. The very mild conditions mean that only the more electrophilic aldehyde carbon serves as the acceptor for aldol chemistry and a high yield of the intermediate (7.35) is formed. Subsequent addition of methanolic potassium hydroxide allows less reactive centres to undergo the aldol reaction, and the preference for formation of five- or six-membered rings ensures that the most favourable course is the desired formation of dienone (7.36) rather than self-condensation of either acetone or ketone (7.35).

7.1.5   The Santonin Rearrangement

Santonin (7.30) occurs in the oil known as Levant Wormseed Oil. This oil is not extracted from the seeds, but from the unopened flower heads of several herbs of the family *Artemisia*, principally from *Artemisia maritima*. Its use in folk medicine for the treatment of intestinal worms stems from the anthelmintic properties of santonin. Santonin undergoes a number of very interesting reactions, three of which are summarised in Figure 7.9.

When treated with acid, santonin (7.30) undergoes what is known as the dienone–phenol rearrangement. Santonin was the first material observed to do this. Protonation of the ketone produces a centre of positive charge at the β-position giving (7.38) and a 1,2-carbon shift then allows elimination of a proton to form an aromatic ring. At some stage, the presence of a positive charge at the other β-position, allows the adjacent alcoholic carbon to epimerise to give the more stable *cis*-fusion of the lactone ring. Thus, the final product is the phenol (7.39).

Treatment of santonin (7.30) with aqueous base also gives a rather unexpected product. Initially, the strained lactone ring is hydrolysed. A series of keto–enol tautomerisations then allows the ene-one-ol system of santonin to rearrange to a dione. Deprotonation of this dione gives an anion, (7.40), which undergoes an internal Michael reaction to give the acid (7.41).

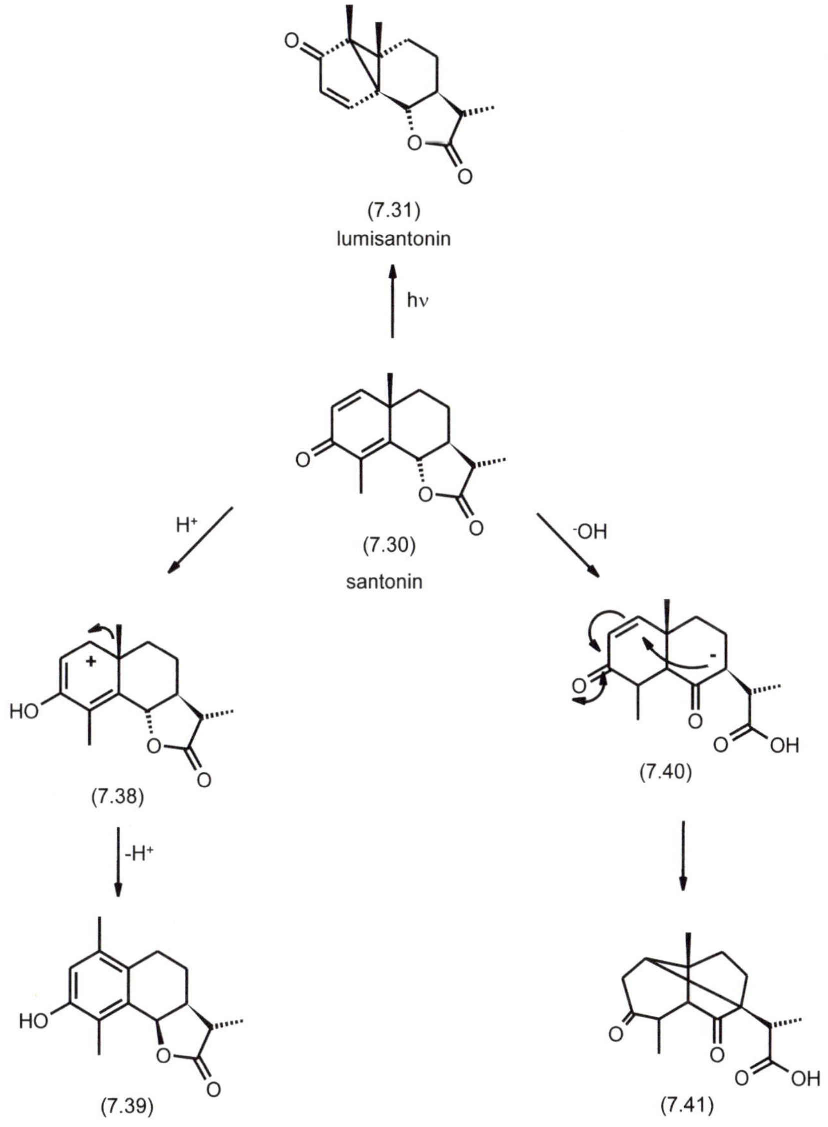

Figure 7.9

The most surprising of all of santonin's reactions is the formation of lumisantonin (7.31) upon irradiation. The mechanism of this reaction is not at all obvious and it was the subject of much research by some of the greatest minds of twentieth-century organic chemistry including Sir Derek Barton and Paul de Mayo. The following description of the mechanism is based on that of Zimmermann and Schuster.[7.7] In order to make it easier to see, we will look at the rearrangement of Bloom's dienone (7.36) rather than santonin itself.

The mechanism is shown in Figure 7.10. This mechanism is not easy to follow and the reader would be well advised to build a molecular model

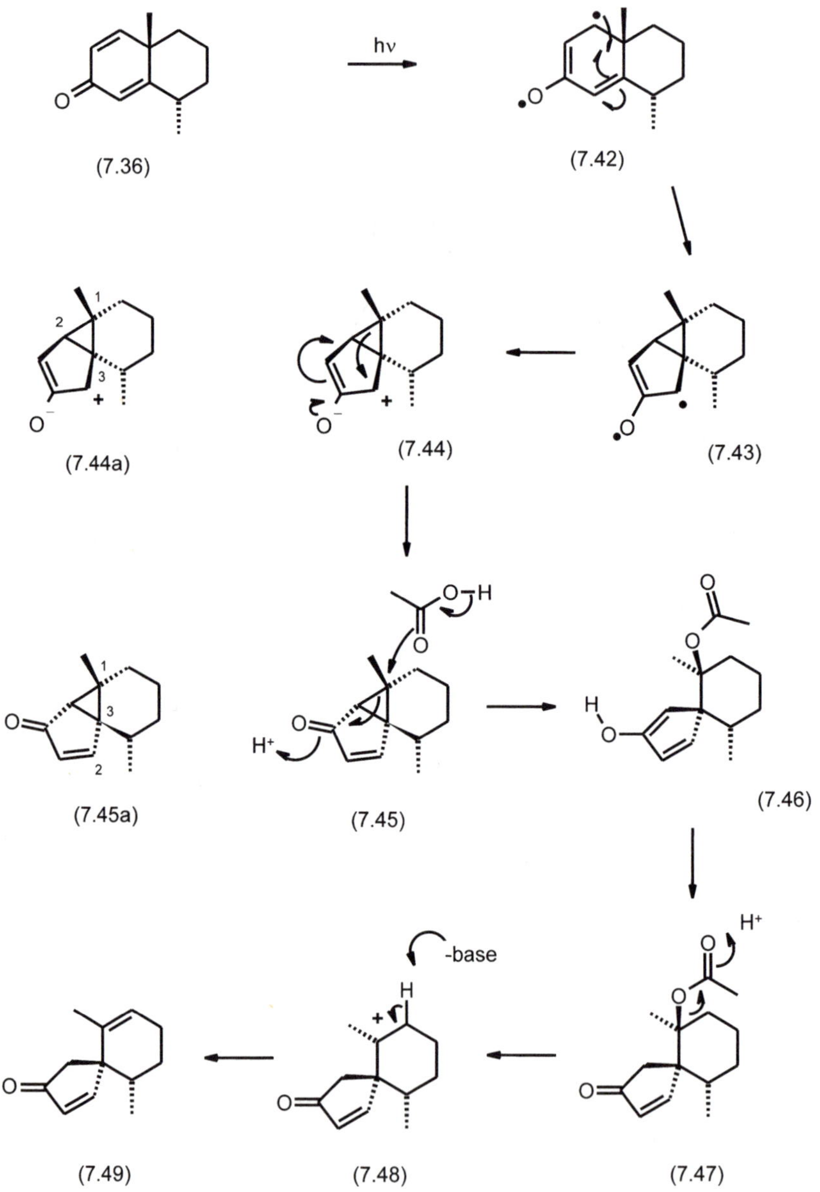

Figure 7.10

of the starting material and then follow the reaction step by step by adjusting the bonds in the model one at a time and thus convincing him/ herself that the drawn structures are correct. For clarity of drawing, the stereochemistry of structures (7.43), (7.44) and (7.45) are defined relative to the cyclopropane ring of each.

The reaction starts with the absorption of ultraviolet light by the dienone system of the starting ketone (7.36). This represents quite an intake of energy by the molecule and the effect is to promote two electrons to higher level orbitals, giving a structure which behaves chemically as the diradical (7.42). The carbon centred radical then adds to the double bond on the opposite side of the ring to produce a new diradical (7.43). In this new diradical, the unpaired electrons are on carbon atoms which are 1,3-relative to each other. This proximity allows a process known as electron demotion to occur. The unpaired electron on the carbon atom moves to the oxygen atom thus generating a full negative charge on the oxygen atom and a positive charge on the carbon atom giving structure (7.44). This process is energetically favourable, as a diradical is a higher energy species than a betaine. Moreover, the molecule can now reduce its free energy even further by allowing the two charges to neutralise each other. This it does through the rearrangement of (7.44) to (7.45). The negative charge on the oxygen atom flows through the double bond towards the cyclopropyl ring, breaking one bond of the latter and forming a new bond, and hence a new cyclopropane ring, by adding to the carbocation. This flow of electrons is shown in structure (7.44) in Figure 7.10. Since the stereochemistry of (7.44) and (7.45) are both defined by the cyclopropane ring of each, it can be somewhat confusing because the cyclopropane rings of each are made up of different carbon atoms. To make this clearer, both structures have been redrawn as (7.44a) and (7.45a), respectively. In these last two structures, the carbon atoms of the cyclopropane ring of (7.44a) have been assigned arbitrary numbers. The same atoms carry the same numbers in (7.45a) and it can be seen that atom no. 2 is not a part of the cyclopropane ring of (7.45a).

This completes the santonin rearrangement and (7.45) is to (7.36) what lumisantonin (7.31) is to santonin (7.30).

The subsequent steps in Figure 7.10 show what happens when (7.45) is treated with acid in the presence of a nucleophile. Acetic acid is used as the nucleophile in Figure 7.10 but other suitable nucleophiles, such as water, will also allow the same conversion to take place.

Protonation of the carbonyl oxygen of (7.45) and simultaneous attack at the rear face of the β-carbon atom (with inversion as in an S_N2 reaction) by the nucleophile allows a reaction, reminiscent of the Michael reaction, to occur with opening of the cyclopropyl ring and hence relief of the associated ring strain. This gives the enol (7.46) which can tautomerise to the ketone (7.47). Elimination of the acetate group generates the carbocation (7.48) and this eliminates a proton to give the final product (7.49). This structure has appeared before as (7.29) but has now been given a new reference number as it is now shown with defined stereochemistry. This introduces us to the fact that an important feature of this overall reaction sequence is that it is stereospecific. The relative stereochemistry of the starting dienone (7.36) is fixed by its synthesis as described earlier and the fact that every intermediate reaction step proceeds with total stereoselectivity, ensures that the relative stereochemistry of (7.49) is exactly as shown in Figure 7.10. This is important as will be seen below.

7.1.6   Marshall and Johnson's Total Synthesis of β-Vetivone

The structure of β-vetivone appeared as (7.21) in Figure 7.6 but this showed no stereochemistry. Figure 7.11 shows the stereochemical detail as (7.50). A brief inspection of the architecture of the molecule reveals two main challenges in its total synthesis. The first is the construction of the spirane system and the second is the introduction of both the methyl and isopropylidene groups in the correct relative stereochemical relationship to each other, *i.e.* both must be on the same side of the plane of the six-membered ring.

Marshall and Johnson used the santonin rearrangement to solve both problems. They started with the decalindienone (7.36) following the synthesis of Bloom. The santonin rearrangement followed by treatment of the product ketone (7.45) with acetic acid, gave the ketone (7.49) which they had used previously as an intermediate in the synthesis of the β-vetivone degradation product (7.25) (see Figure 7.7). Figure 7.11 shows this material (7.49) alongside the target, β-vetivone (7.50).

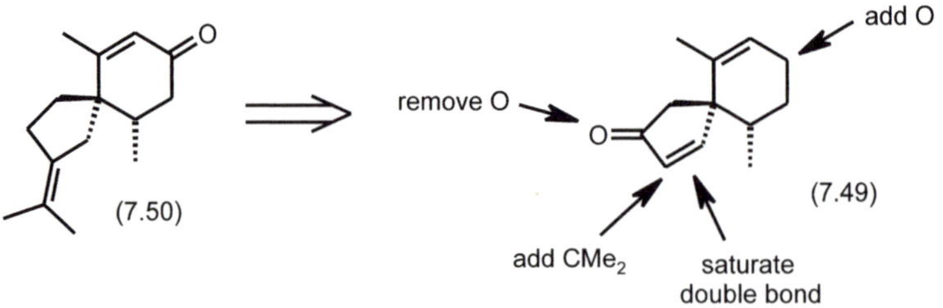

Figure 7.11

The intermediate ketone (7.49) possesses the same spirane system as the target and it remains only to remove one oxygen atom, introduce one other, saturate the double bond and add an isopropylidene group. The two functions to be added must, of course be added at the correct sites and it can be seen that the existing functionalisation of (7.49) will allow this to be done selectively. The methylene group to be oxidised is allylic and so the oxidation will be relatively easy. The isopropylidene group must be added adjacent to the ketone of (7.49) and selectively on one side of it. Steric effects allow the required selectivity. The complete synthesis is outlined in Figure 7.12.

In Figure 7.7, we saw that ketone (7.29) was hydrogenated to the saturated ketone (7.25). However, using controlled hydrogenation conditions, in the presence of sodium hydroxide, it is also possible to achieve selective hydrogenation of the double conjugated to the ketone. This gave ketone (7.51), which could be formylated using ethyl formate and sodium hydride in benzene. This formylation introduces the carbon around which the isopropylidene group can be built. The regioselectivity of the reaction is the result of the so-called neopentyl effect. The "wrong" α-methylene group of (7.51) is adjacent to a quaternary carbon atom. Such positions are known as neopentyl and are notoriously unreactive owing to steric effects. The aldehyde function of the product (7.52) exists in the enol form. This is the more stable tautomer as it allows overlap of the π-electrons of the two double bonds and intramolecular hydrogen bonding. Treatment of (7.52) with butane thiol in the presence of boron trifluoride etherate, caused the replacement of the aldehyde oxygen by the sulfur of the thiol to give the thio-ether (7.53). Sodium borohydride was then used to reduce the ketone function to the corresponding alcohol and thus pave the way for removal of the unwanted oxygen atom, now that it had served its purpose by enabling the introduction of substitution at the desired site. Treatment of (7.54) with mercuric chloride and hydrochloric acid in aqueous acetone, resulted in the hydrolysis of the thio-ether, removal of the sulfur atom by mercury and elimination of the alcohol to give the unsaturated aldehyde (7.55). Treatment of this aldehyde with methyl lithium then introduced one of the methyl groups of the eventual isopropylidene function. Oxidation of the alcohol (7.56) was achieved using dichlorodicyanobenzoquinone (DDQ). Selective reduction of the double bond of the resultant α,β-unsaturated ketone (7.57), was achieved through the Birch reduction using lithium in liquid ammonia, as this does not affect isolated double bonds. However, it did reduce the ketone function to the alcohol (7.58) and so re-oxidation with chromium trioxide was necessary to give the ketone (7.59). Addition of methyl lithium then provided the second methyl group of the eventual

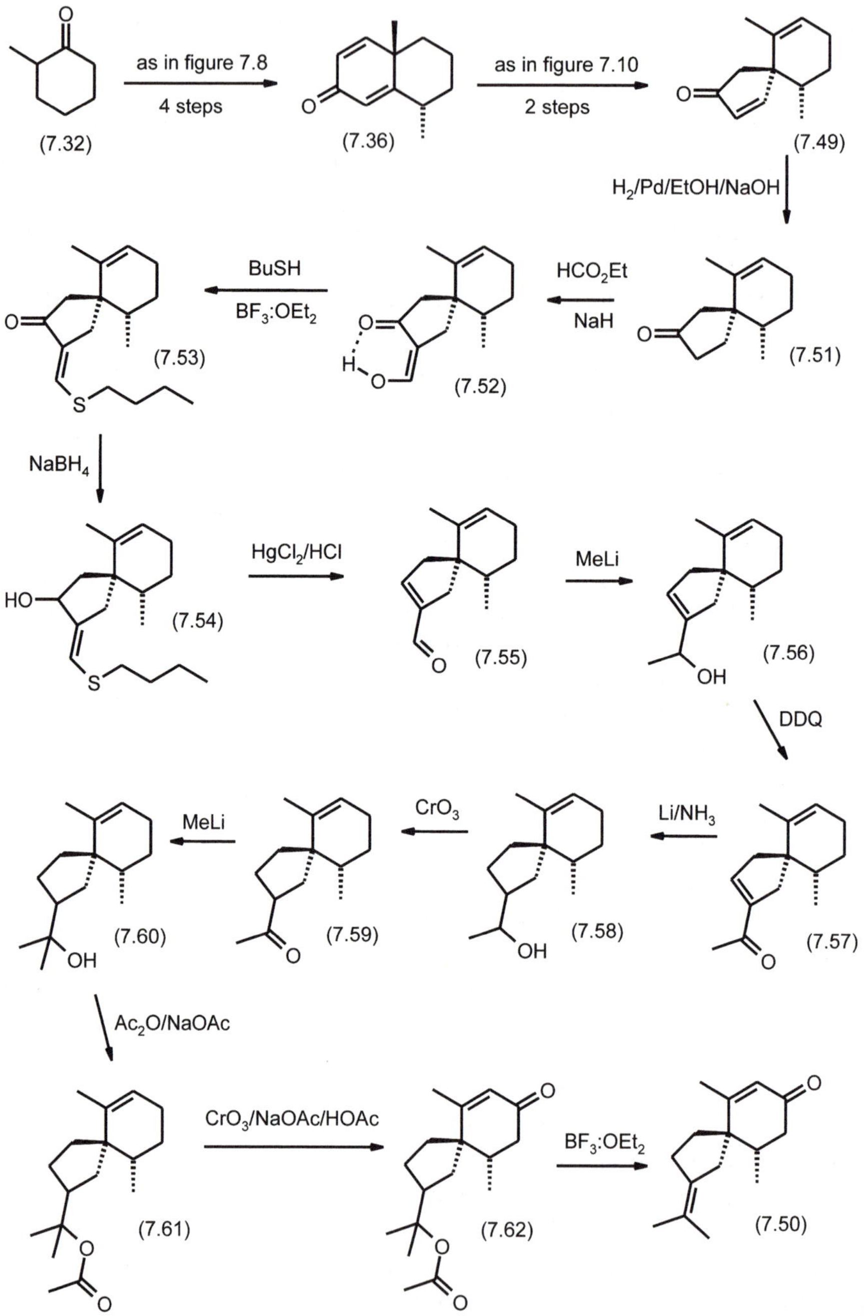
as in figure 7.8
4 steps
as in figure 7.10
2 steps
(7.32)
(7.36)
(7.49)
H2/Pd/EtOH/NaOH
BuSH
BF3:OEt2
HCO2Et
NaH
(7.53)
(7.52)
(7.51)
NaBH4
HgCl2/HCl
MeLi
(7.54)
(7.55)
(7.56)
DDQ
MeLi
CrO3
Li/NH3
(7.60)
(7.59)
(7.58)
(7.57)
Ac2O/NaOAc
CrO3/NaOAc/HOAc
BF3:OEt2
(7.61)
(7.62)
(7.50)

Figure 7.12

isopropylidene function. It now remained only to eliminate the alcohol function of (7.60) and to oxidise the allylic methylene position in the six-membered ring. However, these had to be done in the reverse order, otherwise the allylic methylene positions in the five-membered ring would also be oxidised. The tertiary alcohol function of (7.60) would not survive the conditions of the allylic oxidation and so it had to be protected as the acetate (7.61). The allylic oxidation was achieved using chromium trioxide in acetic acid, buffered by sodium acetate. This system has a preference for secondary allylic positions rather than primary and so only the desired oxidation was effected. It then remained only to eliminate the acetate group of (7.62) using boron trifluoride etherate to give β-vetivone, (7.50). Apart from the fact that the synthetic material was racemic, it was identical to the natural material and thus confirmed Marshall's proposed structure. Taking 2-methylcyclohexanone (7.32) as the starting material, the synthesis comprises a total of 19 steps.

7.1.7 Stork's Total Synthesis of β-Vetivone

There have been many syntheses of β-vetivone since that of Marshall and Johnson but perhaps the one that provides the greatest contrast is that of Gilbert Stork.[7.10] One striking feature of Stork's synthesis is that it is essentially a two-pot synthesis. In one ingenious step, he builds the spirane system and does so with the correct stereochemistry. His reaction scheme, realised in collaboration with his co-workers Rick Danheiser and Bruce Ganem, is shown in Figure 7.13.

The starting material is 5-methylcyclohexa-1,3-dione in which the enol form has been converted to an ether (7.63). This ketone is deprotonated using lithium di-isopropylamide (LDA), a very strong base which results in total and irreversible deprotonation of the ketone. The most reactive proton of the ketone is that on carbon-6, the other acidic proton is that on carbon-4 but this is more hindered and also rendered less acidic by the inductive effect of the ether oxygen. A carbanion is therefore formed completely and selectively on the site α- to the ketone group. To the solution of this carbanion, is added the dihalide (7.64) which is easily prepared from diethyl succinate. The two chlorine atoms in this molecule show very different reactivity because one of them is activated by virtue of being allylic. It is this allylic centre that reacts first and alkylates the carbanion. There is excess LDA present but this is a very poor nucleophile and so does not attack the chloride itself. Instead, it now deprotonates the ketone again, and again selectively in the 6-position to give the anion (7.66). The next step is the crux of the synthesis and one can only admire Stork for his genius in devising it. The detail is shown in Figure 7.14.

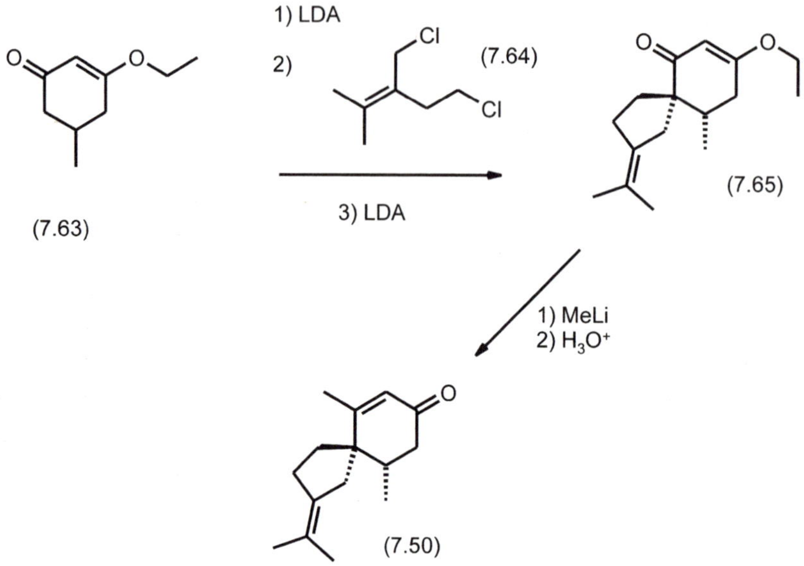

Figure 7.13

Carbanion (7.66) has an almost planar ring and the side chain is now attached to one end of the flattest section. The anion could, in principle, alkylate from either side but, as is clear in the figure, to do so from one side (top in the figure) will result in considerably more steric hindrance than from the other. The reason for this is the methyl group attached to the ring. To make this clearer in the figure, the hydrogen atoms on the methyl group and on the ring carbon to which it is attached, have been drawn in full. The alkylation therefore takes place on the opposite side of the ring from the methyl group and, consequently, the isopropylidene group ends up on the same side of the plane of the six-membered ring as the methyl group, exactly as required for β-vetivone. On work-up, the product is thereby the ketone-enol ether (7.65). It is now necessary to introduce a methyl group on the carbon which currently carries the

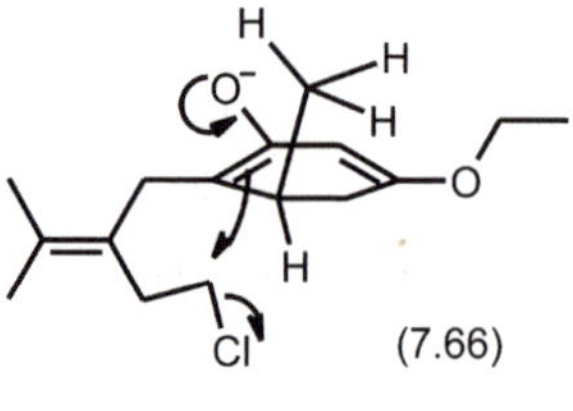

Figure 7.14

ketone function. This is easily achieved by reaction with methyl lithium. In a final piece of elegance, the acidic work-up conditions used at the end of this addition, cause both hydrolysis of the enol ether function, thus freeing up the ketone function, and elimination of the newly formed tertiary alcohol to produce the required olefinic bond, thereby producing racemic β-vetivone.

To say that Stork achieved in two steps what Marshall and Johnson took 19, is something of an exaggeration since neither of Stork's starting materials are available "off the shelf" and therefore have to be synthesised. A fairer comparison can be made if Stork's synthesis is shown in full as in Figure 7.15 which shows how he made the two key components.

The ketone (7.63) was prepared from ethyl acetoacetate (7.66) and ethyl crotonate (7.67) according to the method of Blanchard and Goering. [7.11] In the presence of a basic catalyst, these two molecules undergo a Robinson annulation reaction to give 5-methylcyclohexa-1,3-dione which is shown as the enolic form (7.68) in Figure 7.15. Treatment of (7.68) with ethanol in the presence of an acidic catalyst gave the required ketone (7.63). To prepare the dichloride (7.64), diethyl succinate (7.69) was treated with acetone in the presence of a base to give the isopropylidene derivative (7.70) which was reduced to the diol (7.71) by lithium aluminium ethoxy hydride. Conversion of this diol to the chloride was made difficult because of its tendency to form the cyclic ether. It was achieved by adding methanesulfonyl (mesyl) chloride to the di-lithium salt of the diol and then displacing the mesylate functions by chloride.

We can therefore consider that Stork's synthesis really has seven steps rather than two. However, this is still much shorter than Marshall and Johnson's and also it demonstrates the difference between a linear and a convergent synthesis. A linear synthesis is one in which the product of each reaction is used as the starting material for the next, right from the beginning of the synthesis to the end, as in Marshall and Johnson's synthesis of β-vetivone. A convergent synthesis is one, like Stork's, which starts with two or more raw materials and takes each through to an intermediate which is then added into the main synthesis scheme at some point. It would be nice to always have reactions which deliver 100% of the desired product but this is rare in practice. In fact, a yield of 90% is unusually good in a synthesis of this type. Therefore, every step in the synthesis results in a loss of weight of desired product. Convergent syntheses are better than linear ones in this respect since the weight losses are occurring concurrently rather than sequentially. Thus for the sake of argument, if the yield of each and every step in the schemes shown in Figures 7.12 and 7.15 were 90%, we can compare the effect on the overall

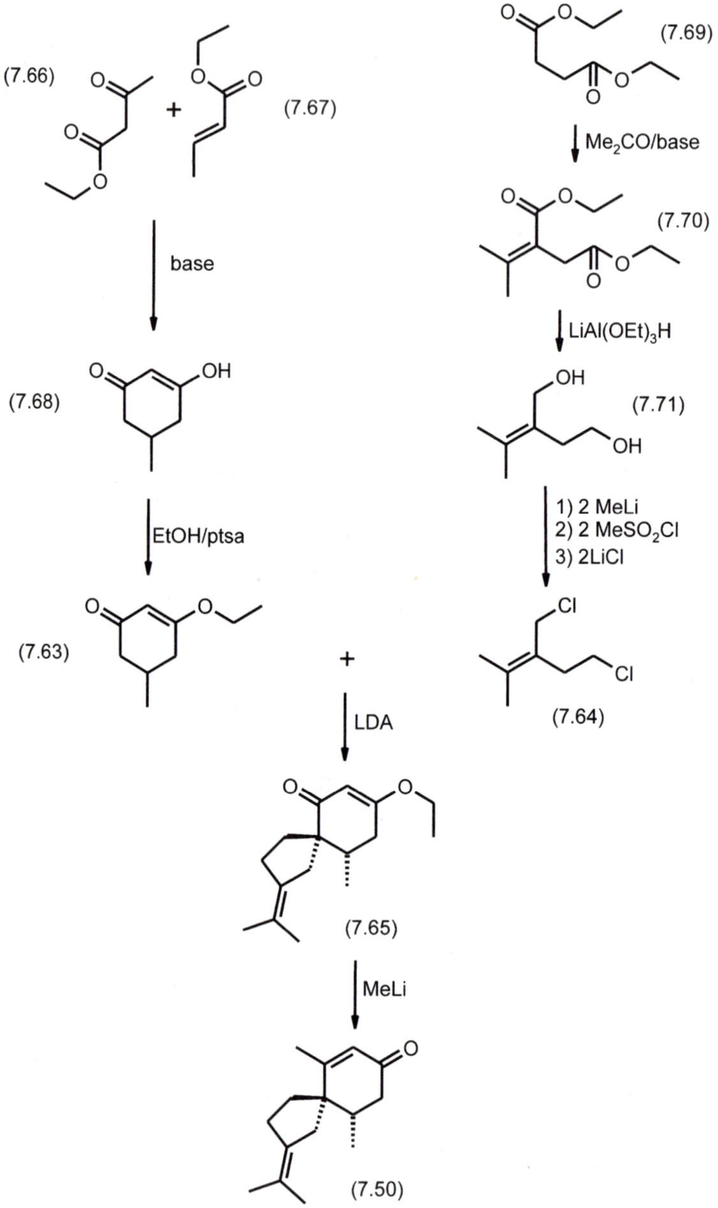

Figure 7.15

yield. Stork's eight steps, running as two and three in parallel and then two, would result in an overall yield of 66% based on starting materials (7.66) and (7.67) or a yield of 59% based on starting material (7.69). For a route in which the seven steps were run sequentially, the overall yield would be 48%. The overall yield of the 19 steps of Figure 7.12 would be 13.5%. If a synthesis is being carried out for commercial production or to provide samples for biological (or other) testing, this loss of material would obviously have serious implications. Even when a synthesis is being carried out purely for confirmation of structure, such loss of material is very undesirable because of the increased volume of starting material required in order to produce manageable amounts of product at the end, and the consequent effects on scale of the early stages. If one of the individual steps has a particularly low yield, this will be much more serious if it occurs towards the end of a synthesis, again because of the effects on scale of the early stages. It is much better to have the first step produce a low yield if all the others are high, than to carry large volumes of material through many stages to lose it all in the last step. Obviously, the higher the conversion and selectivity at every stage, the better. Some basic rules of planning a synthesis are therefore: choose reactions with high conversion; choose reactions with high selectivity; aim for the fewest number of steps possible; try to run as many steps in parallel as possible; and avoid low yielding or unreliable steps late in the synthesis.

7.1.8 Other Vetiver Components

Vetiver has provided terpenoid chemists with a great deal of intellectual stimulation, as is evident from the above discussion. The early work concentrated on the structures of the vetivones, because they are two of the three major components. However, none of the three major components are key contributors to the odour of the oil. It is a handful of the other 150 or so components that are the really important ones as far as fragrance is concerned and it took many years of careful research by various chemists to identify and characterise these trace ingredients. The correct stereostructures of the three major components of vetiver and the seven key odiferous components are shown in Figure 7.16. The relationship of three of these latter components to khusimol (7.67) is obvious. Oxidation of khusimol to the corresponding aldehyde gives zizanal (7.68). The ring position to which the aldehyde is attached is enolisable and hence epimerisation can occur to give epi-zizanal (7.69). Similarly, oxidation to the acid and esterification gives methyl zizanoate (7.71) and its epimer (7.72). In both instances, the epimeric compound has a stronger odour than the one with the stereochemistry of the starting khusimol. The other three organoleptically important components are clearly breakdown products since ketones (7.70) and (7.73) each contain only 12 of the original 15 carbon atoms and khusimone

Components of Vetiver (*Vetivera zizanoides*)

Major components (accounting for 30-35% of the oil)

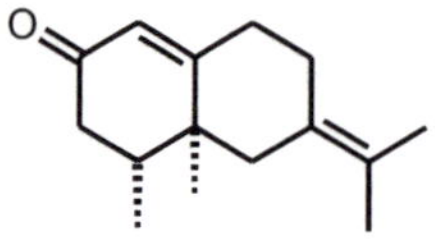

(7.66) α-vetivone 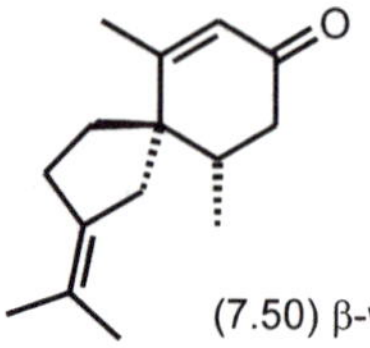   (7.50) β-vetivone

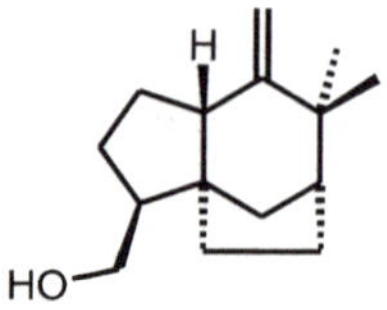

(7.67) khusimol

Minor components of major organoleptic importance

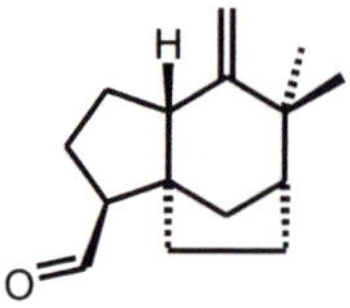

(7.68) zizanal 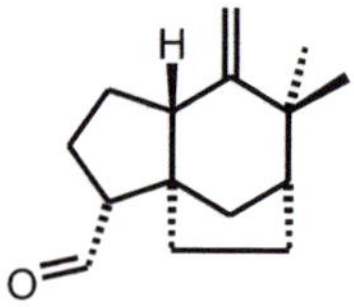   (7.69) *epi*-zizanal 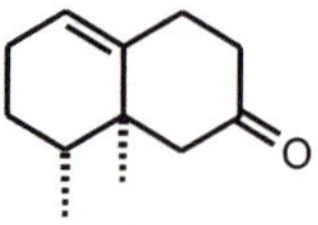   (7.70)

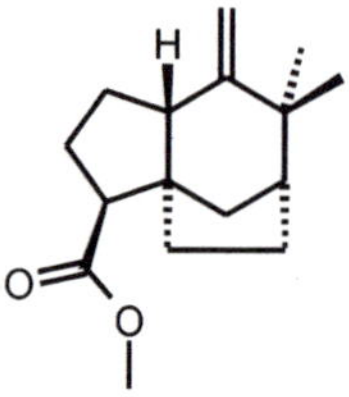

(7.71) methyl zizanoate 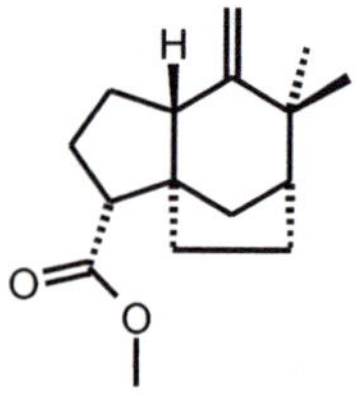   (7.72) methyl *epi*-zizanoate 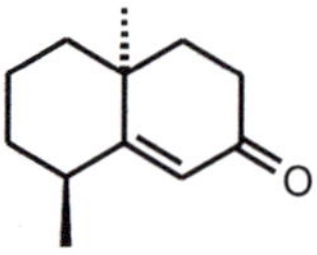   (7.73)

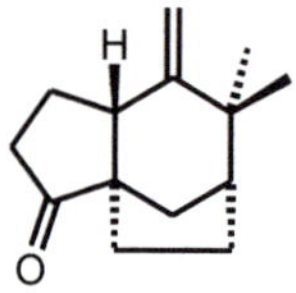

(7.74) (–)-khusimone

Figure 7.16

(7.74) contains only 14. All of the materials shown in Figure 7.16 have been shown to have insecticidal properties.

7.2 PATCHOULI

Patchouli (*Pogostemon cablin*) is a shrub which grows in Indonesia. Its leaves are harvested and allowed to ferment. This results in the formation of an intense and characteristic odour of a somewhat earthy nature. The oil is widely used in perfumery, particularly in after-shaves and colognes for men. Like that of vetiver, the chemistry of patchouli revolves around a large number of sesquiterpenoids with complex cyclic structures. The major component of the oil is patchouli alcohol (7.83) the biogenesis of which is outlined in Figure 7.17.

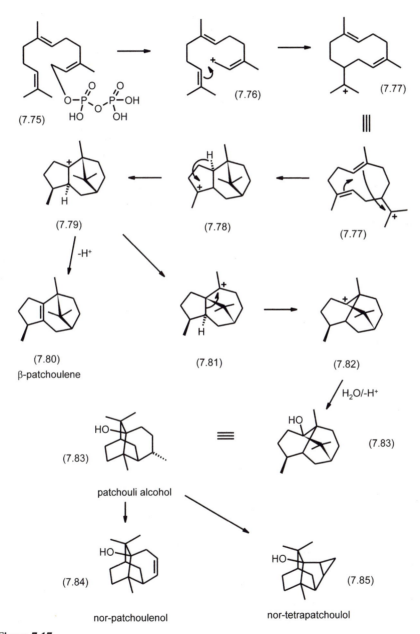

Figure 7.17

The biosynthesis starts with *trans-trans*-farnesyl pyrophosphate. Loss of the pyrophosphate function gives the cation (7.76) which is trapped by the electrons of the remote double bond to give carbocation (7.77) which contains a 10-membered ring. In the figure, this carbocation is drawn in two different ways, the first shows the relationship to the starting pyrophosphate and its geometry and the second helps us to see the ring structure which results from the *trans*-annular cyclisation in the next step. This cyclisation occurs in order to relieve the ring strain involved in the 10-membered ring and produces the bicyclic carbocation (7.78). A 1,3-hydrogen shift then produces the isomeric carbocation (7.79). Elimination of a proton from this intermediate gives β-patchoulene (7.80), a component of patchouli oil. Various 1,2-carbon shifts are also possible. As always, the preference is for the most heavily subsituted carbon atom to move. In this case, it is the gem-dimethyl bridge and the product is the carbocation (7.81). This rearrangement paves the way for a reduction in free energy by setting up the possibility of a further 1,2-carbon shift to convert the skeleton from a bicyclo-[3.2.1]-octane system to the less strained bicyclo-[2.2.2]-octane of (7.82). This carbocation is trapped by water and elimination of a hydroxyl proton gives patchouli alcohol (7.83). This molecule is also drawn in two ways. The first shows the relationship to the intermediate carbocation (7.81) and the second is the easier structure to visualise and hence, the format normally used for drawing patchouli alcohol and related structures.

Another similarity with vetiver is the fact that the odour of Patchouli Oil owes little to the major component. The natural isomer, (−)-patchouli alcohol has some patchouli character whereas its unnatural enantiomer is odourless. However, this odour contribution is a small part of the total and most of the intense character of patchouli comes from minor components such as nor-patchoulenol (7.84) and nor-tetrapatchoulol (7.85). These are both degradation products, probably the result of the fermentation stage and explaining the necessity of including this in the production process. Nor-patchoulenol was identified and characterised by Paul Tesseire of the fragrance company Roure–Bertrand–Dupont and nor-tetrapatchoulol by Braja Mookherjee of International Flavours and Fragrances.

The complex architecture of patchouli alcohol and the dearth of functional groups to serve as handles for manipulation, make the molecule a challenging synthetic target. Figure 7.18 shows the synthetic route of Danishevsky and Duman.[7.12]

The bicyclooctane core of the structure of patchouli alcohol suggests the Diels–Alder reaction as an approach to the synthesis and this was

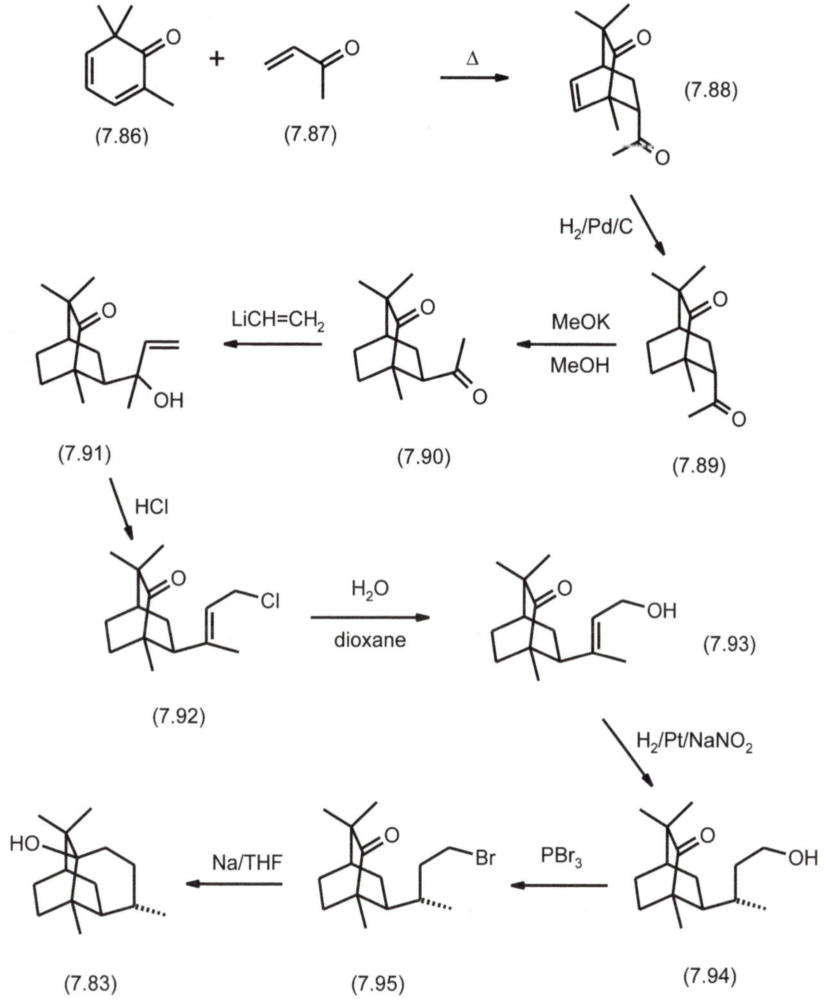

Figure 7.18

the route chosen by Danishevsky. His starting diene was 2,2,6-trimethylcyclohexadienone (7.86) which is easily obtained by exhaustive methylation of phenol. Heating this with methyl vinyl ketone (7.87) gave the bicyclooctenone (7.88). The unwanted double bond was easily saturated by hydrogenation over a palladium catalyst to give (7.89). The acetyl group of (7.89) existed in the *endo*-configuration as a result of the Diels–Alder reaction but, since the position at which it is attached to the ring is enolisable, it was easy to epimerise it to the more thermo-dynamically stable *exo*-configuration by treatment with a base (methox-ide). Thus, in a few steps Danishevsky had constructed a molecule (7.90) which possessed most of the structural features of the target. It remained only to add a two carbon unit between the two carbonyl carbon atoms and selectively remove one of the oxygen atoms. The two carbon unit was introduced in the form of vinyl lithium. Selective addition to the methyl ketone, rather than that of the bridge, was possible because of the high degree of steric hindrance around the latter. This gave the allylic alcohol (7.91). In order to close the tail of the side chain of (7.91) onto the ketone of the bridge, it was necessary to suitably functionalise the terminal carbon atom. It was also necessary to remove the oxygen of the alcohol function. Both of these objectives can be achieved simultaneously by an allylic transposition. We will see few one-step processes for transposition of allylic alcohols in Chapter 9, but for his purposes, it was easier for Danishevsky to carry this out as a two-step process. Firstly, treatment of the alcohol (7.91) with hydrochloric acid caused loss of the hydroxy group to give an allylic carbocation. This cation was trapped at the less hindered end to give the allylic chloride (7.92). Hydrolysis of this chloride with water in dioxane solution, gave the alcohol (7.93). Once again, trapping of the intermediate allylic carbocation occurred at the less hindered end of the system. It was then necessary to remove the olefinic double bond without either reducing the ketone group or inducing hydrogenolysis of the carbon–oxygen bond of the allylic hydroxyl function. This was achieved by the use of Henbest system, *i.e.* hydrogenation using a platinum catalyst in the presence of sodium nitrite. In the 1960s when Danishevsky achieved this synthesis, the chemist had to rely on his own thorough knowledge of the literature to find selective reaction systems such as that of Henbest. Nowadays, life has been made much easier, thanks to sub-structure searchable computer databases of known functional group interconversions. The Henbest reduction of (7.93) was selective in terms of functional groups but not in terms of stereochemistry and so a mixture of epimeric saturated alcohols was obtained. Since the new chiral centre introduced by the hydrogenation was not the only asymmetric centre in the molecule, it meant that the enantiomers produced were actually diastereomers and hence were separable by column chromatography. Thus, Danishevsky was able to obtain racemic (7.94) in which the methyl group had the correct stereochemistry relative to that of the ring system. To complete the synthesis, it was now necessary only to form the bond between the terminus of the side chain and the ketone function on the bridge. Since the required product is a tertiary alcohol, the Grignard reaction (or an equivalent) is the obvious choice. To provide the halide function necessary for formation of the organometallic reagent, the alcohol (7.94) was converted to the bromide (7.95) by reaction with phosphorus tribromide. In bicyclic systems such as (7.95), the ketone function is very hindered and this makes normal Grignard reactions difficult (see the latter part of Chapter 5). In this case, the conversion was achieved by use of sodium, rather than magnesium, as the metal and so patchouli alcohol (7.83) was obtained from (7.95). Organosodium reagents are much closer to being free carbanions than are Grignard reagents and so the steric requirements are less as there is no attached metal with its coordination shell of ethers. The reagents are very much more reactive of course and very short lived, hence they are less useful in general as synthetic reagents. However, in this case, the substrate is held six carbon atoms away from the incipient carbanion and so the latter is trapped by the electrophile as soon as it is formed.

7.3 PINE (HIMALAYAN)

Indian turpentine is obtained from the species *Pinus longifolia*, a tree which grows in the Himalayas. The turpentine obtained from this tree contains about 30% by weight of the sesquiterpene longifolene. The structure of longifolene (7.96) is shown in Figure 7.19. It was first isolated in 1920 by the English terpene chemist Simonsen[7.13] but its structure remained a mystery for over 30 years as chemical degradation gave complex mixtures of highly rearranged products. The structure was finally determined in the 1950s by a combination of X-ray crystallographic[7.14] and chemical[7.15] studies, the latter by the great French natural products chemist, Ourisson.

Longifolene demonstrates very interesting molecular properties. It contains a reactive, polarisable double bond which is susceptible to electrophilic attack. Addition of an electrophile generates a carbocation which is well placed for various Wagner–Meerwein and similar rearrangements. The long bridge is subject to steric overcrowding and the whole ring system is very strained. Both of these features provide plenty of driving force for rearrangements. The overall molecular shape is almost spherical with all of the component atoms in close proximity to each other and this leads to some rather unexpected *trans*-annular reactions.

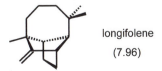

longifolene
(7.96)

Figure 7.19

7.3.1   Synthesis of Longifolene

The total synthesis of longifolene represents a very serious challenge for the natural products chemist. The first synthesis was achieved by one of the greatest masters of the art, E.J. Corey of Harvard University.[7.16,7.17]

His synthesis plan is outlined in Figure 7.20. It is a masterpiece of both strategy and tactics and warrants study in some detail. The basic strategy was to build up the tricyclic ring structure from a suitably functionalised bicyclic precursor by *trans*-annular formation of the final bridging bond. This cyclisation requires the use of a [5.4.0]-bicycloundecane skeleton. These are not readily available and moderately difficult to synthesise. Corey's elegant construction of the required cyclisation substrate therefore adds another admirable dimension to the overall scheme.

The synthesis starts with the Wieland–Mischler ketone (7.97). This ketone is easily prepared from 2-methylcyclohexanone and methyl vinyl ketone and has been used frequently as a starting material for the synthesis of diterpenoids and steroids. The two carbonyl groups have different degrees of reactivity and it is possible to ketalise the more reactive, saturated one leaving the other unaffected. Addition of ethylidenetriphenylphosphorane to the monoprotected ketone (7.98) gives the diene (7.99). Treatment of this with osmium tetroxide led to selective oxidation of the more accessible and more easily polarisable exocyclic double bond to give the diol (7.100). Reaction of the diol with *para*-toluenesulfonyl chloride (*pts*Cl) gave the monotosylate (7.101). This selectivity is not at all surprising as the secondary alcohol is more reactive and much less hindered than the tertiary. The ring expansion on solvolysis of this tosylate is a key feature of the synthesis as it provides the necessary [5.4.0]-bicycloundecane skeleton, perfectly functionalised for the *trans*-annular reaction to form the third ring. The mechanism is shown in Figure 7.21. There are a number of interesting points about this reaction. The choice of reagent, lithium perchlorate, may at first seem rather odd. However, consider the requirements of a reagent for this step. It must be able to induce cleavage of the carbon–oxygen bond by which the tosylate is attached to the ring. At the same time, it must not cause any unwanted reaction of the olefinic bond and it must also leave the ketal function intact to prevent unwanted side reactions. This means that a totally neutral system is required. Lithium perchlorate is neutral but the lithium cation has such an affinity for oxygen that it can, by attaching itself to the tosyl ester function, cause heterolysis of the carbon–oxygen bond. The incipient carbocation is trapped by a 1,2-carbon shift. Interaction with its π-electrons ensures that it is the vinyl carbon which moves rather than either of the other possibilities. The

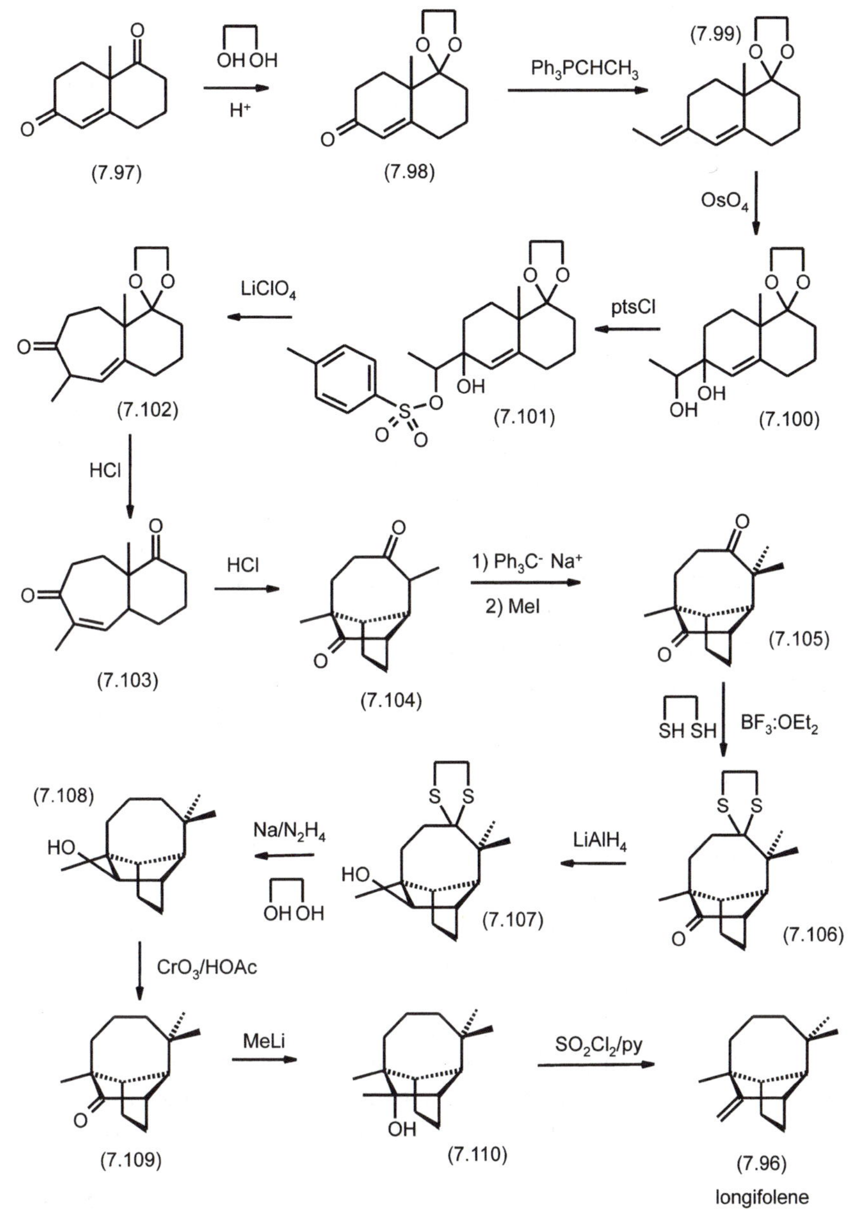

Figure 7.20

product carbocation is stabilised by the pendant hydroxyl group which, essentially becomes a protonated ketone, (7.111). In keeping with the sentiment of the opening quotation of Chapter 5, it is likely that this entire sequence is synchronous and that no free carbocation ever exists.

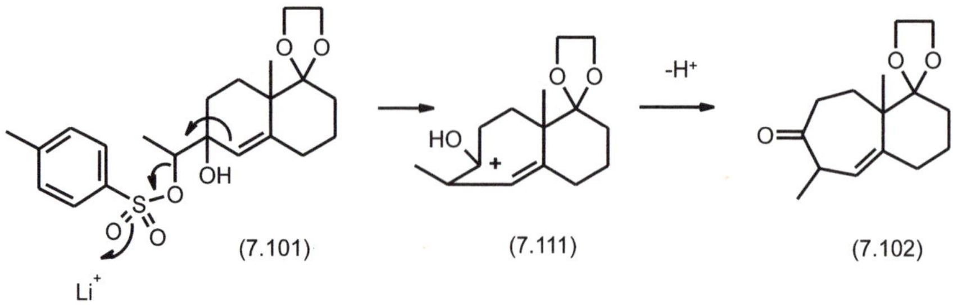

Figure 7.21

Deprotonation of (7.111) gives the ring expanded ketone (7.102). Such are the forces driving selectivity of this reaction that only 5% of the starting material followed other reaction paths.

In order to free-up the functionality required for the cyclisation reaction, it is necessary to hydrolytically remove the ethylene ketal of (7.102) and bring its double bond into conjugation with the other ketone function. Both of these were achieved by treating (7.102) with 2 N hydrochloric acid at room temperature, giving (7.103) as the product. The *trans*-annular cyclisation reaction was a simple Michael reaction. Michael reactions are more usually effected using a basic catalyst but in this case, the hydrochloric acid used in the previous stage was employed. It proved necessary only to raise the temperature to boiling point (100 °C) to result in the formation of the desired diketone (7.104). The detail of this key reaction is shown in Figure 7.22. Protonation of the enol form (7.112) of the starting material occurs on the carbonyl oxygen atom. This induces a flow of electrons across the molecule with elimination of the hydroxyl proton at the opposite end to produce the enol form (7.113) of the product (7.104).

The ring system of the longifolene skeleton is now complete and the synthesis requires only addition of a methyl group, removal of one oxygen atom and conversion of the other to a methylene group. The position at which the methyl group must be added is adjacent to a ketone and so the obvious tactic is to methylate an enolate. This raises the issue of selectivity. There is no problem as far as the other ketone group is concerned since its carbonyl group has only one α-hydrogen atom and that is located on a bridgehead carbon. Bredt's rule therefore dictates that formation of an enolate at this position is not possible. However, there is an issue concerning which side of the other ketone will be alkylated. The tertiary hydrogen which is the one required to be removed, will be the one which is removed fastest to form what is referred to as the kinetic enolate. However, it is essential that no

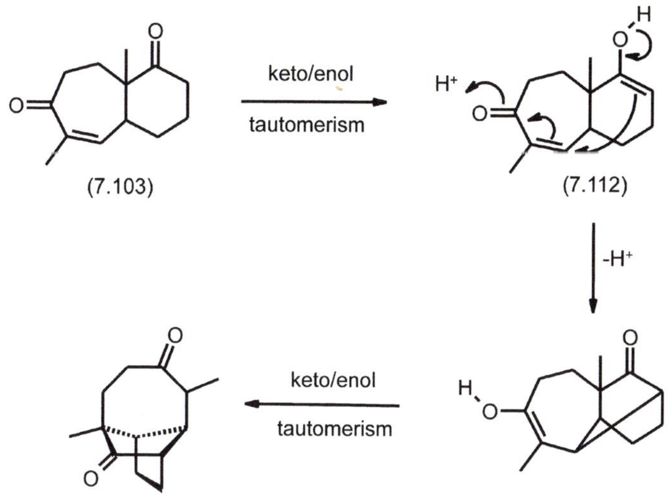

Figure 7.22

equilibration occurs since that would allow the other, more thermo-dynamically favourable, enolate to form. In order to achieve this, Corey used sodium triphenylmethide (trityl sodium) as the base. The trityl anion is a very strong base and the deprotonation of the ketone by it, will be essentially irreversible in an aprotic solvent. Thus, the required diketone (7.105) was obtained in high yield after quenching of the carbanion with methyl iodide.

There now arises the question of selectivity between the two ketone groups. The one on the four carbon bridge must be reduced to a methylene group and the one on the one carbon bridge must be replaced by a methylene group (*i.e.*, by addition of another carbon atom). The former is the more accessible of the two sterically and therefore more likely to react with a bulky reagent such as a Wittig or Grignard reagent. This is the opposite of what is required. Both ketones will react with a reagent with low steric requirements, such as hydrazine, for example, in a Wolff–Kishner reduction. There is, therefore, no simple way of either selectively reducing the ketone group which must be reduced or alkylating that must be alkylated. Corey was thus obliged to use a protecting group in order to achieve the desired conversions. The sulfur atom is larger than the oxygen atom and so ethane dithiol has a larger steric requirement than does ethylene glycol. Thus it was possible to selectively form the thioketal of the more accessible carbonyl group to give the monoprotected compound

(7.106). The exposed carbonyl group was then reduced to the secondary alcohol (7.107) by lithium aluminium hydride. This left that oxygen atom in a state which is unaffected by the conditions of the Wolff–Kishner reduction. Simultaneous deprotection and reduction using hydrazine and sodium in ethylene glycol then produced the alcohol (7.108). This secondary alcohol was oxidised to the ketone by chromium trioxide in acetic acid giving the ketone (7.109). This ketone group is very hindered and will not react with Wittig reagents or Grignard reagents. To introduce the final carbon atom into the molecule, Corey therefore used methyl lithium which has a much lower steric requirement. The alcohol (7.110) produced was then dehydrated using sulfuryl chloride. The conditions employed for this dehydration had to be both mild (0 °C) and basic (pyridine) in order to prevent rearrangement, or other unwanted side reactions, of the sensitive final product, longifolene (7.96).

7.3.2   Acid Catalysed Rearrangement of Longifolene

Treatment of longifolene with even a moderately strong acid, such as formic acid, induces a vigorous and very exothermic rearrangement to give isolongifolene (7.120). This rearrangement involves a series of 1,2-carbon and 1,2-hydrogen shifts some of which appear to be energetically unfavourable. However, it is a feature of carbocation chemistry that most of the rearrangements are reversible and equilibria are set up. In the case of a thermodynamically unfavourable reaction, the equilibrium will lie very much on the side of the starting materials. However, if there is a path by which the product can move to a structure with much lower free energy, then it will do so and eventually, almost all of the material will be converted. This is shown graphically in Figure 7.23.

This is a standard reaction figure showing Gibb's free energy (G) as the dependent variable (Y-axis) and reaction progress (R) as the independent variable (x-axis). If we imagine a carbocation undergoing a series of rearrangements starting from configuration A, then it is easy to see that, with a small energy input, it can cross the transition state barrier to the lower energy configuration, B. It requires more energy to cross the transition state barrier into configuration C and so many fewer molecules will exist in the C configuration than in B at any given time. However, for a relatively small intake of energy, a molecule can move from C, over the final barrier and into the much more energetically favourable configuration D. The total gain in energy by doing so is sufficient to drive a second molecule through the whole sequence. Thus, even though C is a relatively unfavourable and sparsely populated state,

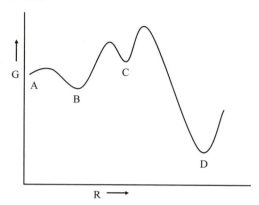

Figure 7.23

essentially all of the molecules will eventually pass through it and into the energy well of configuration D. We see this phenomenon in action in the rearrangement of longifolene to isolongifolene. The mechanism is shown in Figure 7.24.

The rearrangement starts by protonation of the polarisable exocyclic methylene double bond. As expected, this protonates at the less substituted end so as to produce the more stable tertiary carbocation (7.114). The bridgehead methyl group then undergoes a 1,2-shift to produce carbocation (7.115). This cation breaks Bredt's rule as it places an essentially planar carbon atom at the bridgehead. However, the strain is immediately removed by a 1,2-shift of the adjoining bridge to give carbocation (7.116) and a second Wagner–Meerwein rearrangement produces carbocation (7.117). Molecular models will reveal that this series of reactions has significantly reduced the amount of strain energy in the molecule, more than compensating for the energy penalty in going from a tertiary to a secondary carbocation. A 1,2-hydrogen shift to give (7.118) now paves the way for the final Wagner–Meerwein reaction giving the carbocation (7.119). This carbocation can eliminate a proton to give isolongifolene (7.120). The overall loss of strain energy on going from longifolene to isolongifolene is considerable and accounts for the exothermic nature of the reaction.

Complex structures such as that of isolongifolene can be drawn in many ways. Structure (7.120) shows the relationship to the precursor carbocations. The alternative drawing (7.120a) is one which is often used when isolongifolene and its derivatives are shown. Inspection of the two should reveal that the two are the same. One system for doing this is to select an obvious feature of the molecular structure and then follow bond connectivities in every direction around both structural drawings. In the

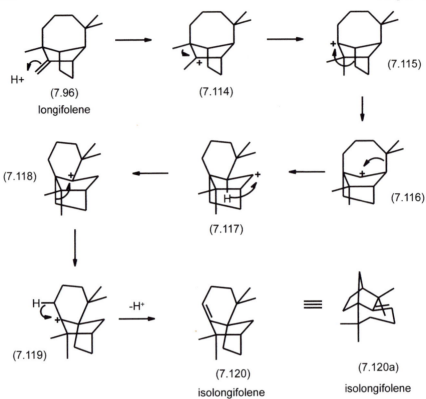

Figure 7.24

case of isolongifolene, one obvious feature to serve as an anchor point is the double bond. So, for example, in either representation, starting at the monosubstituted end of the olefin and progressing away from the double bond, takes us round a six-membered ring and gives the sequence $-CH_2-CH_2-CMe_2-$ and then a bridgehead carbon and back to the olefin. The same thing can then be done working away from the double bond in the opposite direction along both the Z and E legs. Any reader who remains unconvinced that structures (7.120) and (7.120a) represent the same molecule, should resort to molecular models for reassurance.

7.3.3 Reactions of Isolongifolene

Not all of the chemistry of longifolene and related compounds follow the unexpected pathways. Figure 7.25 shows some reactions which proceed as expected and give products which are of value for their woody, amber odours. Epoxidation of isolongifolene gives the normal epoxide (7.121)

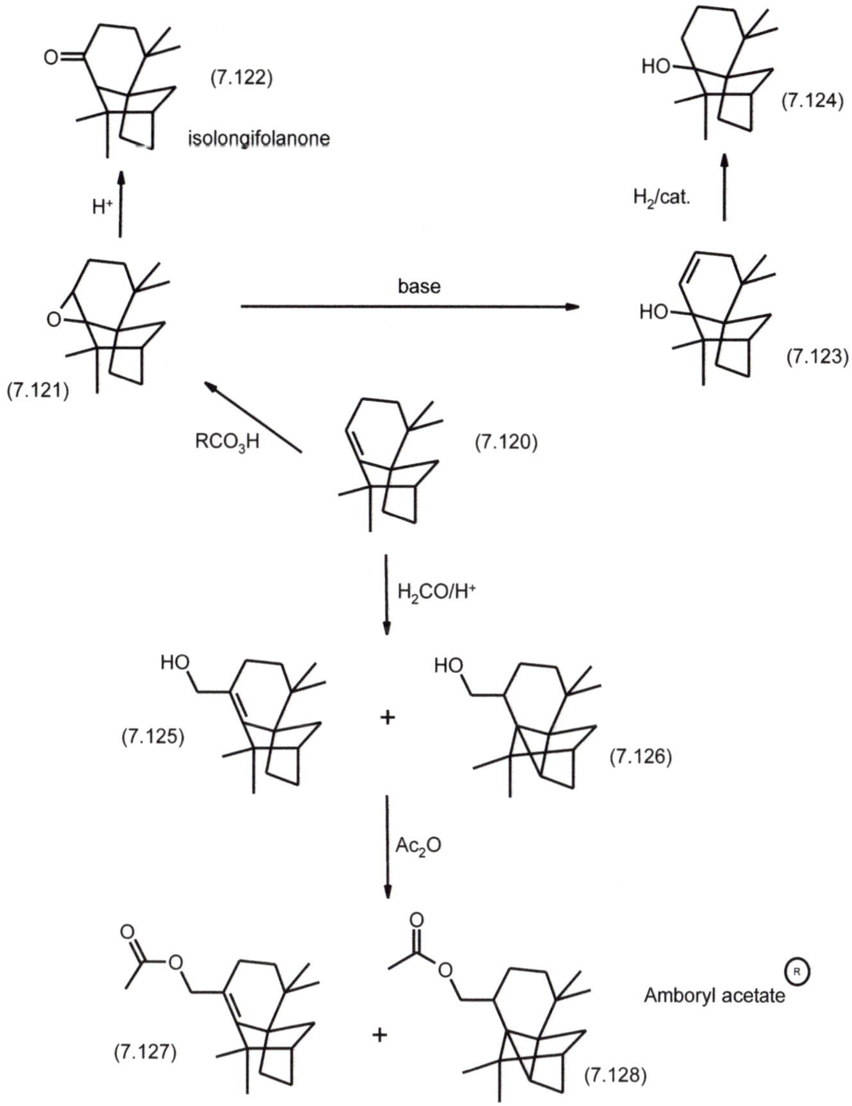

Figure 7.25

and, upon treatment with acid, this rearranges smoothly to the corresponding ketone, isolongifolanone (7.122), without any further skeletal rearrangement. The epoxide (7.121) also undergoes the normal rearrangement to the allylic alcohol (7.123) when treated with a strong, non-nucleophilic base. Hydrogenation of (7.123) also proceeds in a straightforward manner to give the saturated alcohol isolongifolanol (7.124). The Prins Reaction of isolongifolene with formaldehyde gives the expected allylic alcohol (7.125) in admixture with the corresponding tetracyclic product (7.126). The proton eliminated to produce (7.125) is that adjacent to the centre of positive charge whereas the formation of (7.126) requires a *trans*-annular proton elimination. This is not surprising by analogy with the chemistry of the camphene system as discussed in Chapter 5. Esterification of the reaction mixture with acetic acid results in the formation of a mixture of the corresponding acetates, (7.127) and (7.128). This mixture is known as Amboryl acetate®. The fact that all these reactions proceed in a simple manner without any unexpected rearrangements reflects the fact that the isolongifolene skeleton is a less strained, hence lower energy structure, than that of longifolene.

7.3.4   Reaction of Longifolene with Bromotrichloromethane

Before looking at this reaction of longifolene, it will be instructive to see it in the simpler setting of camphene (7.130). The mechanism is shown in Figure 7.26.

Bromotrichloromethane (7.129) contains a very weak carbon–bromine bond which is easily homolysed to give the trichloromethyl radical and a bromine atom. The trichloromethyl radical adds to olefins to give a new carbon-centred radical which then picks up the bromine atom to

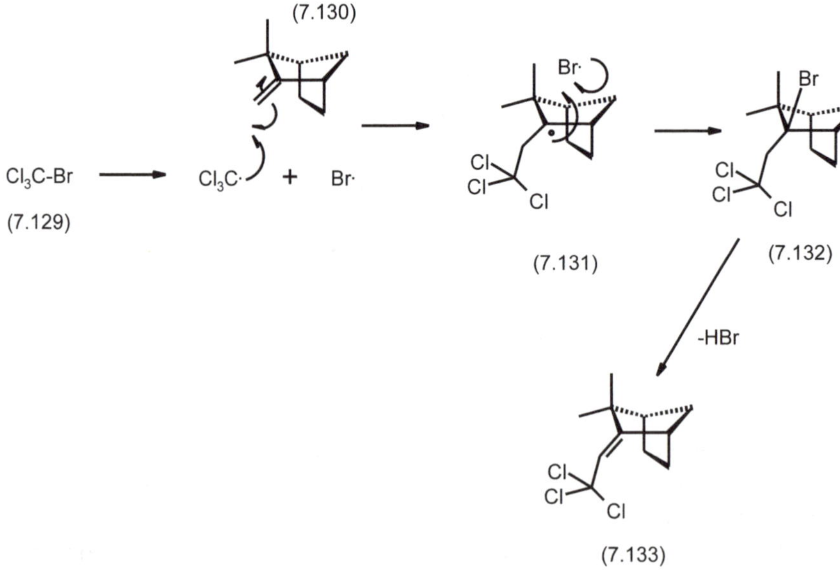

Figure 7.26

form an addition compound. In the case of camphene (7.130), the trichloromethyl radical attacks at the less hindered terminal carbon atom, giving the intermediate radical (7.131) and the bromine then adds to the ring carbon to give (7.132). Because of the acidity of the protons adjacent to the trichloromethyl group (a consequence of the inductive effect of the three chlorine atoms), this material readily eliminates hydrogen bromide to give the olefin (7.133).

In Figure 7.27, we see camphene alongside longifolene and the nearby "isoprene" fragment shows very clearly how closely related the two structures are. To obtain longifolene from camphene it is necessary only to "add" the five carbon fragment across the camphene ring. We might therefore expect the chemistry of longifolene to resemble that of camphene. The reaction with bromotrichloromethane provides us with a very surprising difference between the two. Bromotrichloromethane does indeed add to longifolene and the empirical formula of the product is exactly as we expect, $C_{16}H_{24}BrCl_3$ (*i.e.* $C_{15}H_{24}$ + $CBrCl_3$). However, the first indication that things are not as we might expect is that the product does not readily eliminate hydrogen bromide. The explanation is found in Figure 7.28.

As in the case of camphene, bromotrichloromethane (7.129) homolyses and the trichloromethyl radical adds to the terminal carbon of the methylene group of longifolene (7.96) to give the radical (7.134). The surprise comes at this point. Instead of adding to the bromine atom, the unpaired electron of (7.134) abstracts a hydrogen atom from the long bridge of longifolene to give a second radical, (7.135) and it is this radical which reacts with the bromine atom to give (7.136). The reason why the product does not undergo a facile elimination of hydrogen bromide is now clear; there is no acidic hydrogen atom adjacent to the bromine. Inspection of a model of the longifolene molecule shows that the hydrogen atom which apparently moves across the ring, in fact barely moves at all. Its distance from the olefinic carbon in the ring is very close to the carbon–hydrogen bond distance. Therefore all that is necessary to

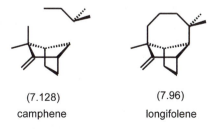

(7.128)   (7.96)
camphene   longifolene

Figure 7.27

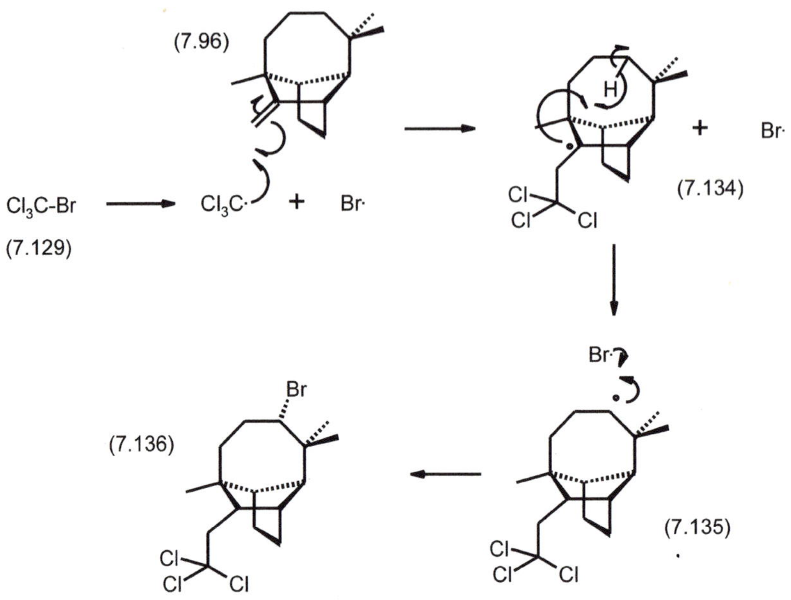

Figure 7.28

change from (7.134) to (7.135) is a reorganisation of the electrons. The bromine atom adds from the opposite face of the molecule to give the stereochemistry shown. This could be the result of the steric effect of the bulky trichloromethyl group or it could indicate essentially synchronous making and breaking of all the bonds involved in the reaction.

7.4   CLOVES AND HOPS

The major terpenoid components of cloves and hops are remarkable in that they contain highly strained medium sized rings. Four of the most important members of the family are shown in Figure 7.29. Of these, the most important are β-caryophyllene (7.137) and α-humulene (7.139). These are the most frequently encountered in essential oils and are often referred to simply as caryophyllene and humulene, respectively. For convenience, these shorter names will be used in this book. For example, α-humulene (7.139) has been found in hundreds of essential oils whereas there are reports of β-humulene (7.140) in only a few. Almost invariably, several members of the family will be found together in oils. For example, all oils which have been found to contain humulene (7.139) also contain caryophyllene (7.137) and isocaryophyllene (7.138).

Caryophyllene (7.137) is the commonest member of the family and is found in a wide variety of plants from herbs such as basil and spices such

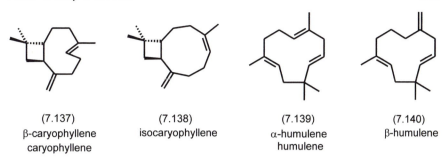

(7.137)	(7.138)	(7.139)	(7.140)
β-caryophyllene	isocaryophyllene	α-humulene	β-humulene
caryophyllene		humulene	

Figure 7.29

as black pepper to patchouli, marihuana and guava. The best known source is clove. The nail like objects used as a spice in cooking, are actually the dried flower buds of the tropical tree *Syzygium aromaticum*, formerly known as *Eugenia caryophyllata*. The oil extracted from these typically contain 10–15% of caryophyllene although eugenol (a shikimic acid metabolite rather than a terpenoid) is the major component. Eugenol is used both in flavours and in dentistry and is separated from the oil either by distillation or by extraction with aqueous base, the latter being possible because eugenol is a phenol. Caryophyllene is therefore a by-product of eugenol extraction. Humulene is also very widespread in occurrence and has been found in such oils as sage, ginseng, litchi and the flower ylang–ylang. Its best-known source is from the flowers of hops (*Humulus lupulus*) which are used in beer making for the characteristic flavour which they add. One of the few sources of β-humulene is pimento or allspice.

The relationship between these materials is clear from their biogenesis which is discussed in Chapter 2 (see Figure 2.11). Cyclisation of farnesyl pyrophosphate produces the 11-membered ring of the humulenes and this then can undergo *trans*-annular cyclisation to give the bicyclic structure of the caryophyllenes.

As mentioned in Chapter 2, these molecules all suffer a high level of strain in the rings. Medium sized rings experience steric strain and the introduction of *trans* double bonds increases this significantly. This stored energy has dramatic effects on the chemistry of the caryophyllenes and humulenes as we shall see.

7.4.1 The Synthesis of Caryophyllene

The synthesis of caryophyllene was another major challenge for the synthetic chemist and again it was E. J. Corey who rose to meet it.[7.18,7.19]

The greatest problem in synthesising caryophyllene lies, not surprisingly, in the construction of the nine-membered ring and its *trans* double bond. This would be exceptionally difficult to achieve by any cyclisation reaction and so Corey chose to tackle it by fragmentation of a bicyclic system. The fragmentation step is the key feature of his synthesis but first let us look at how he constructed the substrate for it. This scheme is shown in Figure 7.30.

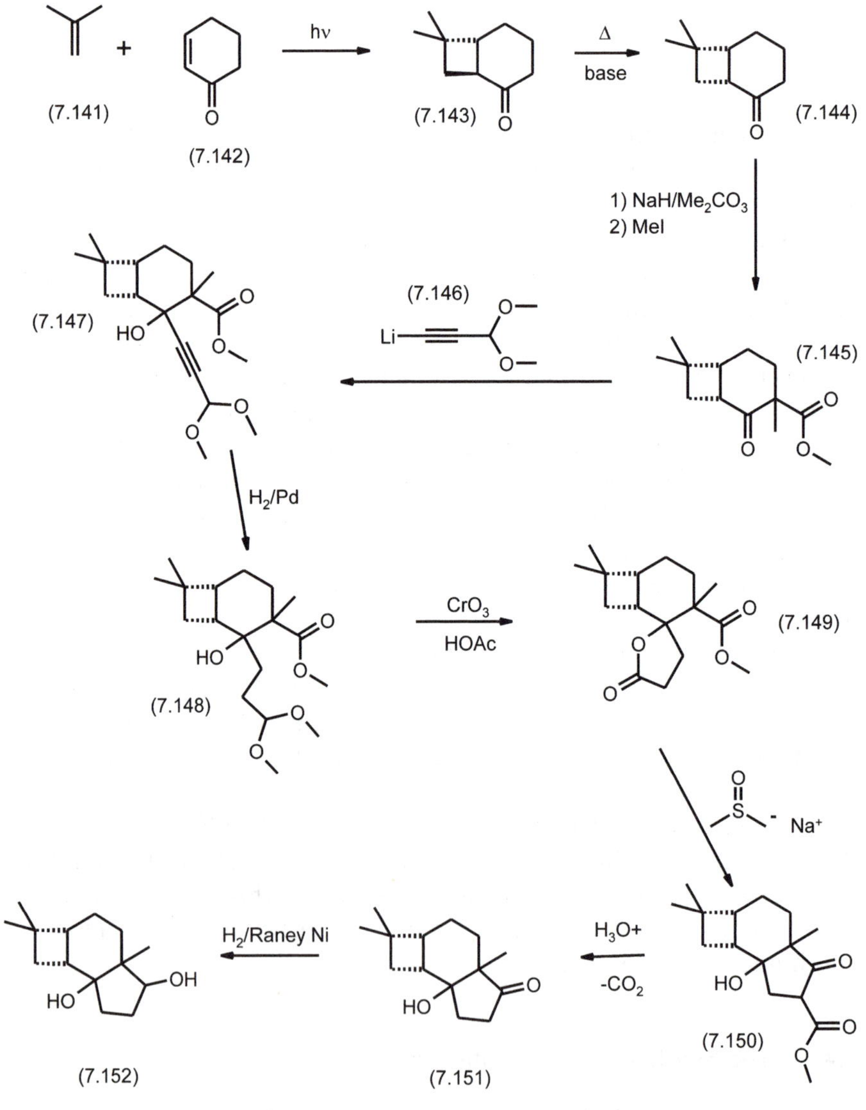

Figure 7.30

Photochemical addition of isobutylene (7.141) to 2-cyclohexenone (7.142) gave the bicyclic ketone (7.143). This 2 + 2 cycloaddition reaction produces the *trans* isomer, in accordance with the Woodward–Hoffmann rules. This is the correct stereochemistry for caryophyllene but, as we will see later, it was necessary to change the stereochemistry temporarily to the *cis* isomer (7.144) by epimerisation using a base. Treatment of this ketone with sodium hydride and dimethyl carbonate resulted in carboxymethylation of the ketone and subsequent formation of the anion of the resultant β-keto-ester. This enolate was not isolated but quenched *in situ* with methyl iodide to give (7.145). The organo-lithium reagent (7.146) is easily prepared from the readily available propargyl alcohol (HCCCH$_2$OH) by oxidation to the aldehyde, ketalisation and then reaction with butyl lithium. Addition of (7.146) to (7.145) gave the acetylenic alcohol (7.147) and subsequent hydrogenation gave (7.148). Treatment of (7.148) with chromium trioxide in acetic acid achieved three transformations at the same time. The protecting acetal function was removed and the aldehyde oxidised to the corresponding acid and this then cyclised to the lactone by esterification with the neighbouring hydroxyl group. The product (7.149) was then treated with the anion of dimethylsulfoxide (dimsyl sodium). This induced a Dieckmann reaction in which the anion α to the lactone carbonyl group attacked the ester carbonyl group displacing methoxide and forming a five-membered ring. The strain created in the lactone ring by this was relieved by methanolysis by the methoxide produced in the Dieckmann reaction. Thus the β-keto-ester (7.150) was produced and was then hydrolysed and decarboxylated to (7.151). Reduction of the ketone function by hydrogenation with Raney nickel gave the key intermediate (7.152).

The diol (7.152) was produced as a mixture of diastereomers and these were separated at this stage. The two diastereomers required for further elaboration were (7.153) and (7.154) as shown in Figure 7.31. Having been isolated in pure form these were converted to their monotosylates (7.155) and (7.156), respectively. This selectivity was easy to achieve because of the greater reactivity of the secondary alcohol functions relative to the tertiary. The stage is now set for the key step of the synthesis. We will look at this for each of the isomers individually, starting with (7.155).

Figure 7.32 shows the anion (7.157) produced by deprotonation of the hydroxyl group of (7.155). The molecule is forced into the configuration shown by the relative geometry of the various substituents, including the *cis*-fused four-membered ring. Now it is clear why the initial stereochemistry of the ring was reversed. The base induced cleavage of such monotosylates is known as the Grob

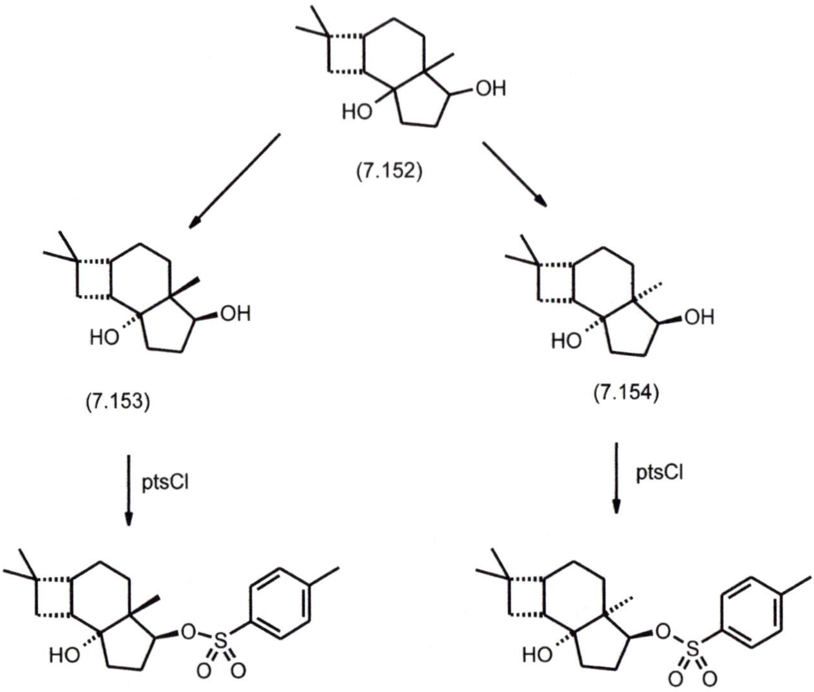

(7.152)

(7.153)

(7.154)

ptsCl

ptsCl

(7.155)

(7.156)

Figure 7.31

fragmentation. As we follow the flow of electrons in structure (7.157), we see that the new double bond will be formed in the *trans*-configuration, as required for caryophyllene. Similarly, in Figure 7.33, we see how the anion (7.158) derived from isomer (7.156), fragments to produce a *cis*-double bond.

The completion of the synthesis is relatively trivial and is shown in Figure 7.34.

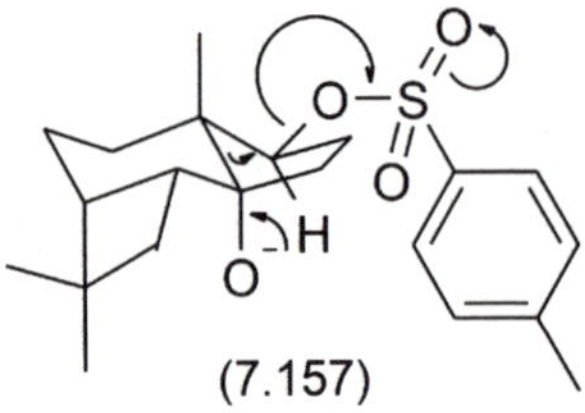

(7.157)

Figure 7.32

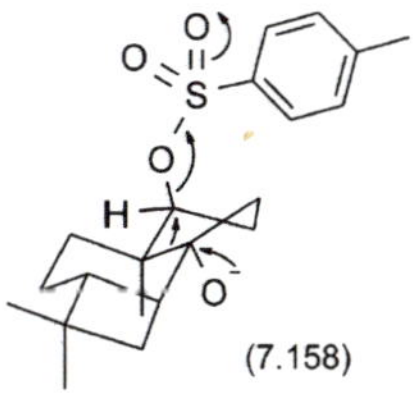

(7.158)

Figure 7.33

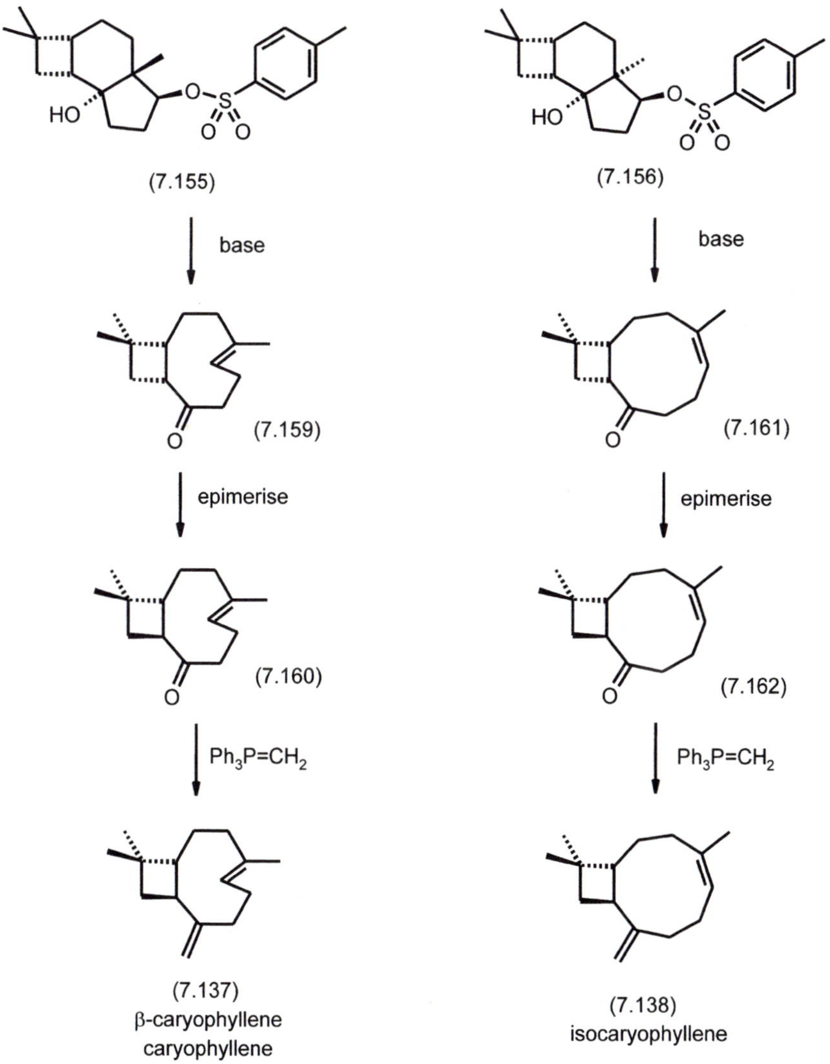

(7.155)

(7.156)

base

base

(7.159)

(7.161)

epimerise

epimerise

(7.160)

(7.162)

$Ph_3P=CH_2$

$Ph_3P=CH_2$

(7.137)
β-caryophyllene
caryophyllene

(7.138)
isocaryophyllene

Figure 7.34

Treatment of hydroxytosylate (7.155) with base leads, as discussed, to the bicyclic ketone (7.159) which contains the nine-membered ring and *trans*-double bond required for caryophyllene. Epimerisation back to the *trans*-ring fusion and subsequent Wittig reaction of the resultant ketone (7.160) with methylenetriphenylphosphorane gave racemic caryophyllene (7.137). In the same way, tosylate (7.156) was converted, *via* ketones (7.161) and (7.162) to isocaryophyllene (7.138).

7.4.2 Acid Catalysed Rearrangement of Caryophyllene

When caryophyllene is treated with acids, it produces a complex mixture of products. The exact composition of the product mixture will depend on the nature of the acid used and the reaction conditions. If water is present, then carbocations can undergo solvolysis to give alcohols. In the absence of nucleophiles, the reaction products are olefins. These olefins are known generically as clovenes. The mechanisms for the formation of the clovenes and the related alcohols demonstrate many of the principles of carbocation chemistry as described in Chapter 5. They also demonstrate the fact that various reaction pathways can operate simultaneously. Furthermore, as with the acid catalysed rearrangement of longifolene, they demonstrate that energetically unfavourable intermediate steps will occur if there is an overall reduction in the free energy of the system. Some of these mechanisms are illustrated in Figures 7.35–7.37. The flat structure (7.137) used above for caryophyllene does not represent its actual preferred configuration very well. To give a better impression of the real shape of the molecule, it has been redrawn as (7.163) in Figures 7.35 and 7.37. The two double bonds lie very close to each other in space and hence, many reactions involve both of them since carbocations formed from one will rapidly react with the electrons of the other.

Normally 1,1-disubstituted double bonds react faster with protons than do 1,1,2-trisubstituted ones because the former are more polarisable, a feature which protons prefer. However, in the case of caryophyllene, the endocyclic *trans* double bond is so strained that the normal reactivity pattern is reversed and it is this bond which is protonated first. In Figure 7.35, protonation of the endocyclic double bond is rapidly followed by addition of the electrons of the exocyclic bond to give the tricyclic carbocation (7.164). As would be expected, in both instances, the electrophile is added to the less substituted end of the olefin. The carbocation can be trapped by water to give the alcohol (7.165) which is known both as caryophyllene alcohol and caryolanol. Alternatively, a 1,2-carbon shift can occur to reduce strain

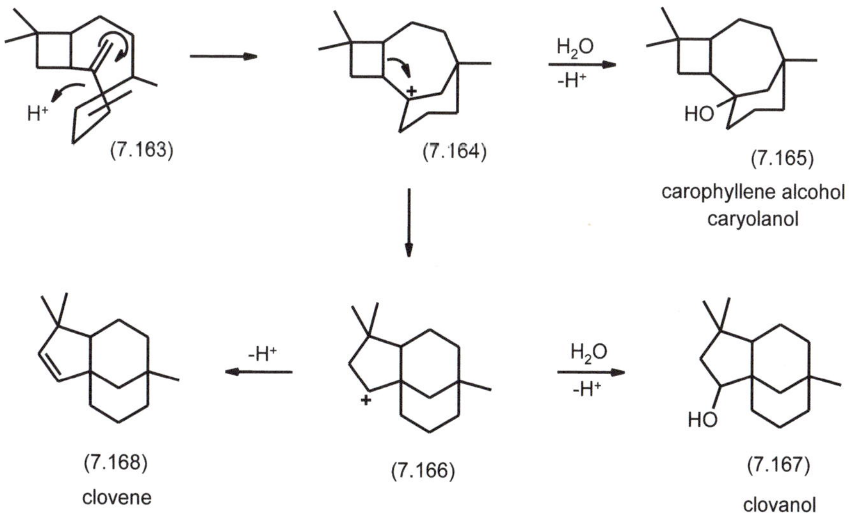

(7.163) (7.164) (7.165) carophyllene alcohol caryolanol

(7.168) clovene (7.166) (7.167) clovanol

Figure 7.35

by breaking open the four-membered ring to give a new tricyclic carbocation (7.166). This carbocation can be trapped by water to give the alcohol clovanol (7.167) or it can eliminate a proton to give clovene (7.168).

Figure 7.36 shows some of the other pathways open to carbocation (7.166) in addition to those shown in Figure 7.35. To save confusion, in those instances where more than one pathway is shown starting from the same intermediate, the intermediate is drawn twice. This is the case with both (7.166) and (7.167). A 1,3-hydrogen shift takes (7.166) down pathway A to carbocation (7.167). This is energetically favourable as it moves from a secondary carbocation to a tertiary. Simple elimination of a proton from this species (pathway D) gives ψ-clovene-A (7.172). Alternatively, a 1,2-carbon shift following pathway C takes us to the primary carbocation (7.168). This is energetically unfavourable, but a subsequent 1,2-carbon shift restores a tertiary carbocation and one which is in a less strained ring system (7.169). A 1,3-hydrogen shift leads to (7.170), again uphill energetically but this paves the way for elimination of a proton to give ψ-clovene-B (7.171).

Instead of the hydrogen shift described above, cation (7.166) can undergo a 1,2-carbon shift along pathway B to give (7.173). This is energetically favourable as the cation centre is now tertiary and the ring system rather less strained. A 1,2-hydrogen shift gives (7.174) and a subsequent 1,2-carbon shift gives (7.175). Elimination of a proton then produces isoclovene (7.176).

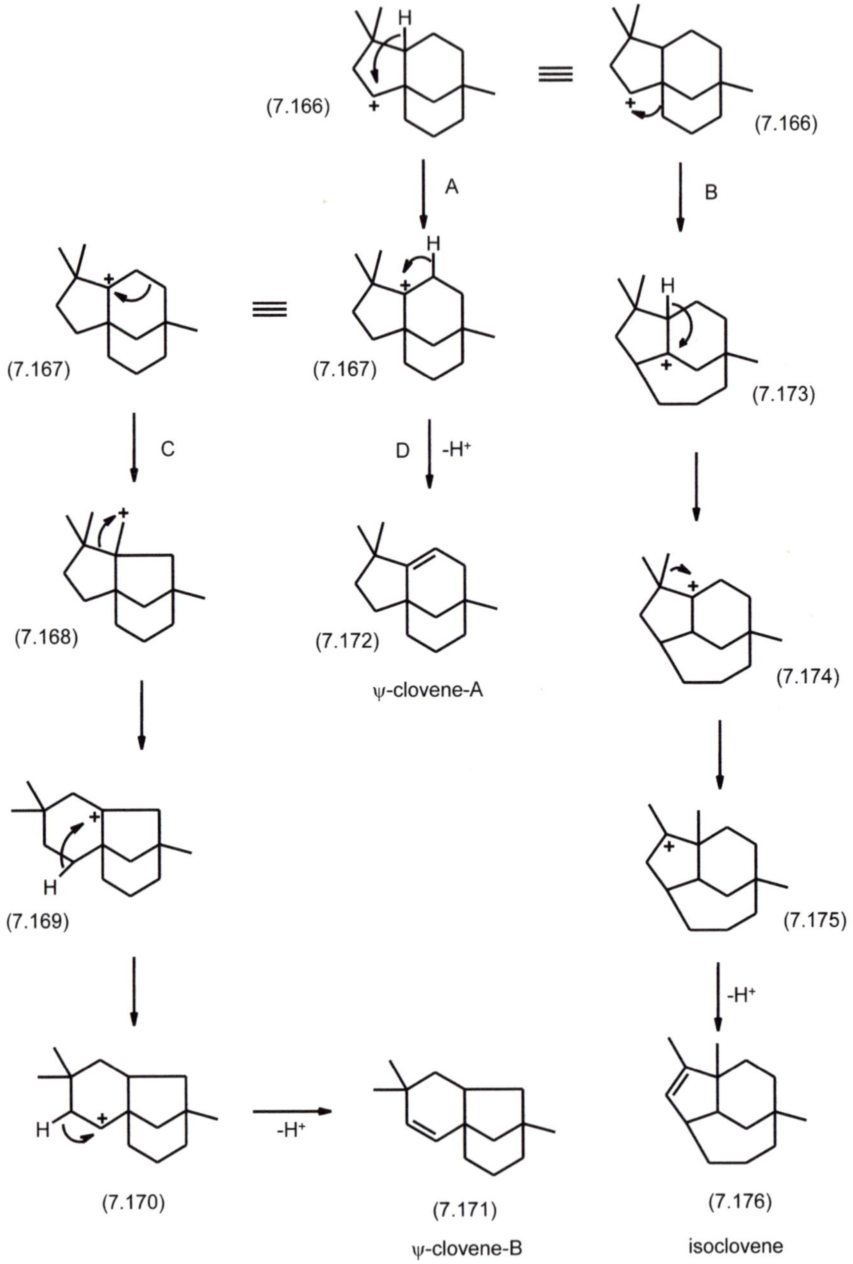

(7.166)

(7.166)

A

B

(7.167)

(7.167)

(7.173)

C

D | -H⁺

(7.168)

(7.172)

ψ-clovene-A

(7.174)

(7.169)

(7.175)

-H⁺

(7.170)

-H⁺

(7.171)

ψ-clovene-B

(7.176)

isoclovene

Figure 7.36

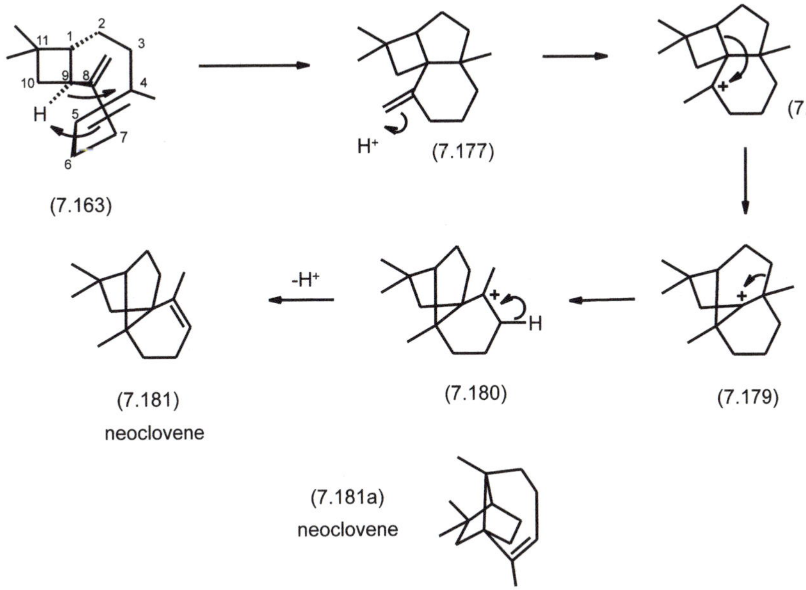

(7.163)

(7.177)

(7.178)

(7.181)
neoclovene

(7.180)

(7.179)

(7.181a)
neoclovene

Figure 7.37

The rearrangement of caryophyllene (7.163) to neoclovene (7.181) starts with a rather different reaction as shown in Figure 7.37. In this figure, the preferred configuration of caryophyllene (7.163) is shown with the ring atoms numbered according to IUPAC rules. The hydrogen atom attached to C9 is held very close to C4. It is thought that, as a proton approaches the strained *trans* double bond from the opposite face of the molecule, the polarisation induced in the 4,5-double bond leads to a *trans*-annular addition of the electrons of the C9–H bond forming a bond between C4 and C9. The hydrogen attached to C9 simultaneously undergoes a 1,5-shift to C5. This produces an isomeric olefin (7.177) which contains a very strained ring system and a highly polarisable *exo*-methylene double bond. Protonation of the latter to give (7.178) allows release of ring strain by a 1,2-carbon shift to give (7.179). This species contains a bridgehead carbocation and is therefore still of relatively high energy. A second 1,2-carbon shift to give (7.180) relieves much of the strain and elimination of a proton gives neoclovene (7.181). This drawing of neoclovene is to allow easy comparison with the intermediates present in its formation. However, it looks rather odd and so a second representation (7.181a), is also given as this shows the structure more clearly as a bicyclo-[2.2.1]-heptane system with an additional four carbon bridge.

7.4.3 Acid Catalysed Rearrangement of Isocaryophyllene

The rearrangements of isocaryophyllene (7.138) are more straightforward than those of caryophyllene (7.163). It does produce neoclovene (7.181) by a mechanism almost identical to that shown in Figure 7.37 for caryophyllene. The formation of the intermediate (7.177) is shown in Figure 7.38 and the rest of the mechanism is exactly as shown in Figure 7.37. Isocaryophyllene (7.138) also gives an isomeric olefin (7.185) by the mechanism shown in Figure 7.38. Protonation of the *exo*-methylene double bond of isocaryophyllene (7.138) gives carbocation (7.182) and a 1,2-carbon shift allows ring expansion to relieve the strain of the four-membered ring. The positive charge of the product (7.183) is then trapped by the electrons of the *endo*-cyclic double bond to give the tricyclic carbocation (7.184). Elimination of a proton from this gives the observed olefin (7.185). In this sequence, we see the more usual order of olefin reactivity towards a Bronsted acid, *i.e.* the 1,1-disubstituted double bond reacts before the 1,1,2-trisubstituted one. This is a clear indication

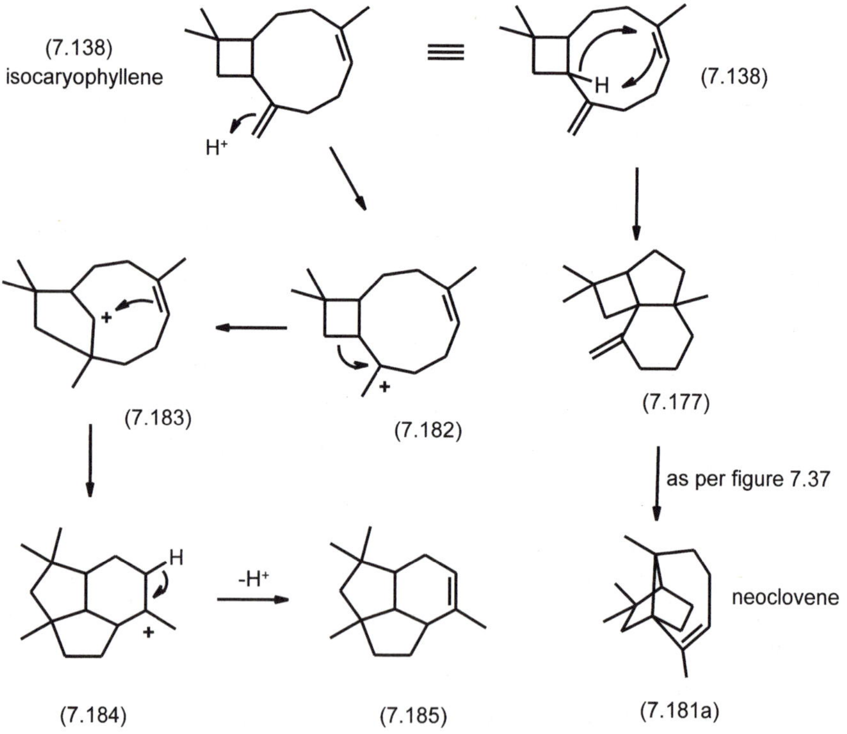

Figure 7.38

of the lower degree of ring strain in isocaryophyllene (7.138) compared to caryophyllene (7.137).

7.4.4 Synthesis of Humulene

One of the difficulties in the synthesis of medium rings from acyclic precursors is that of forcing the two ends of the chain to come close enough together to react with each other. The *trans* double bonds of humulene make this even more difficult. One way of overcoming this problem is to use the template effect of organometallic reagents where both ends of the chain complex with the metal atom and are therefore held in place for the crucial cyclisation step. Figure 7.39 shows the ring forming steps of two such reactions.

Allylic acetates become electrophilic on forming complexes with palladium (0) and will therefore alkylate stabilised enolates. Kitagawa *et al.*[7.20] used this to good effect in the synthesis of (7.187) from (7.186) *en route* to humulene (7.139). Similarly, Corey and Hamanka[7.21] used nickel carbonyl to intramolecularly couple the two allylic bromide functions of (7.188). The *E*-double bond in the substrate (7.188) was

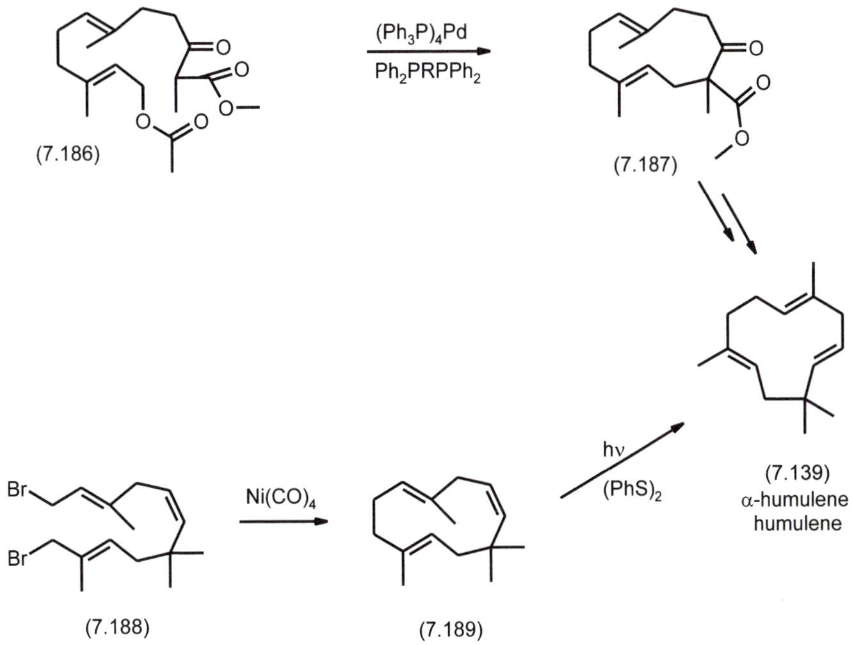

Figure 7.39

important in helping the cyclisation reaction to proceed by bringing the ends of the chain closer together. However, this meant that the cyclisation product (7.189) had to be isomerised by ultraviolet irradiation in the presence of diphenyl disulphide in order to produce all *trans* humulene (7.139).

REFERENCES

7.1.   Y.R. Naves and E. Perrottet, *Helv. Chim. Acta*, 1941, **23**, 768.

7.2.   A.S. Pfau and P.A. Plattner, *Helv. Chim. Acta*, 1939, **22**, 640.

7.3.   A.S. Pfau and P.A. Plattner, *Helv. Chim. Acta*, 1940, **23**, 768.

7.4.   J.A. Marshall, N.H. Andersen and P.C. Johnson, *J. Am. Chem. Soc.*, 1967, **89**, 2748.

7.5.   J.A. Marshall and P.C. Johnson, *J. Am. Chem. Soc.*, 1967, **89**, 2750.

7.6.   S.M. Bloom, *J. Am. Chem. Soc.*, 1958, **80**, 6280.

7.7.   H.F. Zimmermann and P.L. Schuster, *J. Am. Chem. Soc.*, 1962, **84**, 4527.

7.8.   J.A. Marshall and P.C. Johnson, *J. Chem. Soc. Chem. Commun.*, 1968, **21**, 391.

7.9.   J.A. Marshall and P.C. Johnson, *J. Org. Chem.*, 1970, **35**, 192.

7.10.   G. Stork, R.L. Danheiser and B. Ganem, *J. Am. Chem. Soc.*, 1973, **95**, 3414.

7.11.   J.P. Blanchard and H.L. Goering, *J. Am. Chem. Soc.*, 1951, **73**, 5863.

7.12.   S. Danishevsky and D. Duman, *J. Chem. Soc. Chem. Commun.*, 1968, **21**, 1287.

7.13.   J. Simonsen, *J. Chem. Soc.*, 1920, **117**, 570.

7.14.   R.H. Moffett and D. Rogers, *Chem. Ind.*, 1953, 916.

7.15.   G. Dupont, R. Dulou, P. Naffa and G. Ourisson, *Bull. Soc. Chim. Fr.*, 1954, 1075.

7.16.   E.J. Corey, M. Ohno, P.A. Vatakencherry and R.B. Mitra, *J. Am. Chem. Soc.*, 1961, **83**, 1251.

7.17.   E.J. Corey, M. Ohno, R.B. Mitra and P.A. Vatakencherry, *J. Am. Chem. Soc.*, 1963, **86**, 478.

7.18.   E.J. Corey, R.B. Mitra and H. Uda, *J. Am. Chem. Soc.*, 1963, **85**, 362.

7.19.   E.J. Corey, R.B. Mitra and H. Uda, *J. Am. Chem. Soc.*, 1964, **86**, 485.

7.20.   Y. Kitagawa, A. Itoh, S. Hashimoto, H. Yamamoto and H. Nozaki, *J. Am. Chem. Soc.*, 1977, **99**, 3864.

7.21.   E.J. Corey and E. Hamanaka, *J. Am. Chem. Soc.*, 1964, **86**, 1641.

© Jaqui Hurst/CORBIS

CHAPTER 8

Degradation Products

The characteristic scent of violet flowers is
due to carotenoid degradation products

Change and decay in all around I see

Henry F Lyte

Once an organic compound has been biosynthesised it comes under chemical attack from biochemical processes in living organisms and chemical attack from oxygen in the air, light and other chemicals present in the environment such as the salts in sea water. Eventually all organic materials return to carbon dioxide and water. However, there are some very exciting intermediates on the way. In this chapter, we will look at two very important groups of terpenoids which arise, in nature, from the degradation of larger units.

KEY POINTS

Chemical compounds isolated from nature may be the products of degradation of other materials.
Complex products are sometimes produced most economically from available natural feedstocks.
Terpenoids are essential for the sense of sight.

8.1 AMBERGRIS

The reader will no doubt be familiar with amber, the fossilised form of rosin, which is a sought after material for use in jewellery. The main source of amber is the coastline of the Baltic where it is often found washed up on beaches. Since amber is derived from rosin, we can consider it to be terpenoid in nature. There is another terpenoid material which resembles amber to some extent and is also found on beaches, but which is much less known. This second product is called ambergris, from the French *ambre gris* – grey amber. This name was coined to distinguish it from the fossilised rosin, amber or, as it is also known, *ambre brun* – brown amber. Ambergris is a complex mixture formed by degradation of a whale metabolite.

The sperm whale (*Physeter macrocephalus*) produces, in its intestinal tract, a mixture containing mostly a triterpenoid called ambreine (8.1) and a steroid, (+)-epi-coprosterine (8.2). Triterpenoids usually comprise 25–45% of the total mass and steroids about 45%.[8.1] It is not known exactly why the whale produces this mixture, but it is probably in response to some irritation of its gastro-intestinal tract. The whale excretes lumps of ambergris, usually up to 20 cm in diameter, though sometimes there are larger ones which can weigh up to 100 kg. The largest recorded piece of ambergris was taken from a whale which had been slaughtered and weighed no less than 400 kg.[8.2] Once excreted into the sea, the presence of salt water, air and sunlight starts a chain of degradation reactions eventually producing a complex mixture of breakdown products. Ambergris is usually dark brown when first excreted, but turns grey on aging. The aged material possesses a very characteristic, powerful and persistent odour and it has been prized as an ingredient in perfumes for millennia. Abu'l Kasim Obaidallah (d. 912 AD) wrote about its importation into Spain from the Malay archipelago in the ninth century. The odour is complex and contains aspects of wood, incense, tobacco, brine, ozone, musk and animalic notes. Perhaps it is because of these last two that ambergris gained a reputation as an aphrodisiac. It has also been used as a spice in cooking and pharmaceutically as a restorative.[8.3,8.4]

Some of the more organoleptically important of the degradation products are shown in Figure 8.1, along with the structures of the two major components of fresh ambergris. Inspection of the structures of the degradation products (8.3)–(8.10) will quickly reveal that they are derived from ambreine (8.1) by cleavage around the central double bond. Structures (8.3)–(8.7) are derived from the monocyclic part of the molecule whilst (8.8)–(8.10) are from the bicyclic part. The aldehyde (8.3) contributes both briny and amber notes whereas (8.4) and (8.5) are predominantly briny and ozone-like. The ketone (8.6) is a major contributor to the tobacco aspects of ambergris and the alcohol (8.7) adds earthy and faecal characteristics. As far as odour is concerned, the most important of all is the perhydronaphthofuran (8.10) which possesses the characteristic animalic note of ambergris.[8.5,8.6,8.7] It is also interesting to note that one of the breakdown products (8.4) contains a chlorine atom. This originates from the chloride ions present in sea water. Many terpenoid components of marine algae (seaweed) contain chlorine, bromine or iodine, all of which are present in sea water.

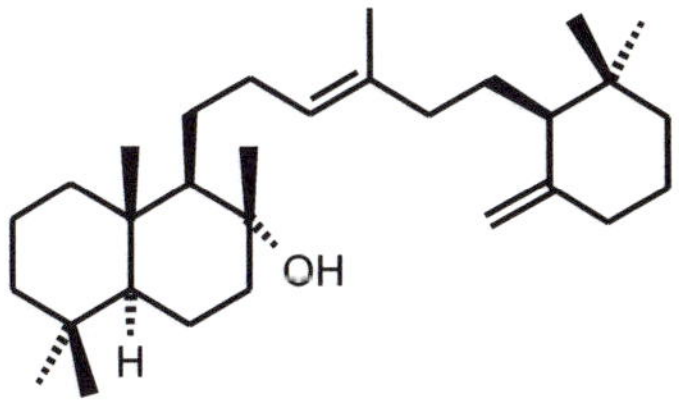

(8.1)

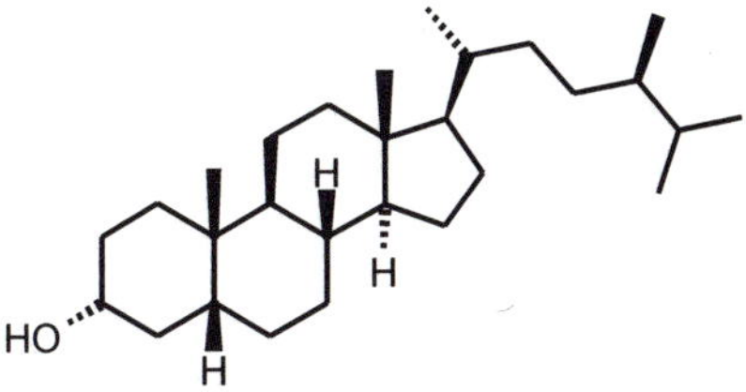

(8.2)

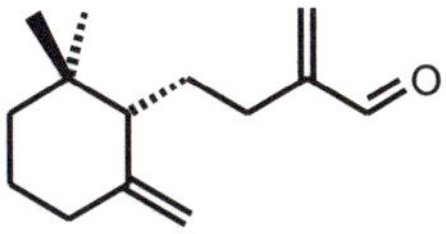

(8.3)

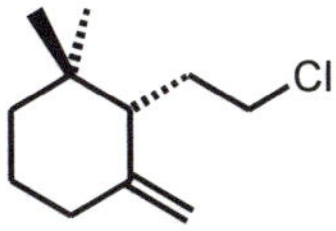

(8.4)

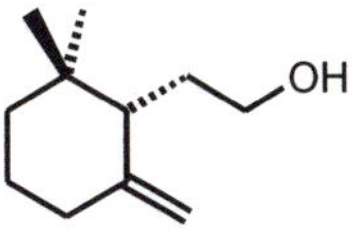

(8.5)

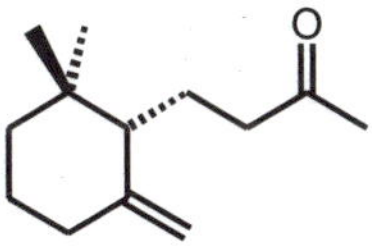

(8.6)

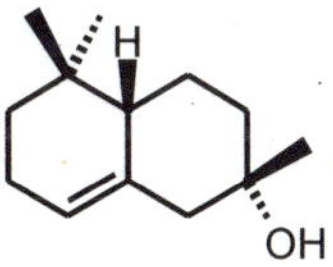

(8.7)

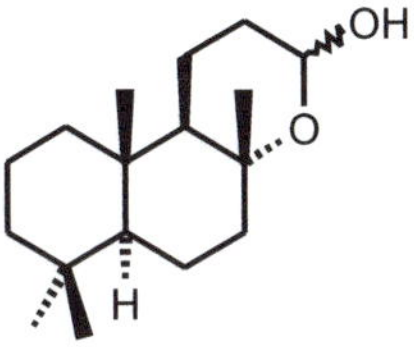

(8.8)

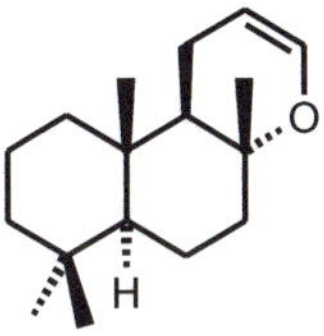

(8.9)

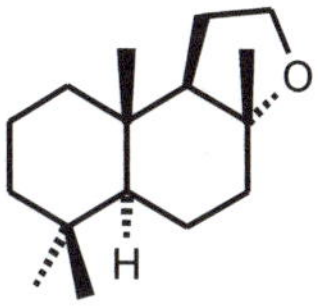

(8.10)

Figure 8.1

This is in contrast to terpenoids produced by terrestrial organisms, in which halogenation is rare.

8.1.1 Degradation of Ambreine

Mechanisms have been proposed for the reactions leading to all these compounds and experimental support for them has been obtained through replication in the laboratory using pure ambreine and/or model substrates.[8.8,8.9,8.10,8.11] The key reaction which starts the break down is photo-oxidation of the double bond by singlet oxygen. The mechanism of this reaction is shown in Figure 8.2.

The molecular structure of oxygen (O_2) is such that it has two unpaired electrons in its outermost occupied molecular orbitals. In the ground state, these two electrons have parallel spins. This means that the electron spin resonance (ESR) spectrum of oxygen shows a triplet pattern. The ground state of O_2 is therefore referred to as triplet oxygen. Irradiation with ultraviolet light can promote oxygen to a higher energy state in which the two unpaired electrons have anti-parallel spins. This gives a single line in the ESR spectrum and hence this state of O_2 is known as singlet oxygen. (For a fuller account of this, see for example Coulson's book on Valence.[8.12]) Oxygen does not absorb ultraviolet light of the frequencies reaching sea level and, therefore, another species must be involved in the process. It is known that many dyes can absorb ultraviolet light and transfer the energy they receive by doing so to oxygen. The dye is referred to as a sensitiser and the overall process is known as dye-sensitised photo-oxidation. One natural dye which has been identified in ambergris is the porphyrin-based compound haemocyanin. This compound contains

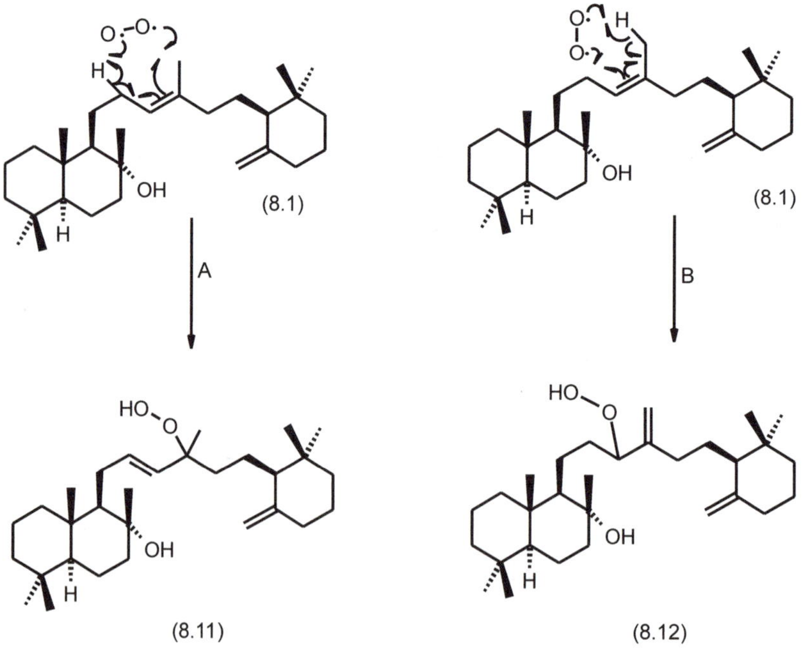

Figure 8.2

copper and either the dye or copper from it, could act as a sensitiser for this photochemical reaction. Singlet oxygen is a very reactive species and behaves as if it were a di-radical. The reaction of this "di-radical" with the electron-rich double bond in the middle of ambreine proceeds *via* a six-membered ring transition state. There are three possible six-membered rings. One of these involves the double bond and one of the methylene hydrogens on the decalin ring side of the olefin. Reaction in this way follows pathway A of Figure 8.2 to give (8.11) containing a tertiary hydroperoxide group. Alternatively, a six-membered ring involving the olefin and one of the methyl hydrogens adjacent to it, follows pathway B to give (8.12) in which the hydroperoxide group is secondary. The third possibility, *i.e.* involving the double bond and the methylene group on the cyclohexane ring side of it, does not seem to be important in the chemistry leading to odorous ambergris materials. Hydroperoxides are unstable because of the high energy oxygen–oxygen single bond and they, therefore, tend to break down fairly easily and quickly. Some of the possible decomposition reactions of (8.11) and (8.12) are shown in Figures 8.3–8.5.

Figure 8.3 shows two pathways for acid-catalysed breakdown of the tertiary hydroperoxide (8.11). Both start with protonation of the terminal oxygen atom. In pathway A, a nucleophile attacks the methylene carbon α to the hydroperoxide function. The resultant flow of electrons from the nucleophile, through the molecule to the positively charged oxygen atom, leads to the elimination of a neutral water molecule and fragmentation of the ambreine skeleton to give the unsaturated ketone (8.13) and a cyclohexyl fragment to which the nucleophile is now attached. If the nucleophile is chloride, then the product is the chlorinated odorant (8.4) whereas, nucleophilic attack by water will give the alcohol (8.5). Fragmentation to the other side of the hydroperoxide (pathway B), gives the tobacco-scented ketone (8.6) and a cationic fragment which will undergo further reaction. Ketone (8.6) is the precursor for ambrinol (8.7), by means of an acid-catalysed Prins reaction.

Similar breakdown pathways operate in the case of (8.12) as for (8.11) and two of these are shown in Figure 8.4. Attack by a nucleophile α to the hydroperoxide function, following pathway A, gives the marine aldehyde (8.3) whereas fragmentation in the opposite direction (pathway B) gives the aldehyde (8.14) and a cationic species. The aldehyde will form the observed hemi-acetal (8.8) by intramolecular acetalisation. Dehydration of (8.8) will give the dihydropyran (8.9). Figure 8.5 shows a special example of the fragmentation of pathway A in Figure 8.4. In this case, the nucleophile is the alcohol function of ambreine and this

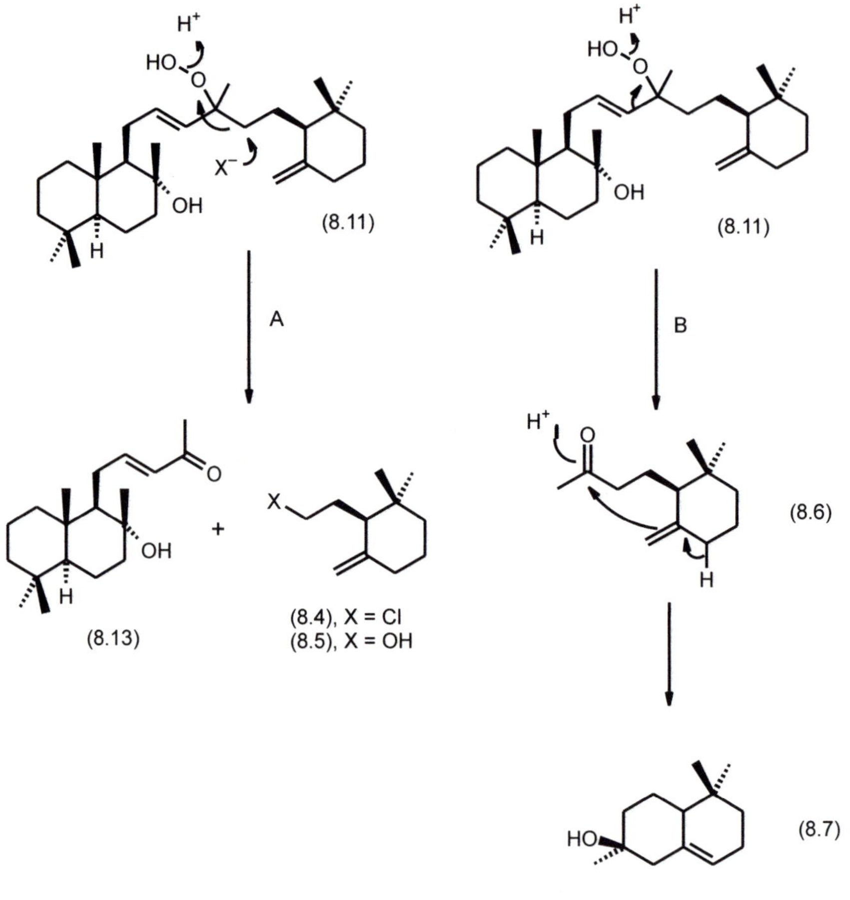

Figure 8.3

intramolecular reaction accounts for the key odorous principle (8.10), in addition to the marine aldehyde (8.3).

8.1.2 Ambergris Materials from Other Natural Products

Ambergris has always been very expensive and supply is precarious since it essentially relies on beachcombing. The decline in the whale population as a consequence of the whaling industry in the nineteenth century has exacerbated the situation. The price and availability of the natural material preclude its use in all but occasional fine fragrances and hence, much work has been done on synthetic substitutes. Obviously, the naphthofuran (8.10) has received most attention. Quite a number of total syntheses of this material have been published but the most important

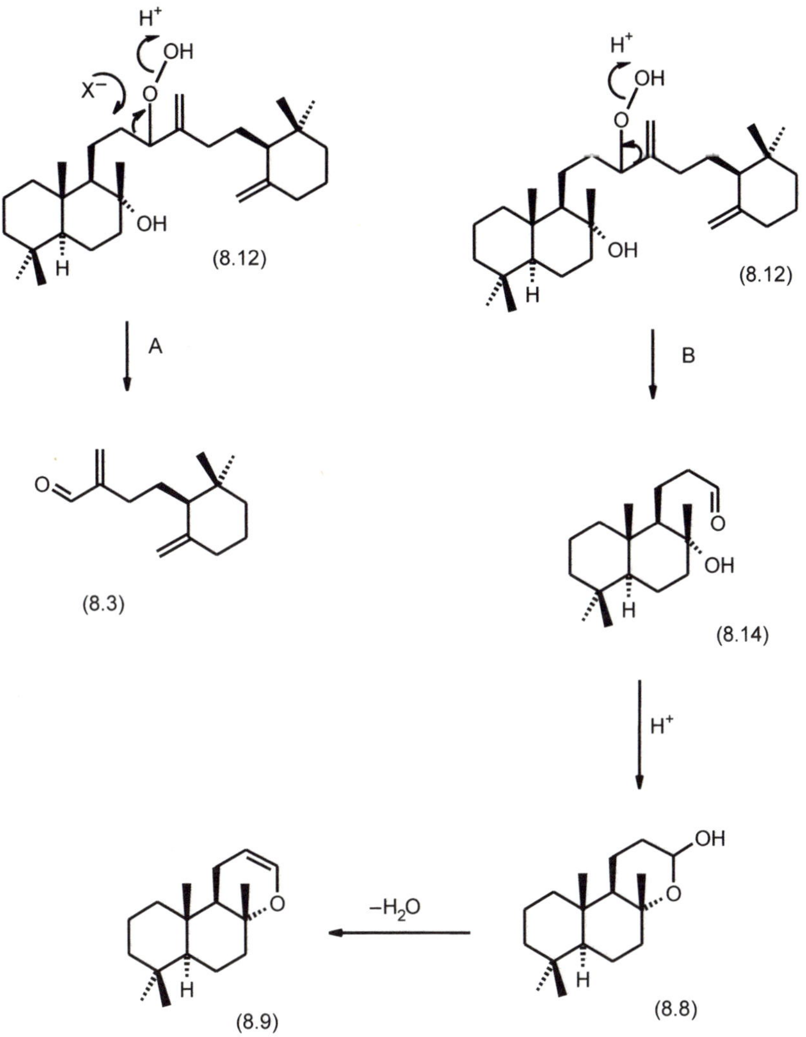

Figure 8.4

routes for commercial synthesis take advantage of the fact that the labdane family of diterpenes share the same stereochemical features in the decalin ring as (8.10). Furthermore, the most readily available labdanoid starting materials also share the same absolute stereochemistry as (8.10). This is important since the enantiomer of (8.10) is very much weak in odour and so the product of a non-enantioselective synthesis possesses only about half the odour impact of the natural compound.

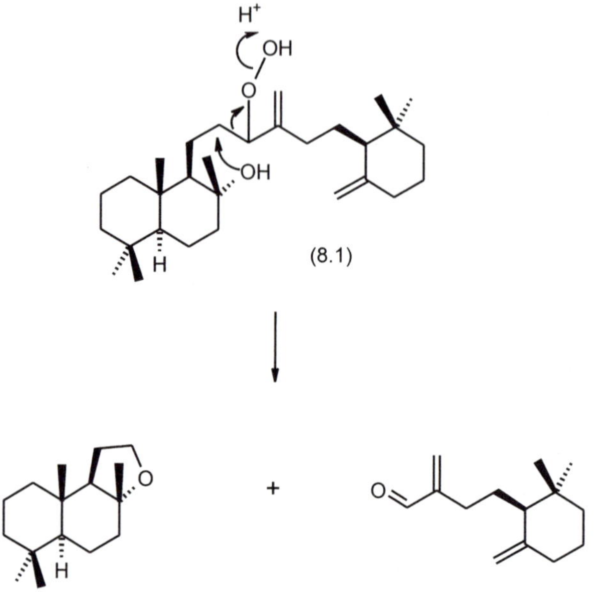

Figure 8.5

8.1.2.1 Clary Sage.   Clary sage (*Salvia sclarea*) is a relative of sage. It grows in Mediterranean countries and is harvested for its oil which contains mostly linalool and linalyl acetate. After the oil has been removed by steam distillation, extraction of the spent herb produces a concrete which contains about 50% by weight of the labdane diterpenoid sclareol (8.15). It is obvious from the structure of sclareol that suitable degradation of the side chain followed by intramolecular cyclisation will give the naphthofuran (8.10). The oldest, and still the most commercially important, routes for achieving this are shown in Figure 8.6.

The key intermediate is sclareolide (8.16) which is also important in its own right as a flavouring material in tobacco. Degradation of sclareol (8.15) with chromium trioxide in acetic acid gives sclareolide (8.16) directly. It is also possible to use a two-stage process in which permanganate oxidises the sclareol to the vinyl ether (8.17) which is a methylated analogue of one of the natural ambergris components (8.9). Ozonolysis of this gives (8.18) which can be hydrolysed and lactonised to sclareolide (8.16). Reduction of sclareolide using either lithium aluminium hydride (LAH) or borane gives the diol (8.19). Treatment of (8.19) with acid causes cyclisation, with loss of water, to give the naphthofuran (8.10). This

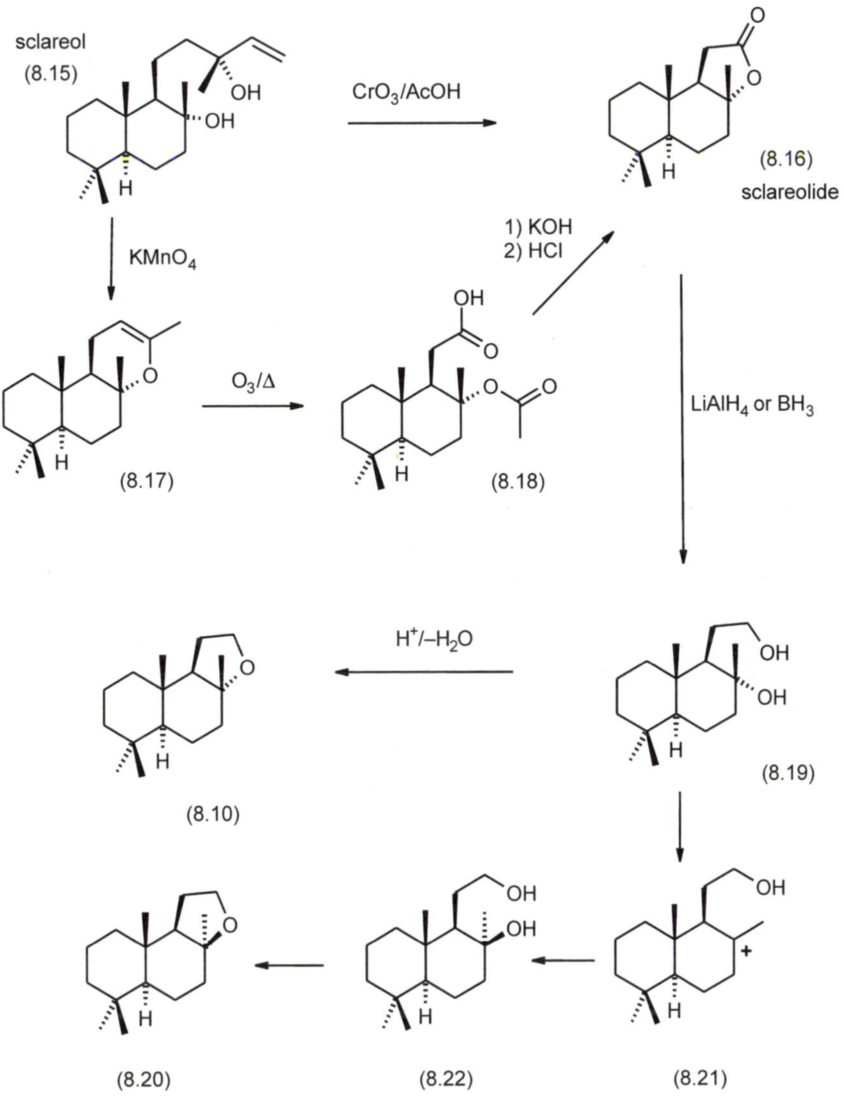

Figure 8.6

material is known under various trade names such as Amberlyn® (Quest), Ambrox® (Firmenich) and Ambroxan® (Cognis).

The advantages of this synthetic approach lie in the availability of the feedstock and the fact that enantio-pure material is produced. However, the processes also have some serious disadvantages. The oxidation of sclareol (8.15) requires the stoichiometric use of either chromium or manganese salts. This leads to a serious effluent problem. The oxidation leaves the intermediate, sclareolide (8.16), at too high an oxidation level and a vigorous reducing agent is required to take the lactone down to an alcohol. Both of the successful reductants are difficult to handle leading to high process costs. The final cyclisation requires acidic conditions and this leads to formation of an isomeric ether (8.20). This happens through epimerisation of the ring alcohol. This alcohol group is acid sensitive since it is tertiary. Protonation and subsequent loss of water gives the cation (8.21) which can add water from either side thus leading to the isomeric (8.22) as well as the starting diol (8.19). Cyclisation of the isomeric diol (8.22) will, obviously, lead to the isomeric ether, (8.20). Equally, direct ether formation by intramolecular cyclisation of (8.21) to (8.20), could occur from either side. A further problem which occurs is loss of material through elimination of a proton from the carbocation (8.21) to give unwanted unsaturated derivatives. The flattening of the ring through the introduction of a double bond into these materials, reduces the steric crowding of the ring and so re-hydration or ether formation are not readily achievable. In view of the value of the naphthofuran, it is not surprising that a great deal of research has been invested in attempts to improve individual stages of this approach or to find alternative routes from sclareol to the target. For example, some more modern processes recycle the heavy metal used in the oxidation of sclareol or avoid it entirely by using micro-organisms to carry out the oxidation enzymatically.

Figure 8.7 shows an idealised scheme for the conversion of sclareol to sclareolide. The concept involves formation of an alkoxy radical on the side chain of sclareol giving (8.23). This radical then reacts with another radical, cleaving the four unwanted carbon atoms from the side chain to produce (8.24). If the group X, which originated from the second radical, is a good leaving group, then it could be displaced to give the desired naphthofuran (8.10). Thus, it might be possible to achieve a one-pot conversion. One team of researchers (Farbood, Willis and Christenson of the fragrance company Fritsche, Dodge and Olcott) discovered a hitherto unknown fungus which could carry out something like this.[8.13] When they fed sclareol (8.15) to *Hyphozyma roseoniger*, they obtained the diol (8.19). The cyclisation was then effected in the usual way to give the final product. Another team (Decorzant, Vial and Näf of the fragrance company Firmenich and Whitesides of Harvard University) succeeded in realising a very similar concept in a purely chemical system.[8.14] They first prepared the isomeric hydroperoxides (8.25) and (8.26) from sclareol by treatment with hydrogen peroxide in the presence of an acid catalyst. Decomposition of the hydroperoxides using a mixture of copper(I) and iron(II) salts, gave the naphthofuran

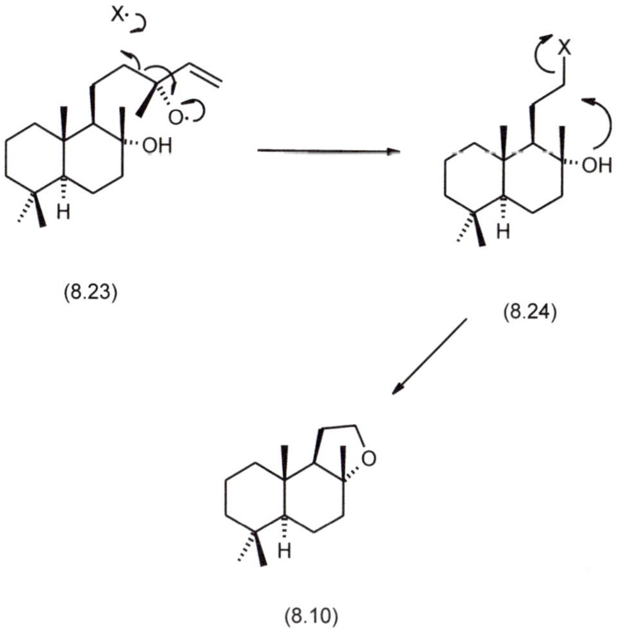

(8.23)

(8.24)

(8.10)

Figure 8.7

(8.10) by a radical process, albeit in rather low yield. This is shown in Figure 8.8.

8.1.2.2 Labdanum. Around 450 BC, when the Greek historian Herodatus was writing his "Histories"[8.15] he included a chapter on the perfumes of Arabia. He asked the Arabian perfumers about their sources of ingredients and was told some rather tall stories. In those days, sources of raw materials and perfume formulae were invaluable trade secrets since anyone who had access to them would be able to compete with the perfumers for their business. Modern analytical chemistry has destroyed the value of such secrecy since, with gas chromatography–mass spectrometry (GC–MS) and a good library of reference samples, the chemical composition of a perfume can be determined in an hour or less. However, Herodatus' interlocutors, anxious to protect their livelihood, told him one lie. When he believed that, they told him another, bigger one and so on. When they reached the limits of his credulity, they told him something close to the truth and this he refused to believe. This last story concerned the source of labdanum, a resinous gum with an amber-like odour. The Arabians told Herodatus that they collected it from the beards of goats which had been browsing

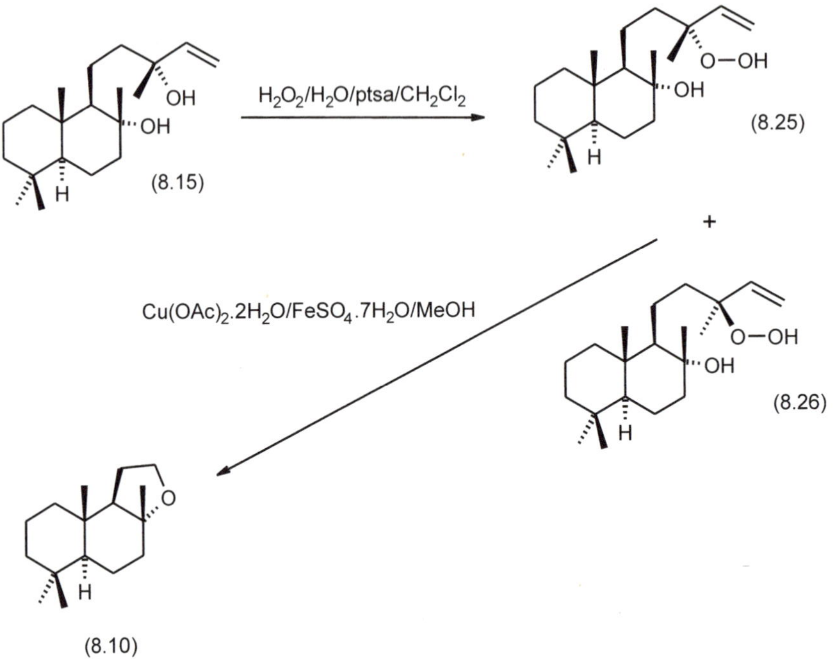

Figure 8.8

among bushes. Labdanum is a resin produced by the shrub *Cistus ladaniferous* which grows all over the Mediterranean region. Cistus is harvested for its essential oil which comprises a variety of mono- and sesquiterpenoid hydrocarbons, alcohols and ketones. The resin from the plant contains, among other things, a diterpenoid acid known as labdanolic acid (8.27). It can be seen from its structure that this acid, like sclareol, shares many structural features in common with the ambergris naphthofuran, (8.10). Since labdanolic acid is readily available as a by-product of cistus production, it is therefore another potential starting material for the naphthofuran. One recent example of the synthesis of (8.10) from labdanolic acid (8.27) is shown in Figure 8.9.

To convert labdanolic acid (8.27) to the naphthofuran (8.10), it is necessary to shorten the side chain and then cyclise the residue onto the alcohol to form the furan ring. In this synthesis, Bolster and de Groot carried out the chain shortening in two steps.[8.16] Firstly, after protection of the ring hydroxyl group by conversion to the acetate, the terminal acidic carbon atom was removed by photo-decarboxylation in the presence of iodine and iodobenzene diacetate. This gave the iodinated material (8.29). Treatment of (8.29) with potassium *tert*-butoxide

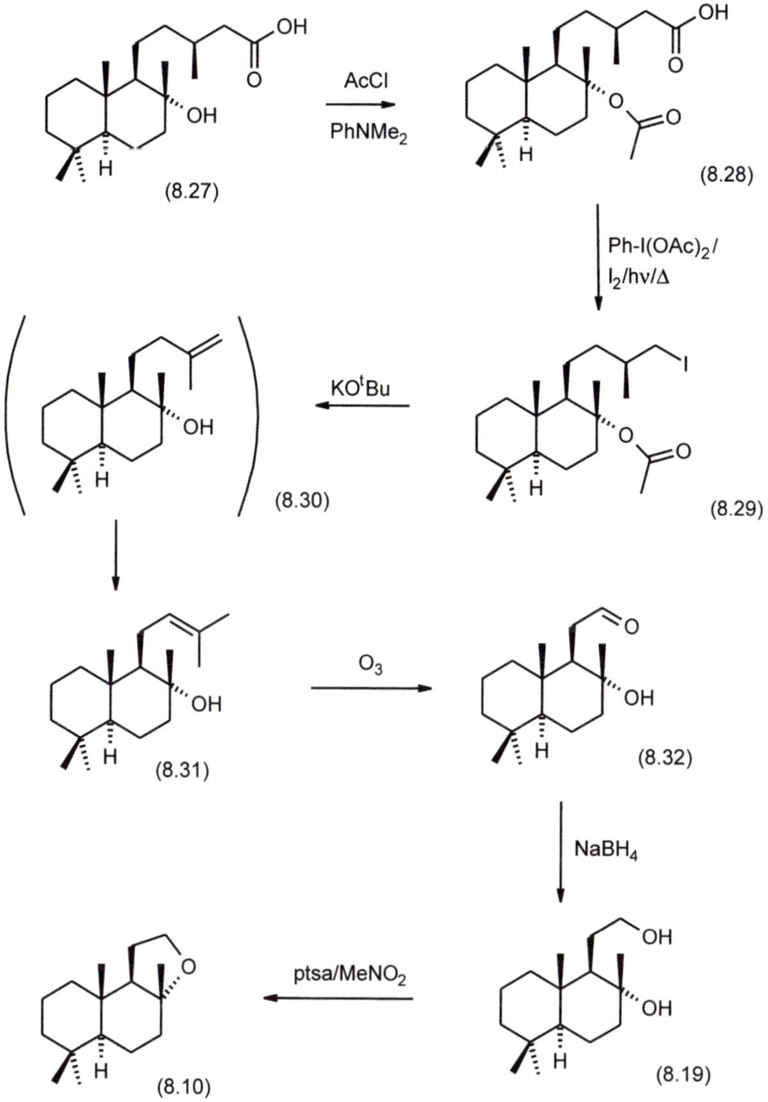

Figure 8.9

caused elimination of hydrogen iodide to give the olefin (8.30) which rearranged under the conditions employed to the thermodynamically more stable olefin (8.31). The protecting acetate group was also removed during this step, presumably by hydrolysis or *trans*-esterification. Ozonolysis then reduced the chain to the required length giving the hydroxy-aldehyde (8.32). It then remained only to reduce this to the known diol (8.19) and cyclise this to the target (8.10).

8.1.2.3 Total Synthesis. There are many total syntheses of the naphthofuran reported in the literature. This is not surprising in view of its commercial importance and the intrinsic challenge for the synthetic chemist. The strategic approaches employed fall into two main groups. The first group are those which employ conventional ring construction methodology and the second are those which emulate nature by forming both carbocyclic rings in one acid catalysed reaction. The term bio-mimetic is used to describe reactions which copy those of nature.

Figure 8.10 shows the synthesis of Büchi and Wuest of the Massachusetts Institute of Technology.[8.17,8.18] The starting point was dihydro-β-ionone (8.33). (The synthesis of ionones will be covered later in this chapter.) Treatment of this with sodium hydride and diethyl carbonate gave the β-keto-ester (8.34). Cyclisation of this using stannic chloride as a catalyst gave (8.35) which could be O-alkylated by allyl bromide in the presence of sodium hydride. The product, (8.36) was set up for a Claisen rearrangement to give (8.37). Hydrolysis of (8.37) was achieved by the use of hydrate of calcium chloride in dimethyl sulfoxide (DMSO). Reaction with methyl Grignard reagent occurred from the less hindered face of the decalin ring system to give the alcohol (8.39) possessing the correct stereochemistry for the final product. Ozonolysis of (8.39) gave the same hydroxy-aldehyde (8.32) that we saw in Figure 8.9 and the synthesis was completed in exactly the same way as that of Bolster and de Groot.

The synthesis shown in Figure 8.11 is an example of a bio-mimetic approach. It is the work of Vlad and co-workers at the Moldavian Academy of Science.[8.19] The starting material is homofarnesol (8.40). Treatment of this with fluorosulfonic acid induced cyclisation forming both carbocyclic rings of (8.10) and, at the same time, also the furan ring. The fusion between the two carbocyclic rings proceeded selectively but the fusion between these and the furan ring gave a mixture of products. Thus the isolated products were racemic (8.10) and racemic (8.41) in 40 and 35% yield, respectively. The mechanism is shown in the lower part of Figure 8.1. The flow of electrons in this reaction should be compared with that in the cyclisation of squalene epoxide in Figure 2.15 in order to appreciate the bio-mimicry involved in Vlad's synthesis.

8.1.2.4 Jeger's Ketal from Manool. Manool (8.42) is another labdane terpenoid which has found use as a precursor for fragrance ingredients with an ambergris odour. It differs from sclareol in that it has an

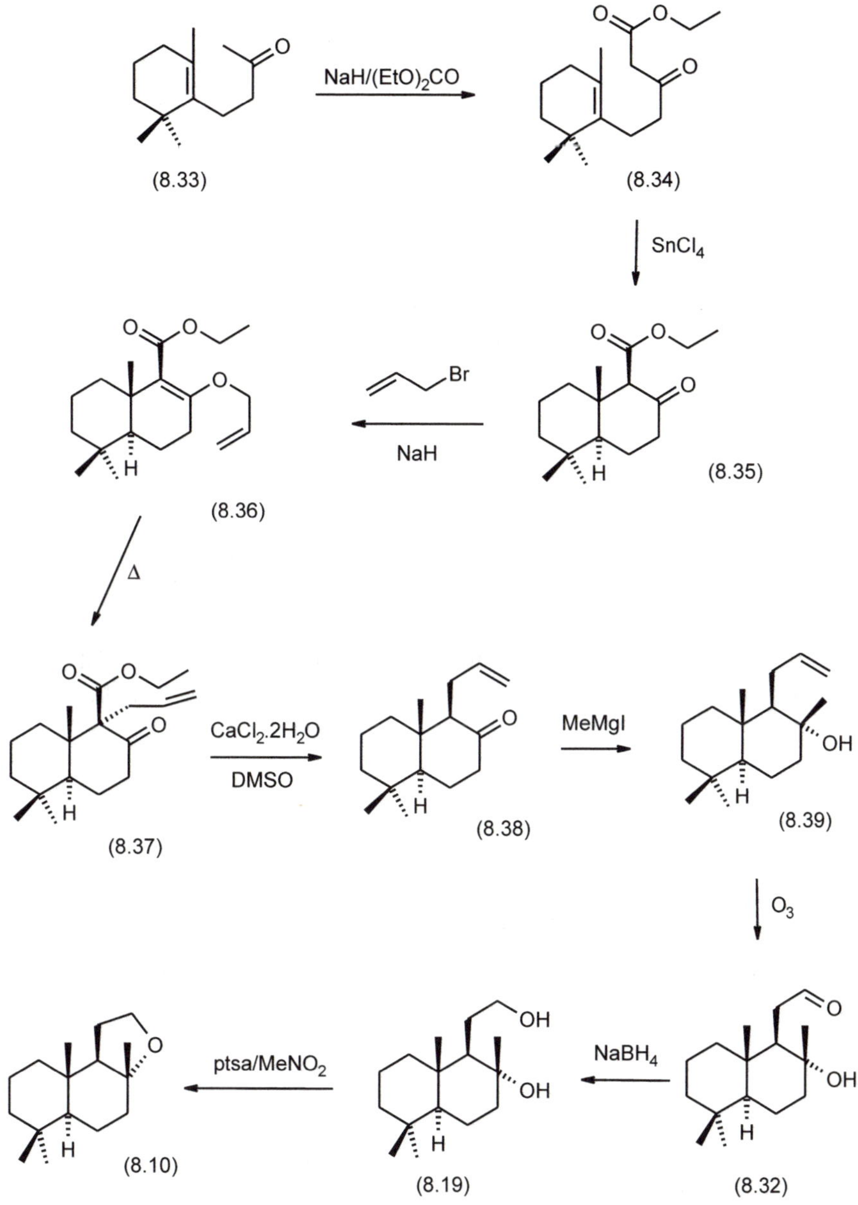

Figure 8.10

exo-cyclic methylene group in the place of sclareol's tertiary alcohol. It occurs in the wood of the tree *Halocarpus biformis* (formerly known as *Dacrydium biforme*). This tree is endogenous to New Zealand and is a protected species, thus limiting the supply of manool.

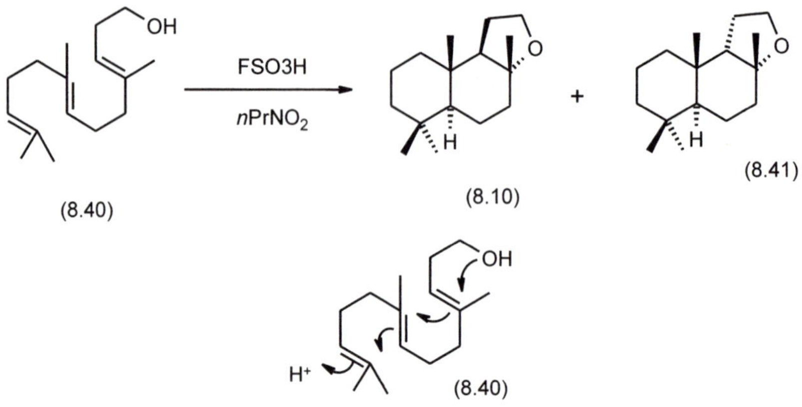

(8.40)   FSO3H
                *n*PrNO₂   (8.10)   +   (8.41)

H⁺   (8.40)

Figure 8.11

Treatment of manool with permanganate cleaves the side chain in the same way that it cleaves the side chain of sclareol. This gives the unsaturated ketone (8.43), as shown in Figure 8.12. Epoxidation of (8.43) gives the epoxide (8.44) as expected. Treatment of this epoxide with acid produces the ketal (8.45). This ketal is known as Jeger's ketal, after

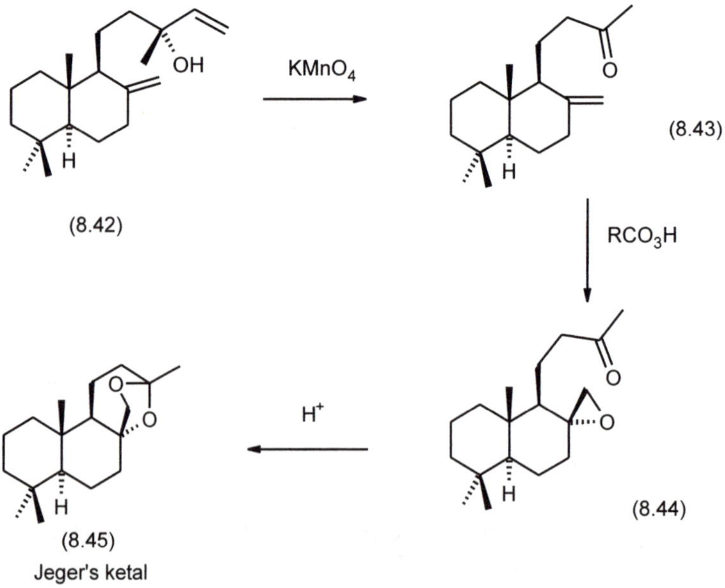

(8.42)   KMnO₄   (8.43)

RCO₃H

(8.45)   H⁺   (8.44)
Jeger's ketal

Figure 8.12

Otto Jeger who first synthesised it. It has a powerful ambergris odour and is highly prized in perfumery.

8.2 CAROTENOIDS

My first ever practical experience of chromatography was at school when we extracted pigments from grass and put a spot of the extract in the centre of a piece of filter paper. I can still remember my excitement when, as I dropped solvent onto the spot, I saw three concentric circles appear and grow outwards, at different rates, following the movement of the solvent front. Two of these circles were green in colour and one was orange. The greens were, of course, chlorophylls a and b and the orange one was α- and β-carotene. I had repeated the work of Pliny the Elder, whose work on the separation of plant pigments using paper chromatography was brought to an abrupt end in AD79 as his laboratory was destroyed, along with the rest of Pompeii, when Vesuvius buried the town under metres of volcanic ash. The three pigments I had separated were all terpenoids. The carotenes are produced, not only by grass, but also by a vast range of plants. The best known source is carrots, from which these terpenoids take their name. The notion that eating carrots helps one to see in the dark, has some basis in fact. Breakdown products of the carotenes are essential for the sense of sight and a shortage of them will result in impaired vision. This would be particularly noticeable when the light is poor.

8.2.1 Vitamin A – The Chemistry of Sight

Figure 8.13 shows the structures of α-carotene (8.46) and β-carotene (8.47) and two of the most important breakdown products of these, namely, vitamin-A or retinol (8.48) and 11-*cis*-retinal (8.49). Retinol and retinal are both quite unstable molecules and also moderately toxic. Therefore they are not easily stored in the body. The carotenes, on the other hand, can be kept in fatty tissue and, when required, removed and metabolised to retinal by enzymes in the liver. Enzyme catalysed oxidative cleavage in the centre of β-carotene produces vitamin-A and this is oxidised to the corresponding aldehyde by the enzyme retinal dehydrogenase. The all *trans*-retinal thus produced is isomerised to 11-*cis*-retinal by another enzyme, retinal isomerase. Combination of 11-*cis*-retinal with opsin gives the light sensitive pigment rhodopsin which enables our eyes to detect incident light. In order to understand the role of rhodopsin, it will be necessary first to look at the structure of a group of proteins known as 7-transmembrane G-coupled receptor proteins and their role in cellular communication.

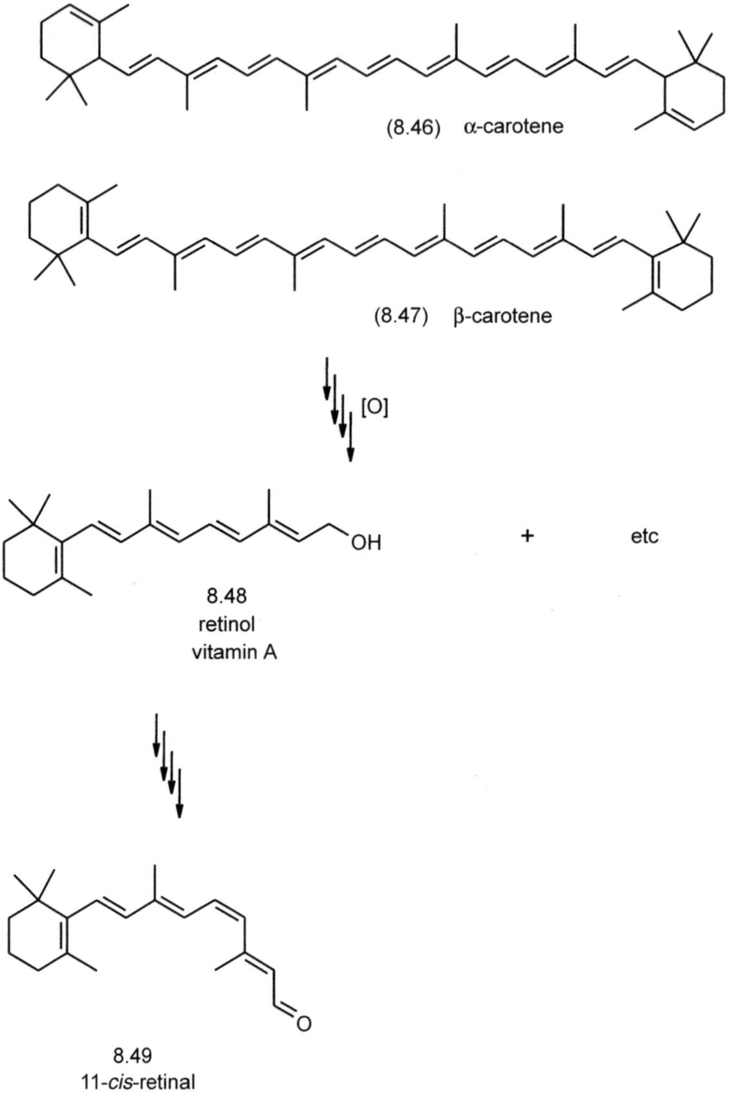

(8.46)   α-carotene

(8.47)   β-carotene

[O]

8.48
retinol
vitamin A

+   etc

8.49
11-*cis*-retinal

Figure 8.13

8.2.1.2 7-Transmembrane G-coupled Receptor Proteins. Mammalian cell walls are made up of lipid bilayers. The building blocks of these bilayers are molecules containing a polar head group and two long fatty tails. Such lipid derived materials are referred to as lecithin. A typical example, phosphatidylcholine (8.50) is shown in Figure 8.14. There is a glycerol unit in the centre of the molecule and two of its hydroxy groups

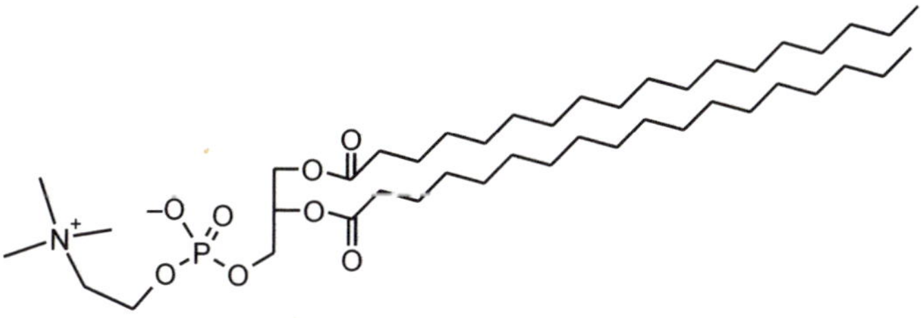

(8.50) phosphatidylcholine

Figure 8.14

are esterified with a fatty acid (stearic acid in the figure). The third hydroxy group is esterified to a molecule of phosphoric acid and this is also esterified to choline ($Me_3N^+CH_2CH_2OH$). The fact that one end of the molecule is hydrophilic and the other hydrophobic, makes phosphatidylcholine surface active and hence capable of forming micelles and bilayers.

Figure 8.15 shows a schematic cross-section of a cell wall, and it can be seen how the individual surfactant molecules (each depicted as a circle to represent the polar head group and two lines to represent the fatty chains) organise themselves into a double layer with polar groups along both outer faces and a lipophilic region between the two. This bilayer extends both in the plane of the paper and at right angles to it. Thus it can form a skin around a pocket of hydrophilic material, in other words, the contents of the cell. Both the liquid surrounding the cell and its contents are aqueous in nature and separated from each

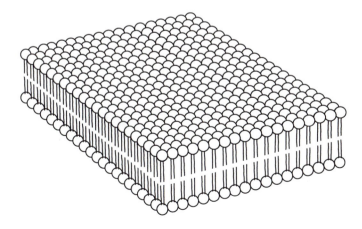

Figure 8.15

other by a fatty membrane. Each cell is therefore isolated from its environment and from other cells. However, it is advantageous for cells to be able to obtain information about changes in their environment and, in multi-cellular organisms, it is necessary for the cells to be able to communicate with each other chemically. Both of these needs can be achieved through proteins which are built into the cell wall. A protein which passes through the cell wall can either serve as a channel through which hydrophilic material can pass or as a means of sending messages from one side of the cell wall to the other through changes in tertiary structure of the protein. The former are known as ion channels and they have gates built into them which can open or close as necessary. The latter type are known as receptor proteins. If another molecule interacts with the extracellular face of a receptor protein in such a way as to alter its tertiary structure, the effects of this will be seen at the intracellular face also and this change can be used to initiate chemical changes within the cell. The molecule which sets this entire chain in action by recognising the receptor protein, is known as the substrate, ligand or agonist of the latter.

One such receptor protein is opsin, the structure of which is shown schematically in Figure 8.16. As with all proteins of its class, the amino-acid chain of opsin begins with the carboxylate terminus inside the cell wall. The chain then passes through the cell wall, loops round inside the cell and back through the wall to the outside again. This is repeated until there are seven points at which the backbone passes through the cell wall, leaving the amino terminus outside the cell. Each *trans*-membrane section of the protein is folded as an α-helix. Opsin belongs to the class of proteins known as 7-transmembrane G-coupled receptor proteins though

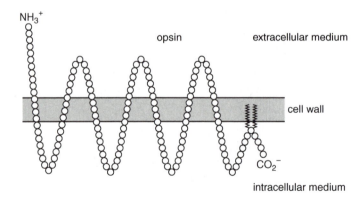

Figure 8.16

it is unusual in two major aspects. Firstly, and most importantly, its substrate is not another molecule, but a photon. Secondly, its carboxylate end is also locked back into the cell wall by means of two fatty acid residues attached to cysteine units in the chain. The seven *trans*-membrane helices come together to form a channel through the centre of the protein. This is also typical of the class but, since the substrate is a photon, opsin, which is transparent to visible light, needs to employ retinal in the role of a co-factor. The retinal is bonded to the protein chain by forming the carbon–nitrogen double bond of a Schiff's base to the free amino group of a lysine unit in the protein backbone. When the pigment is present, the protein is known as rhodopsin (from the Greek word *rhodos* meaning red). Figure 8.17 shows a schematic cross-section of the *trans*-membrane part of rhodopsin with retinal bonded in place. This adaptation of the receptor allows rhodopsin to detect light falling on it.

Receptor cells are nerve cells and they initiate electrochemical nerve signals. The chain of species involved in this process is shown in Figure 8.18.

The change induced in the receptor protein by interaction with its substrate, triggers a change in another protein, from a class known as G-proteins (hence the name G-coupled receptor) which is present at the inner face of the cell wall. In the case of rhodopsin, the G-protein involved is transducin. The modification of transducin in turn induces a

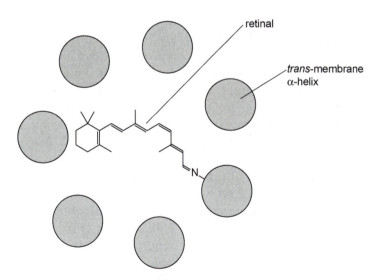

Figure 8.17

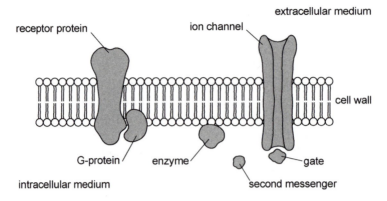

Figure 8.18

change in a third protein, this time an intracellular enzyme. There are two enzymes which are associated with this receptor train, *viz.* adenylyl cyclase and phospholipase-C. Once activated by the G-protein, they catalyse some chemical changes in the cell. Adenylyl cyclase catalyses, amongst other reaction, the formation of cyclic adenosine monophosphate (c-AMP) and similarly, phosphospholipase-C produces inositol triphosphate (IP_3). These small molecules (c-AMP and IP_3) are known as second messengers and their role is to act on the ion channel, causing it to open or close the gate which allows ions to pass through the cell membrane and thus generate the electrical current which constitutes the nerve signal. It has been estimated that the interaction of one photon with one molecule of rhodopsin is sufficient to stop one million sodium ions from passing through the ion channel. This leads to an electrical imbalance between the cell and its environment, causing the nerve synapse to fire and generate an electrochemical signal which then proceeds on to the brain where it is interpreted.

The role of retinal in detection of light is outlined in Figure 8.19. An amino group of opsin (8.54) forms a Schiff's base with 11-*cis*-retinal (8.49) giving rhodopsin (8.51). The extended conjugation across this part of the molecule makes it an absorber of visible light. There are two types of light detecting cells in the retina. These are known as rods and cones because of their shapes. The rhodopsin in the rod cells absorbs light over the range of wavelengths from 400 to 600 nm, that is from blue to red visible light. The exact maximum frequency of absorption of the rhodopsin in the cone cells is determined by the distance between the positive charge on the protonated Schiff's base and the counter ion associated with it. By engineering three different discrete distances, the

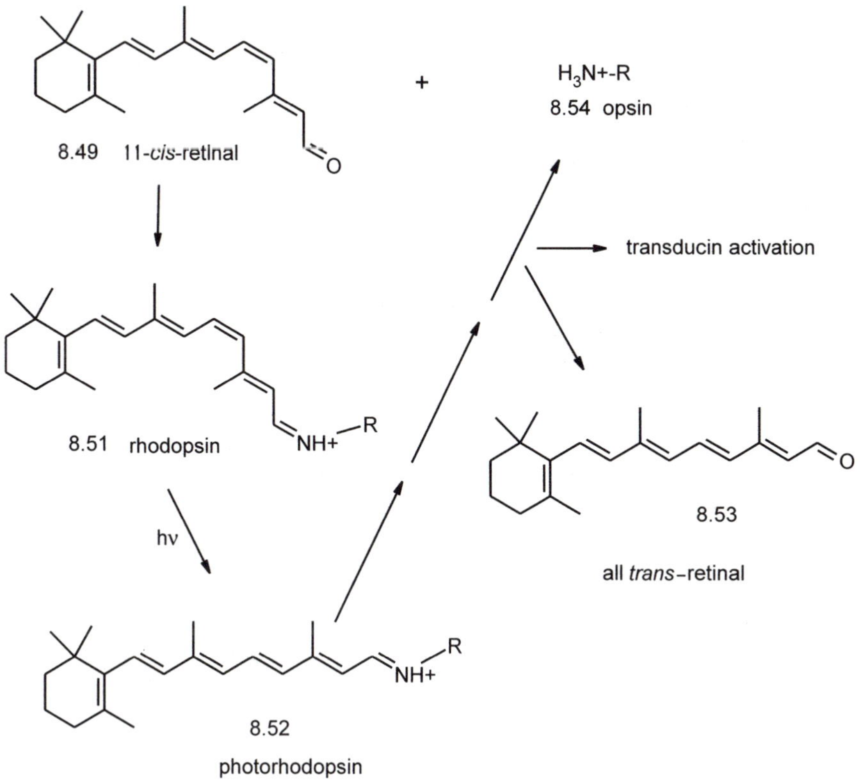

Figure 8.19

cone cell receptors are tuned into three distinct wavelengths corresponding to red (570 nm), green (530 nm) and blue (440 nm) light. The vast
array of colours which we can perceive is thus made up of different
combinations of the three basic receptor types in the cone cells. When a
photon hits rhodopsin, it promotes the chromophore into an excited
state. The energy from the photon is used to isomerise the *cis*-double
bond to the *trans*-configuration to give a new material known as
photorhodopsin in an almost instantaneous process. The change in
configuration results in a change in the binding energy of the protein
since the much more linear form of photorhodopsin no longer fits the
binding pocket in the same way. This induces a series of changes in the
detail of the retinal binding and each intermediate state has a different
maximum frequency of ultraviolet absorption, allowing accurate
measurement of reaction times by standard photochemical techniques.
The change from photorhodopsin to bathorhodopsin takes 45 ps;
from bathorhodopsin to lumirhodopsin, 30 ns; from lumirhodopsin to

μ-rhodopsin I, 75 μs; and from μ-rhodopsin I to μ-rhodopsin II, 10 ms. It is at this point that all *trans*-retinal dissociates to regenerate opsin and transducin is activated. The all *trans*-retinal is recycled back to 11-*cis*-retinal by retinal isomerase. Any reader who wishes more detail of this amazing process should read the excellent article on the subject by Hideki Kandori.[8.20]

8.2.2 Violets, Roses, Orris, Osmanthus, Geranium, Grapes, Vanilla, Raspberries, Passionfruit and Tea – The Chemistry of Ionones and Related Compounds

Degradation of carotenes also leads to the formation of three related groups of compounds: the ionones, damascones and theaspiranes. The generic structures for these materials are shown in Figure 8.20. In structures (8.55), (8.56) and (8.57), the dotted lines can be either single or

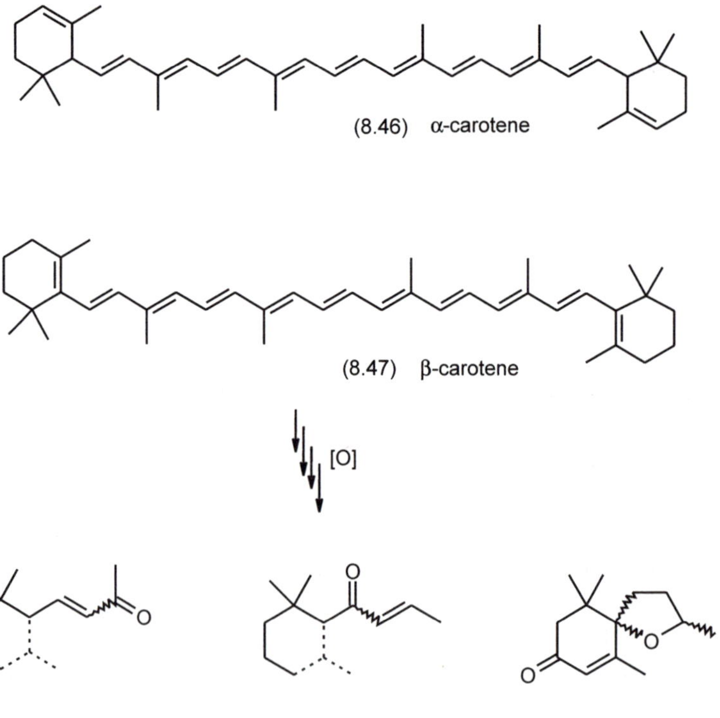

(8.46) α-carotene

(8.47) β-carotene

[O]

(8.55)
ionones

(8.56)
damascones

(8.57)
theaspiranes

Figure 8.20

double bonds and the wavy lines indicate that either isomer is included. It is obvious from the figure that all three families are derived by oxidative degradation of carotenes, with each six-membered ring carrying only four of the carbons which once constituted the long bridge between it and its opposite number at the far end of the carotene molecule. The differences between the three families lie in the oxidation patterns. In ionones, the third carbon atom along the chain carries the oxygen atom: in damascones it is the first and in the theaspiranes, there is a spiro fusion to a tetrahydrofuran ring and further oxidation at C-4 of the six-membered ring.

The system used for nomenclature of the ionones and damascones is illustrated in Figure 8.21. The position of the double bond is indicated by a Greek letter and the prefixes *n-* and *iso-* are used to indicate the position of substitution. Only two substitution positions are named, for reasons that will become clear later. The prefixes *Z-/E-* and *R-/S-* are used, in the usual manner, to indicate the geometry of the side-chain double bond and the absolute stereochemistry of the stereochemical centre, where one exists. To illustrate the application of these rules, the structure of *Z-α-*isomethylionone is shown as structure (8.58).

8.2.2.1 Ionones. Violet flowers were used in perfumery in the past, but the oil was always very expensive because of the size of the flowers and the difficulty in extracting it. Iris roots possess a similar scent to that of violet flowers and the early work on their composition concentrated on the more readily available orris (an extract from iris rhizomes). It was assumed, incorrectly, that the key odour components were isomers of each other. The key components of orris are the irones (*vide infra*) and it

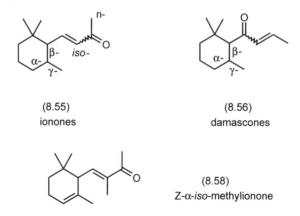

(8.55)
ionones

(8.56)
damascones

(8.58)
Z-α-*iso*-methylionone

Figure 8.21

was wrongly concluded that their empirical formula was $C_{13}H_{20}O$. This second mistake was fortuitous in that the target structures for synthesis became those of the ionones. Tiemann succeeded in synthesising (and patenting) α-ionone in 1893 but it was not until 1927 that an ionone was first identified in nature when Penfold discovered β-ionone in the Australian shrub *Boronia megastigma*. In 1937, Ruzicka reported isolating a ketone, which he called Parmone, from violets and, in 1972, Ohloff showed that Parmone was, in fact, (+)-α-ionone. The air above an odorous material is known as the headspace and it has been shown that α- and β-ionone account for 57% of the organic materials present in the headspace of violet flowers (*Viola odorata*). Tiemann's synthesis of α-ionone is shown in Figure 8.22.

Condensation of citral (8.59) with acetone in the presence of base gives the aldol condensation product (8.60), which is known as Ψ-ionone. Treatment of this with an acid leads to protonation of the isolated double bond. This protonation occurs selectively at the less substituted end, as expected, to give the tertiary carbocation (8.61). There is now the possibility of forming a six-membered ring by trapping of the cation by

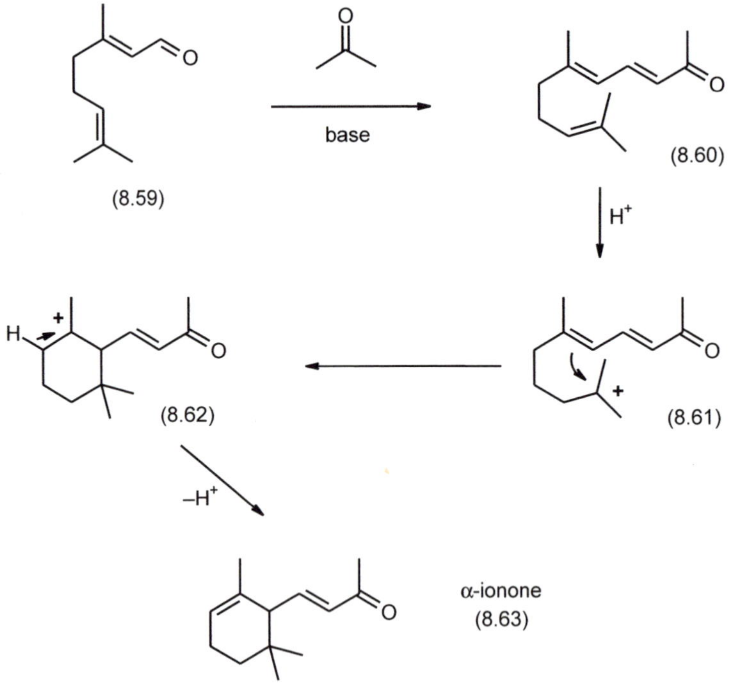

Figure 8.22

the electrons of the double bond in the middle of the chain. This gives another tertiary carbocation (8.62) which eliminates a proton to give α-ionone (8.63). Clearly, there are many other possible reactions which can take place during this process, and they do. There are other possible aldol reactions, for example, ψ-ionone can condense with another molecule of citral to give a C23 material. Citral can also undergo a retro-aldol reaction to give 2-methylhept-2-en-6-one and acetaldehyde and these components can then undergo forward aldol reaction. Both *E*- and *Z*-isomers can be formed, though the former would be expected to predominate (see the section on aldol stereochemistry in Chapter 6). The elimination of a proton from (8.62) can lead to the α-, β- or γ-isomer. If the acid used for cyclisation is phosphoric acid, the α-isomer predominates whereas sulfuric acid gives mainly the β-isomer. The use of boron trifluoride etherate in a polar aprotic solvent gives mostly the γ-isomer. Higher ketones can be used in place of acetone to produce higher homologues of the ionones, the commonest being 2-butanone which leads to methylionones. Use of an unsymmetrical ketone such as 2-butanone, of course, further complicates the issue since the initial aldol can occur at either side of the carbonyl function to give the *n*-alkylionones or *iso*-alkylionones, hence the development of a specific nomenclature system for the two possible resultant sites of alkylation (*vide supra*).

The ionones possess odours which are reminiscent of violet, sometimes also with woody notes. Each isomer has its own combination of violet and wood character and small variations in the composition can have relatively large effects on the odour. The most valued odour is that of (*Z*)-α-*iso*-methylionone (8.58). The exact composition of the product mixture from any ionone synthesis of this type depends very much on the nature of the catalysts and reaction conditions employed. A vast amount of development work has gone into each step of the synthesis in order to optimise yield and product isomer ratio. All the companies which produce ionones have their own signature blends of isomers and the mixtures are available under many different trade names. For obvious reasons, much of the detail of development work and reaction and distillation conditions are kept as company secrets. Citral and the ionones are very important commercially, not just in their own right but also as intermediates in the synthesis of vitamins. Consequently, there is a large volume of published academic and patent literature on their synthesis. This will be considered in more detail in Chapter 9, in the context of factors concerning synthesis on commercial production scale.

Figure 8.23 shows another possible product of the acid catalysed cyclisation of ψ-ionone (8.60). In this case, the protonation occurs on the oxygen atom and the positive charge is trapped by the isolated double

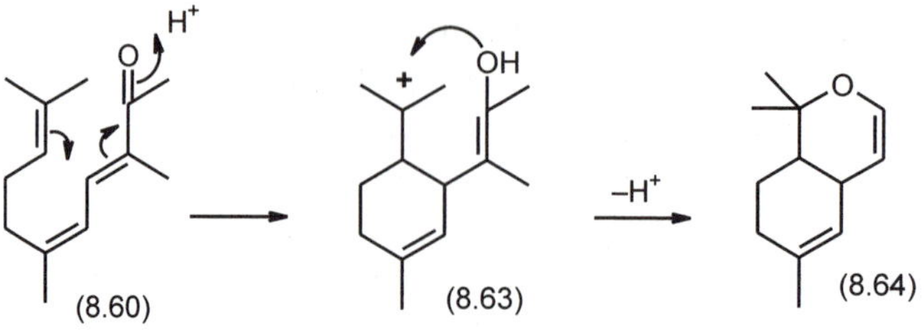

Figure 8.23

bond at the far end of the chain to give the cation (8.63). In Figure 8.23, ψ-ionone is drawn differently from before so as to show how this can occur easily to form a different six-membered ring. The oxygen atom can now trap the positive charge of (8.63) and eliminate a proton to give the pyran ring of the product (8.64).

8.2.2.2 Damascones. The damascones are isomers of the corresponding ionones in which the α,β-unsaturated function of the side chain has been transposed. They occur in rose oils and their name is derived from the Damask rose (*Rosa damascena*) in which they were first detected. They have powerful fruity odours with a hint of florality and rose character.

The transposition of the functional group might seem like a relatively small structural change but it has a major effect on the ease of synthesis as the simple condensation and cyclisation sequence of Tiemann's ionone synthesis is no longer possible. One approach to α-damascone is shown in Figure 8.24. The starting point is methyl cyclogeranate (8.65) which reacts with allyl magnesium chloride to give the ketone (8.66). Normally ketones are more reactive towards Grignard reagents than are esters and so it would be expected that a second mole of allyl magnesium chloride would add to ketone (8.66) to give the tertiary alcohol (8.68). However, by carrying out the addition in the presence of a strong base (lithium di-isopropylamide, LDA), it is possible to arrest the addition at the ketone stage with a reasonable degree of selectivity. The terminal double bond of the initial product (8.66) is brought into conjugation with the ketone by the base, giving α-damascone (8.67). Alternatively, it is possible to use excess Grignard reagent and allow the tertiary alcohol to form. On pyrolysis, (8.68) undergoes a retro-ene reaction with the elimination of propylene to give (8.66) and hence, after isomerisation, α-damascone (8.67).

The ready availability of ionones as potential starting materials, was exploited by Büchi in an elegant synthesis of β-damascone (8.69) and

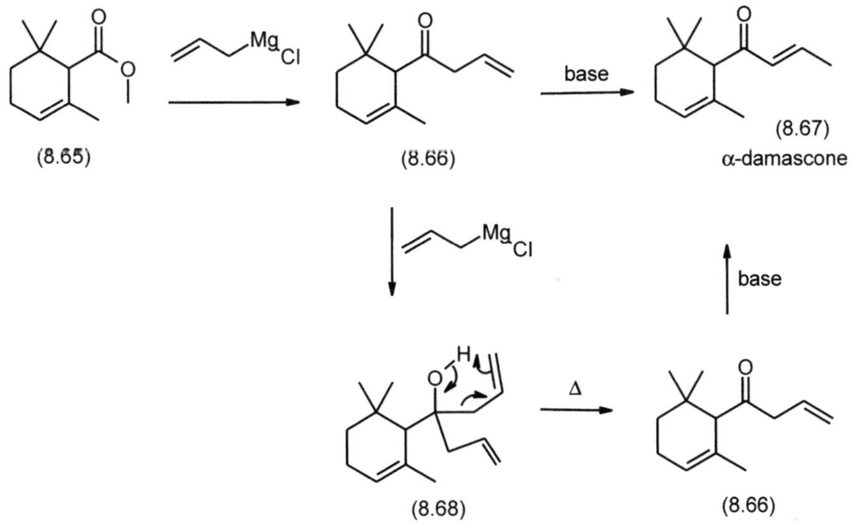

Figure 8.24

damascenone (8.70), another analogue highly prized for its odour. His synthetic approach involved an ingenious transposition of the functional group by means of an oxazole intermediate. The scheme is shown in Figure 8.25.

The synthesis started with β-ionone (8.71) which was converted to its oxime by reaction with hydroxylamine. Oxidation of the oxime with iodine in the presence of potassium iodide gave the oxazole (8.73) and thus achieved the introduction of an oxygen atom at the desired position in the side chain. Birch reduction of the oxazole using sodium and tertiary butanol in liquid ammonia, gave β-damascone (8.69). Alternatively, epoxidation of the oxazole (8.73) gave (8.74), selective hydrogenation of which using Raney nickel as catalyst gave the enaminoketone (8.75). Birch reduction of this gave epoxy-β-damascone (8.76). Treatment of (8.76) with aqueous acid led to hydrolysis of the epoxide and elimination of both hydroxyl functions of the resultant diol, to give damascenone (8.77).

8.2.2.3 Theaspiranes and Vitispiranes. The shrub *Osmanthus fragrans* is a member of the *oleaceae* family which grows in eastern Asia. It is a rich source of carotenoid degradation products, many of which have also been found in other plant sources. Some of these are shown in Figure 8.26 and it is evident in the figure how they are related to each other and to the ionones.

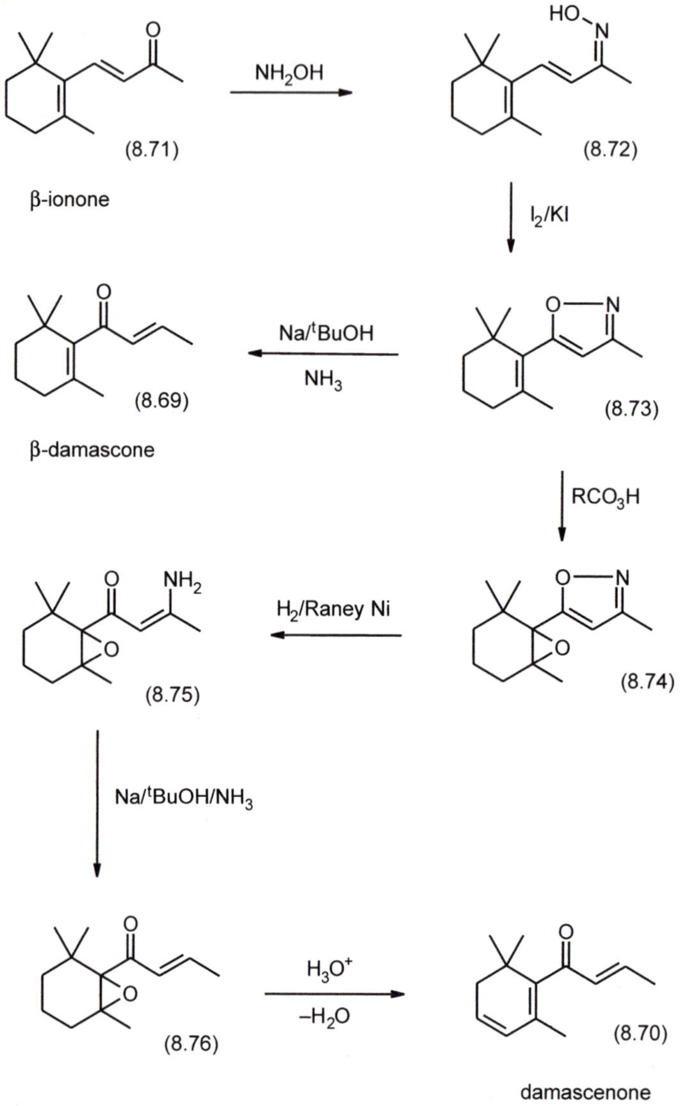

Figure 8.25

Both theaspiranes are found in osmanthus: theaspirane-A (8.77) has also been identified in raspberry and passionfruit. The theaspirones (8.79) and (8.80) are found in geraniums and tea, in addition to osmanthus. The vitispiranes (8.81) and (8.82) are found in osmanthus though they were first identified in grape must and vanilla.

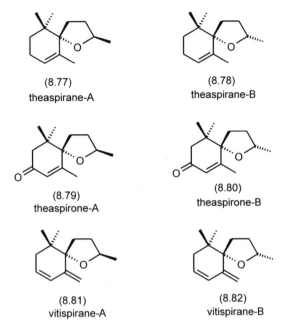

(8.77)
theaspirane-A

(8.78)
theaspirane-B

(8.79)
theaspirone-A

(8.80)
theaspirone-B

(8.81)
vitispirane-A

(8.82)
vitispirane-B

Figure 8.26

The theaspiranes are accessible synthetically from β-ionone (8.71) as shown in Figure 8.27. The diol (8.83) is easily prepared from β-ionone (8.71) and treatment of it with acid results in selective dehydration to give a mixture of the isomeric alcohols (8.84) and (8.85). Acid catalysed cyclisation of these leads to a mixture of the two isomeric theaspiranes (8.77) and (8.78). The theaspirones and vitispiranes can be prepared from these by relatively straightforward functional group manipulation.

8.2.2.4 Irones – The Chemistry of Iris. At first glance, the irones might seem to be simple analogues of the ionones since they differ from the latter only by the addition of one methyl group in the six-membered ring. However, consideration of the biosynthetic pathway which leads to the ionones, reveals that this is not a position which is likely to be methylated in any carotenoid or degradation product thereof. In fact, the irones are derived from oxidative degradation of a triterpenoid called iripallidal (8.86) which has a somewhat unusual and clearly rearranged skeleton. It also contains 31 carbon atoms rather than the 30, which one would expect from a triterpenoid, indicating further modification of the original triterpenoid backbone. The structure of iripallidal and a generic structure of the irones are shown in Figure 8.28. As in Figure 8.20, the dotted lines can be either single or double bonds and the wavy lines

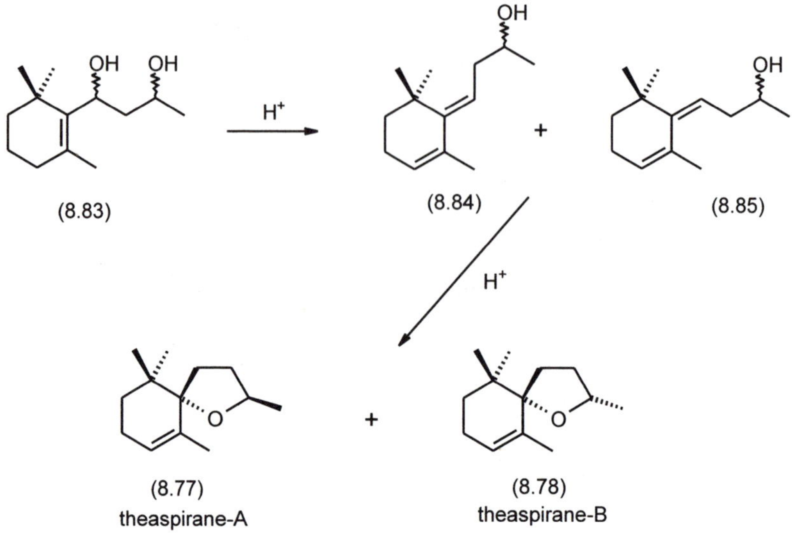
OH
OH
OH
H+
+
OH
(8.83)
(8.84)
(8.85)
H+
O
+
O
(8.77)
theaspirane-A
(8.78)
theaspirane-B

Figure 8.27

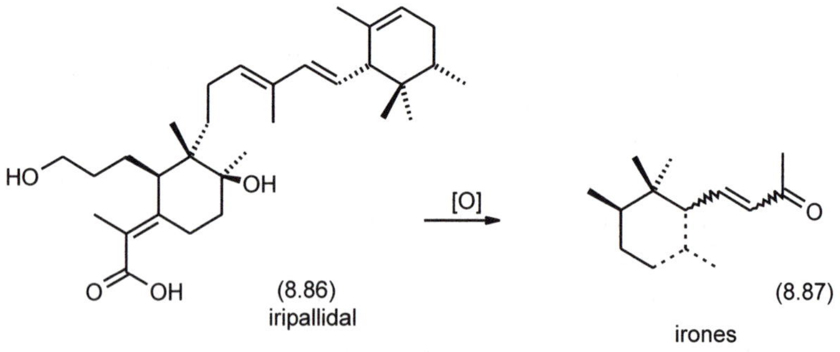
HO
OH
O
OH
(8.86)
iripallidal
[O]
O
irones
(8.87)

Figure 8.28

indicate that either geometrical isomer is included. The Greek prefixes α, β and γ have the same significance as they do for the ionones. In order to produce orris butter, as the extract from the rhizomes is known, the rhizomes are stored, allowed to age and subjected to aerial oxidation and the action of a species of grub which feed on them. It is during this ageing process that the degradation of iripallidal to the irones takes place.

The presence of the additional methyl group in the irone structure makes them more difficult synthetic targets than the ionones. Figure 8.29 shows Barton's [8.21] modification of Tiemann's ionone synthesis, in

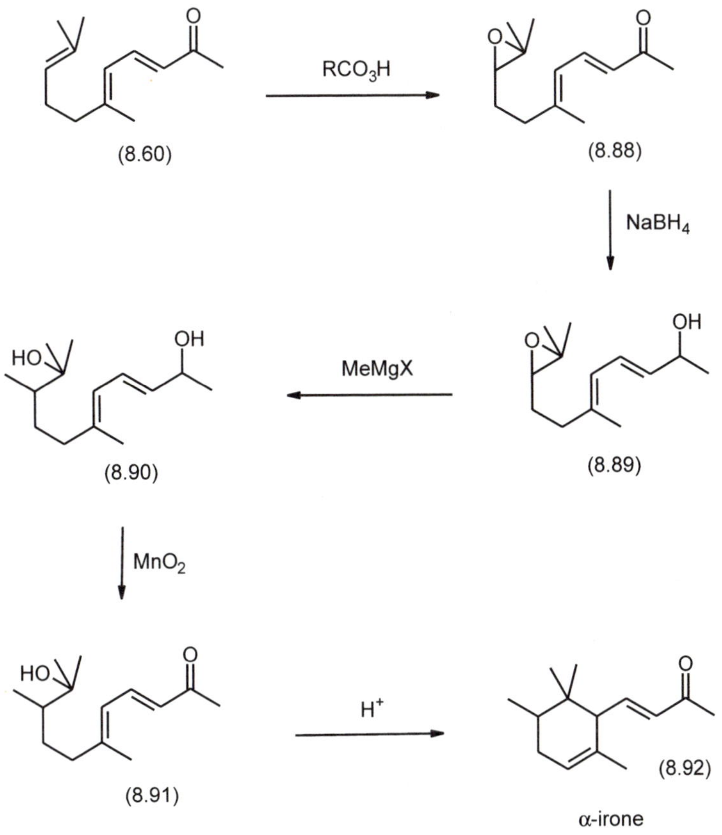

Figure 8.29

which he introduces the extra methyl group by means of addition of methyl Grignard reagent to epoxy-ψ-ionone and then uses the resultant tertiary alcohol function to serve as the source of the carbocation for cyclisation.

Epoxidation of ψ-ionone (8.60) can be achieved selectively at the most electron rich double bond, *i.e.* the isolated one at the end of the chain. It was necessary to protect the ketone group of the product (8.88) before carrying out the Grignard reaction and this was done by reducing it to the corresponding alcohol (8.89). The Grignard reaction then gave the diol (8.90) and the allylic alcohol was oxidised, using manganese dioxide, back to the ketone level of oxidation giving (8.91). Treatment of this with acid caused loss of the tertiary alcohol function and the resultant carbocation cyclised, as in Tiemann's ionone synthesis (see Figure 8.22), to give α-irone (8.92) as a mixture of isomers.

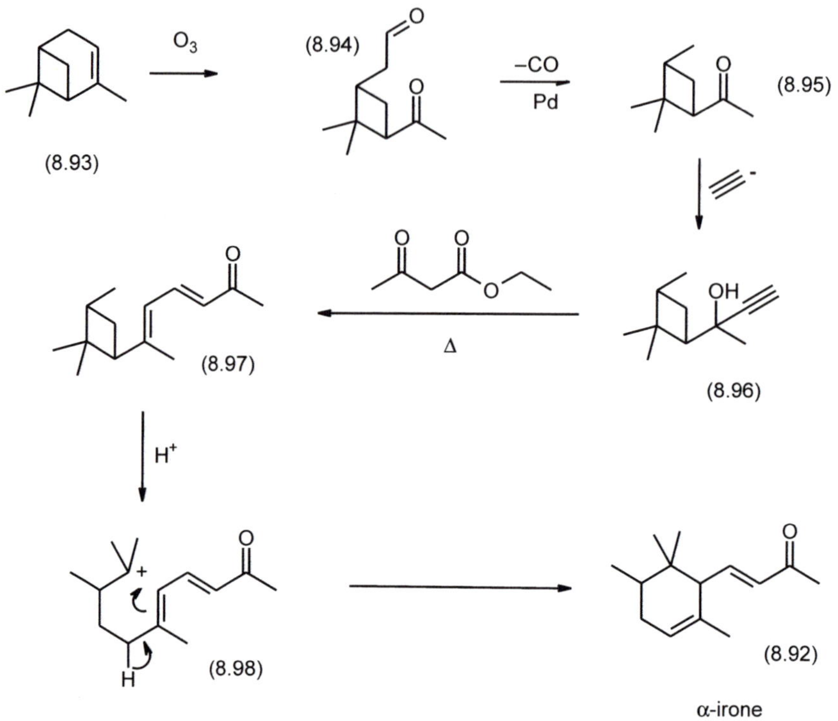

Figure 8.30

A less obvious way of introducing the extra methyl group is shown in Figure 8.30. This is a synthesis by Eschinazi.[8.22] He started with α-pinene (8.93) and ozonolysed it to give the keto-aldehyde (8.94). He then decarbonylated this over a palladium catalyst. Treatment of the resultant ketone (8.95) with acetylene in the presence of a base gave the acetylenic alcohol (8.96). This was then subjected to a Carroll reaction by heating it with ethyl acetoacetate to give the α,β,γ,δ-unsaturated ketone (8.97). The Carroll reaction and its mechanism will be discussed in more detail in Chapter 9. Treatment of (8.97) with a Bronsted acid led to opening of the cyclobutane ring to give the cation (8.98) which is of course, identical to that produced in Barton's synthesis by protonation of the alcohol (8.91) with subsequent loss of water. As in Barton's synthesis, the carbocation (8.98) cyclised to give a mixture of isomers of α-irone.

In order to illustrate the significant increase in difficulties involved in the synthesis of single enantiomers rather than mixtures of stereo-isomers, we will now look at a more recent irone synthesis. This synthesis was carried out by Brenna and co-workers[8.23] and also illustrates the use of an enzyme to achieve a stereospecific synthesis. This is a tactic which

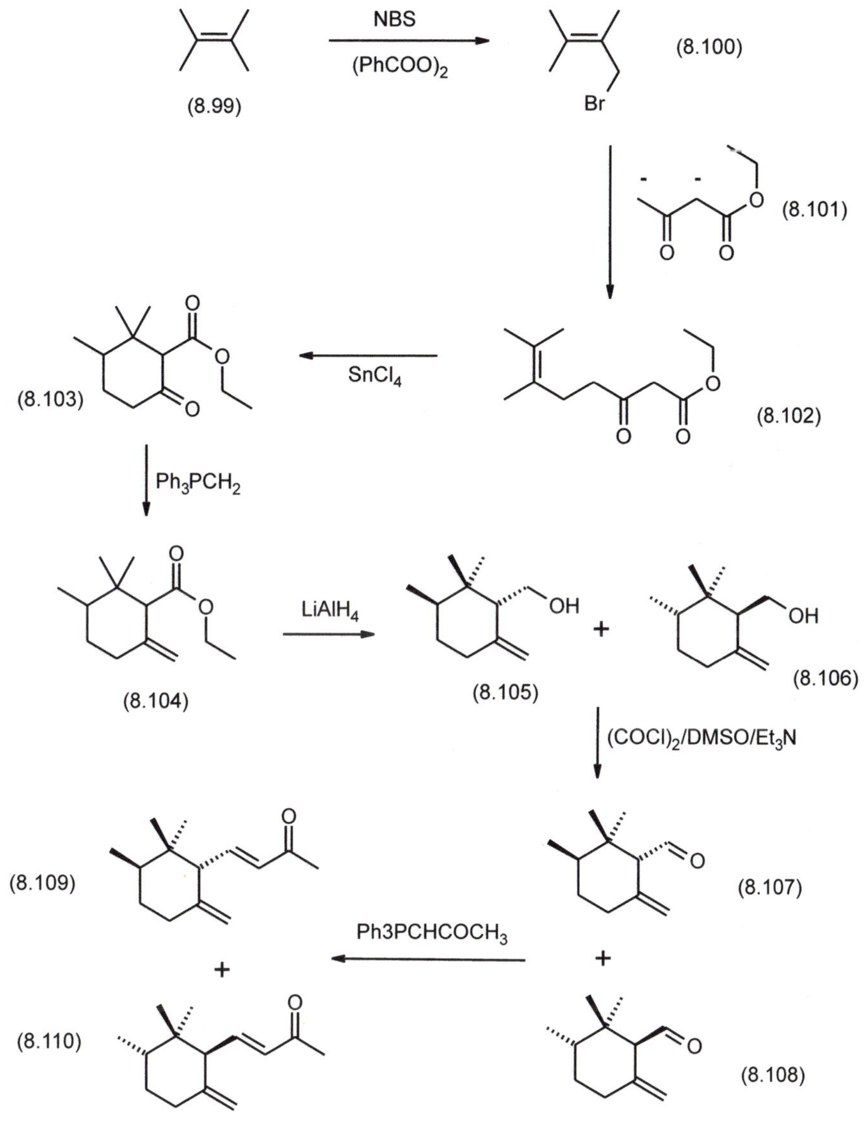

Figure 8.31

has become increasingly popular in recent years. The first part of the synthesis involves the preparation of racemic *trans*-γ-irone (8.109) + (8.110) and is shown in Figure 8.31.

The synthesis started with the bromination of tetramethylethane (8.99) using *N*-bromosuccinimide (NBS) and dibenzoyl peroxide. This gave the monobromide (8.100) which was used to alkylate the dianion (8.101) of ethyl acetoacetate. This dianion can be generated by first treating ethyl acetoacetate with sodium hydride to remove the more acidic proton between the two carbonyl groups and then subsequently with butyl lithium to remove one of the methyl protons. The more reactive terminal anion reacts faster and so monoalkylation, as shown, is achievable. Acid catalysed cyclisation of the product (8.102), using stannic chloride, led to the β-keto-ester (8.103). The Wittig reaction was then used to convert this to the ester (8.104) which was reduced with LAH to the *trans*-alcohols (8.105) and (8.106). The ester (8.104) was a mixture of both *cis*- and *trans*-isomers, the basic nature of the LAH reduction served to equilibrate the *cis*-isomers to the thermodynamically more stable *trans*-forms during the reaction. Oxidation of this mixture with oxalyl chloride in the presence of DMSO and triethylamine gave the corresponding racemic aldehydes (8.107) and (8.108) and these were converted to racemic *trans*-γ-irone (8.109) and (8.110) by reaction with 2-oxopropylidenetriphenylphosphorane. It was then necessary to resolve this mixture in order to obtain the pure enantiomers. This was achieved as shown in Figure 8.32.

The racemic mixture of *trans*-γ-irone (8.109) and (8.110) was reduced using sodium borohydride. This introduced another chiral centre into the molecule (the carbon atom of the newly formed alcohol function) and thus doubled the number of stereoisomers to four. These are shown as structures (8.111) – (8.114) in Figure 8.32. The reduction was used to give a "handle" on the molecule to enable an enzyme to distinguish between enantiomers later in the process. In this case, unfortunately, the further introduction of a chiral centre led to increased loss of material. There was also the need for a second additional process step since re-oxidation to the ketone level was necessary. Treatment of the mixture of four diastereomeric alcohols (8.111) – (8.114) with *p*-nitrobenzoyl chloride in pyridine converted it to a mixture of the four corresponding stereoisomeric esters (8.115) – (8.118). In Figure 8.32, the symbol Ar refers to the *p*-nitrobenzoate radical. When this mixture was crystallised from hexane, the less soluble pair of diastereomers, (8.116) and (8.117), crystallised first and could be removed by filtration. After three successive recrystallisations, the mother liquor contained only the other diastereomeric pair, (8.115) and (8.118). Hydrolysis of this pair with potassium hydroxide in methanol gave the corresponding racemic alcohols (8.111) and (8.114). This racemate was then resolved by treating it with vinyl acetate in the presence of a lipase, Lipase PS. The active site of the lipase is chiral and only catalyses esterification of one of the enantiomeric alcohols. Thus, alcohol (8.114) was converted to acetate (8.119) whilst its antipode, (8.111) was left unchanged. Alcohol (8.111)

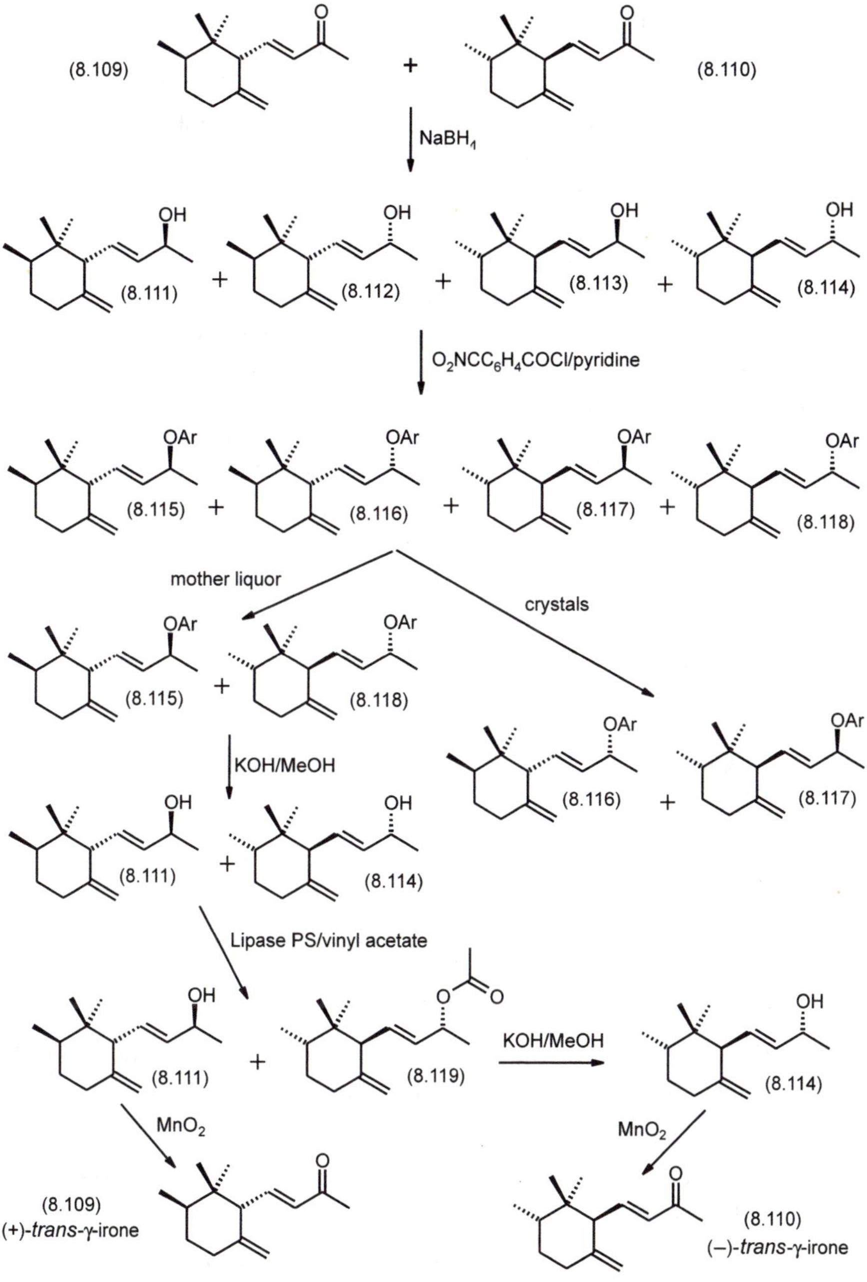
(8.109)
(8.110)
NaBH$_4$
(8.111)
(8.112)
(8.113)
(8.114)
$O_2NCC_6H_4COCl$/pyridine
(8.115)
(8.116)
(8.117)
(8.118)
mother liquor
crystals
KOH/MeOH
Lipase PS/vinyl acetate
(8.119)
MnO$_2$
(+)-*trans*-γ-irone
(−)-*trans*-γ-irone

Figure 8.32

and acetate (8.119) were easily separated by column chromatography. Oxidation of (8.111) with manganese dioxide gave pure (+)-*trans*-γ-irone (8.109). Pure (−)-*trans*-γ-irone (8.110) was obtained by hydrolysis of the acetate (8.119) and oxidation of the resultant alcohol (8.114).

Resolutions such as this are inevitably lengthy and require numerous stages with handling and purification issues at each one. This makes them expensive to carry out on a large scale. Thus, chiral synthesis or synthesis from a homochiral (only one enantiomer) precursor are always preferable, if they can be found. It is not surprising therefore that the Nobel Prize for chemistry in 2001 was awarded to William S. Knowles, Ryoji Noyori and K. Barry Sharpless for their pioneering contributions to research on chiral catalysts which can be used to introduce chirality into a synthesis starting from achiral precursors. We saw some of the fruits of Noyori's work in the chiral synthesis of menthol described in Chapter 4.

REFERENCES

8.1.   E. Lederer, *J. Chem. Soc.*, 1949, 2122.

8.2.   R. Clarke, *Nature*, 1954, **174**, 155.

8.3.   R. Cornon, *Ind. Perfum.*, 1955, **10**, 291, 351.

8.4.   E.W. Bovill, *Dragoco Rep.*, 1973, **1**, 3 and **2**, 23.

8.5.   G. Ohloff, *The Fragrance of Ambergris* in *Fragrance Chemistry* E.T. Theimer (ed.), Academic Press, New York, 1982.

8.6.   E.-J. Brunke, *Dragoco Report*, 1980, **1**, 9.

8.7.   C.S. Sell, *Chem. Ind.*, 1990, **16**, 520.

8.8.   B.D. Mookherjee and R.R. Patel, Proceedings of the Seventh International Conference of Essential Oils, Kyoto, 1977, paper 137, 479.

8.9.   E. Lederer, in *Fortschritte Der Chemie Organischer Naturstoffe*, Springer, Berlin, 1950, Vol. 6, pp. 120.

8.10.   G. Ohloff, K.-H. Schulte Elte and B.L. Muller, *Helv. Chim. Acta.*, 1977, **60**, 2763.

8.11.   K.-H. Schulte Elte, B.L. Muller and G. Ohloff, *Nouv. J. Chim.*, 1978, **2**, 247.

8.12.   C.A. Coulson, *Valence*, Oxford University Press, Oxford, 1961.

8.13. M.I. Farbood, B.J. Willis and P.J. Christenson, South African Patent 8504306, 1985.

8.14. R. Decorzant, C. Vial, F. Näf and G.M. Whitesides, *Tetrahedron*, 1987, **43** (8), 1871.

8.15. Herodatus, *The Histories*, translated by Aubrey de Selincourt, Penguin Classics, Harmondsworth, England, 1968.

8.16. M.G. Bolster, PhD. Thesis, University of Wageningen, 2002.

8.17. G. Büchi and H. Wuest, *Helv. Chim. Acta*, 1989, **72**, 996.

8.18. G. Büchi and H. Wuest, 1984, US Patent 4 613 710.

8.19. P.F. Vlad, N.D. Ungur and V.B. Perutskii, *Khim. Geterotslic. Soedin*, 1990, 896.

8.20. H. Kandori, *Chem. Ind.*, 1995, 735.

8.21. D.H.R. Barton and M. Mousseron-Canet, *J. Chem. Soc.*, 1960, 271.

8.22. H.A. Eschinazi, *J. Am. Chem. Soc.*, 1959, **81**, 2905.

8.23. E. Brenna, C. Fuganti, S. Ronzani and S. Serra, *Helv. Chim. Acta*, 2001, **84**, 3650.

CHAPTER 9

Commercial
Production
of Terpenoids

Taking chemical reactions up from laboratory scale to commercial production is an
intellectually challenging task

*Nowadays, the synthetic organic chemist can make any target molecule. The real
question is "At what cost?"*

Prof. George Buchi

In previous chapters, we have studied academic syntheses of terpenoids.
When designing a synthesis for large-scale commercial production,
factors such as safety, environment, feedstock availability and raw
material and process costs come into play. The development chemist
must find a satisfactory balance between them all. In principle, every
production hurdle can be overcome, but the solution will always involve
a cost. This cost, whether it relates to the cost of feedstocks or reagents,
waste treatment and disposal, engineering, logistics and even R&D, will
become part of the total production cost. (The cost of capital investment
and R&D must be recovered and this is usually done by adding on a
percentage of it to the selling price of the product.) The role of the
development chemist is, therefore, to find the lowest possible overall
production cost. This is a difficult task and requires skills in both
chemistry and economics together with knowledge of the fine chemicals
market, available process technology and legislation affecting chemical
production.

KEY POINTS

Seven crucial parameters in a commercial-scale chemical process are
safety, environment, purity, reproducibility, capacity, cost and
sustainability.
Different routes to the same product will have different balances of
positive and negative points in each of the seven parameters.

Statistical methods of experimental design are employed to determine the best overall balance of reaction parameters.

Factors other than technical ones may determine the choice of route (*e.g.*, safety, environment, feedstock availability).

External factors (such as changes in legislation, new technology and changes in feedstock supply) may force change in process routes.

Different routes may operate simultaneously with each producer taking advantage of opportunities open to him.

Constraints on commercial-scale chemistry can lead to excellence and elegance, as demonstrated by the BASF synthesis of citral.

Companies often develop "technology trees" of products.

No route is perfect and improvement is always possible.

9.1 DIFFERENCE BETWEEN ACADEMIC AND COMMERCIAL SYNTHESES

When carrying out a synthesis for the purpose of confirmation of a proposed structure, the prime consideration is that each step of the scheme must proceed in an unambiguous way so that there will be no doubt about the molecular structure of the final product. The most important parameters when carrying out preparation of a sample for applications research purposes (usually for screening of biological, chemical or physical properties) on the other hand are usually speed and cost. The faster the sample is prepared, the sooner the results of its evaluation will be known. In this context, cost will be the cost of producing the requisite amount of material and, especially as it will be the work of a skilled, relatively highly paid bench chemist, the principal contributor to the total cost will be the chemist's time. Therefore, expensive starting materials, reagents and conditions will be used if their use will result in a faster synthesis. Design of a synthesis route for commercial production of a chemical is a very different matter. There are many factors to be taken into consideration and an optimum balance will have to be sought since in many, if not most, cases, the constraints imposed on the synthesis will be pulling in different directions.

9.2 CONSTRAINTS ON COMMERCIAL PROCESSES

The reason for commercial production of chemicals is to provide a benefit to the consumer. That benefit could be one of a drug to cure illness, a fragrance to enhance the pleasure and benefit of using a bar of soap, an improved construction material and so on. In every case, there will be a limit to the amount that the consumer is prepared to pay for this benefit and also a limit to what society is prepared to tolerate in terms of side effects of the production and use of the product. Seven crucial parameters in a commercial-scale chemical process are safety, environment, purity, reproducibility, capacity, cost and sustainability.

Each of these is described briefly in the following seven sections.

9.2.1 Safety

By far the most important consideration is that of safety. No company would wish to endanger either its employees or the public and any responsible company will invest considerable time, effort and money into ensuring that it does not do so.

Chemical process safety relates largely to handling of hazardous materials and to exothermic reactions. Handling of hazardous materials, which may be starting materials, reagents, catalysts, intermediates or products present a danger to the process operators and anyone else within range of the plant in the case of a discharge. Exothermic reactions present a danger in that if they are not controlled adequately and the heat produced removed efficiently, they can lead to escape of material from the reactor possibly with explosive violence. In the event of an explosion, fragments of the reactor body and associated plant, are likely to become dangerous missiles.

Good practice dictates that a detailed analysis of the thermodynamics of the process will be carried out before scale up. The ratio between surface area and volume decreases as the size of an object increases. Thus, the larger the reactor, the smaller is its surface area relative to its volume and so the less able it is to dissipate heat. An exotherm which is too small to be noticeable on a laboratory scale, might be sufficient to present a serious hazard on manufacturing scale. The effects can also be magnified if, as is usually the case, the heat produced leads to an increase in the reaction rate. In turn, this increase in the reaction rate leads to an increase in the rate of release of heat which, in turn, leads to a further increase in rate and so on. Such a situation is known as a thermal runaway.

The materials of construction of a chemical plant must be resistant to their contents over a long period of time and testing of samples of the construction materials in the reaction media will be carried out to ensure that corrosion does not lead to escape of reaction materials or failure of plant components. This work is usually carried out in collaboration between development chemists and chemical engineers.

Before a process can be introduced into a factory, it is usual to carry out what is referred to as a hazard and operability study, or HAZOP.

In this study, all possible issues of safety will be considered and checked to see that safety precautions are adequate. This will include consideration of the possible consequences of accidental omission of one or more reagents or catalysts, accidental addition of excess of any reagent, wrong order of addition, mixing of any combination of the reagents outside the reactor (*e.g.* leaks on storage), incorrect profile of reaction temperature and so on.

In the United Kingdom, inspectors from the Health and Safety Executive (HSE) have the power to close down any operation which they deem unsafe and to fine and imprison companies and individuals who carry out, authorise the operation of or allow unsafe practices to be carried out. Corresponding government bodies have similar powers in almost all other countries in the world. Thus, the law provides strong motivation for ensuring the safety of chemical processes, in addition to the ethical and financial (loss avoidance) reasons for doing the same.

9.2.2   Environment

Any of the starting materials, reagents, catalysts, products and by-products of a chemical process could possibly find their way into the environment. The chemicals company must ensure that precautions are taken to ensure that there are no accidental releases into the environment. Nonetheless, it will also have contingency plans for dealing with any escape of material if such precautions should fail. All excess materials, spent catalysts, *etc.,* and other effluent must either be recycled or disposed of in an acceptable manner. Obviously, for environmental reasons, hazardous feedstocks, reagents and catalysts are best avoided. Every raw material must be transported to the production site, and they and all intermediates produced must be stored and handled safely there. Hazardous materials, therefore, present issues in terms of the capital cost of the transport systems and plants which will be necessary to enable their safe use. Everything can be handled safely but the more intrinsically dangerous it is, the higher will be the cost of safety precautions.

As far as disposal of materials is concerned, the environmental effects of materials are handled in a similar way to product safety. The environmental fate of chemicals must satisfy certain criteria for their use to be desirable or even legal. For organic materials, one of the most important parameters is the fate of a material in a typical sewage treatment plant. The ease with which sewage bacteria will convert the material to carbon dioxide and water can be measured in laboratory equipment which replicates the first stage of sewage treatment and this has now become a standard test for materials undergoing development. Inorganic waste is usually disposed of either to water treatment plants or to landfill. There are strict rules concerning what may be disposed of in such ways and how much any one company can dispose of in a given time. It is much better, therefore, if inorganic waste can be recycled or sold for a different use. For example, aluminium waste from Friedel – Crafts reactions can be passed on to a company which will be able to convert it into materials for water treatment. In many countries, such as the Netherlands, the production site will have a total limit on inorganic salts which can be removed from it. This means that the use of mineral acids or bases as catalysts is undesirable. For example, if a process uses sulfuric acid as a catalyst and it is neutralised after the reaction by addition of sodium hydroxide, then sodium sulfate becomes an inorganic effluent. Although sodium sulfate is environmentally benign, the volume limit becomes an issue. Therefore, it is better to use an acid catalyst, such as an ion exchange resin or a zeolite, which can be filtered off and reused.

The government body with responsibility for prevention of environmental pollution in the UK is The Environment Agency (EA). The EA has considerable powers, similar to those of the inspectors from the HSE; powers which far exceed those of the police, for example, in their ability to enter and search premises without a search warrant. As with safety, there are many very good reasons for a chemicals company to put every effort into avoiding polluting the environment.

Much work has been done in recent years around the environmental impact of chemical processes. The question is, "How do we measure the environmental impact of a process?" Almost simultaneously, in the early 1990s, Professors Barry Trost and Roger Sheldon came up with proposals for doing this.

Trost[9.1] proposed a factor called "atom utilisation" or "atom efficiency". Atom utilisation is defined as the molecular weight of the desired product of a reaction divided by the total of the molecular weights of all species produced in the reaction, multiplied by 100 to give a percentage figure. This concentrates on the stoichiometry of the reaction. Thus, a reaction of the type A + B = C is likely to be better than one of the type A + B = C + D. For example, a Diels – Alder reaction will have a higher atom utilisation than hydrolysis and decarboxylation of a β-keto ester, as illustrated in Figure 9.1. Here two reactions are shown, both of which produce intermediates of potential use in terpenoid synthesis. In the first, isoprene (9.1) adds to methyl vinyl ketone (9.2) in a Diels – Alder reaction to give 4-acetyl-1-methylcyclohex-1-ene (9.3) with 100% atom utilisation. On the other hand, the hydrolysis and

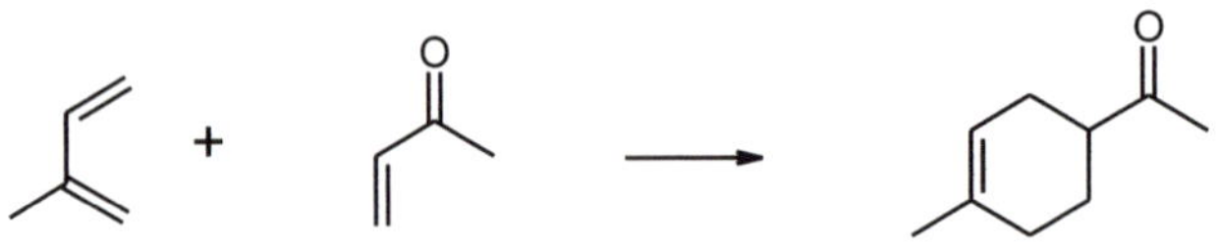

(9.1)   (9.2)   (9.3)

Atom utilisation = 138 / 138 x 100 = 100%

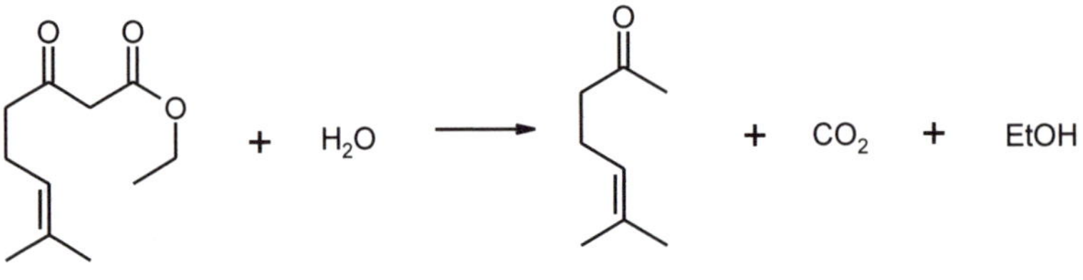

(9.4)   (9.5)

Atom utilisation = 126 / (126 + 44 + 46) x 100 = 58%

Figure 9.1

decarboxylation of ethyl-7-methyloct-6-en-3-onoate (9.4) to give 2-methylhept-2-ene-6-one (9.5) proceeds with only 58% atom efficiency.

Sheldon's E-factor, $E^{9.2,9.3}$ considers not only the starting materials and products of a reaction, but also all of the materials consumed by the process. It is defined as the ratio of desired product to the sum of all by-products as shown in Equation (9.1). An alternative way of expressing this is shown in Equation (9.2). Thus, for example, solvent and catalyst handling losses are included. Energy can also be included by adding the weight of fuel oil burnt to provide it. The energy consumption is often very difficult to measure in practice because a steam generator, for example, will be used to heat a number of reactors and also other buildings, such as laboratories and offices, on the site. Therefore, determining exactly how much steam is consumed by any individual reactor is not straightforward. Multiplication of the E-factor by an "unfriendliness quotient", Q, (*e.g.* 1 for an environmentally benign salt such as sodium sulfate and 1000 for a highly toxic heavy metal) we obtain an Environmental Quotient (EQ), as shown in Equation (9.3). The lower E and EQ are, the more environmentally friendly is the process.

$$E = \text{kg waste/kg product} \qquad (9.1)$$

$$E = \text{(total kg material used and not recycled}$$
$$- \text{kg product)/kg product} \qquad (9.2)$$

$$EQ = E \times Q \qquad (9.3)$$

9.2.3 Purity

Product purity may seem an obvious factor in development of a chemical process. However, there are a few subtleties which need to be considered.

There will always be a number attached to statements of purity. For example, there may be a claim that a chemical is 99% pure or contains less than 1% of other components. The other components may have specified upper limits. For example, a product may be labelled as α-pinene containing no more than 0.1% β-pinene. The ultimate purity limit will always be determined by the limit on the ability to analyse the product. If, for the sake of argument, β-pinene could not be determined at concentrations below 1 part per billion (ppb) in α-pinene, then the purest grade of α-pinene could not claim any better than "less than 1 ppb β-pinene".

In many cases the nature of individual impurities may be more important than the total level of all of them. For instance, in a pharmaceutical product, the presence of 1 part per million (ppm) of a potent neurotoxin is likely to be more of an issue than 1% (1000 ppm) of sodium citrate. In the fragrance industry, where the materials used are all of low toxicity, the main issue in purity terms is the presence of ingredients with very low odour thresholds since traces of these can seriously affect the odour of the sample. For example, 1% of citronellol in a sample of geraniol will have less effect than would 1 ppm of thioterpineol as the latter would be more than enough to add a very noticeable grapefruit note to the otherwise rosy odour. In any fine chemicals industry, a sound understanding of all aspects of the product is very beneficial when considering specifications on product purity.

Another important factor to be considered is the effects of starting material purity on product purity. Minor components in starting materials can make their way through processes, with or without being modified structurally by the process chemistry, to appear as minor components of the product. It is best to have several suppliers of each starting material (for the sake of security of supply) but each must be checked by putting it through the reaction process to determine what, if any, effects on product quality will arise from the changes of feedstock supply.

One parameter which has become increasingly important over the last few decades is that of enantiomeric purity. This issue came into

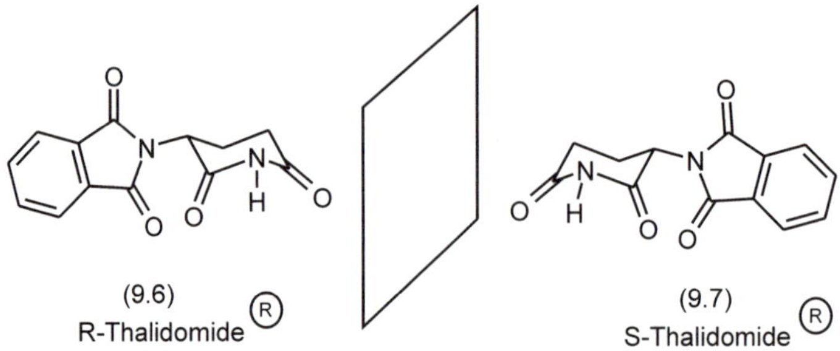

(9.6) R-Thalidomide® sedative

(9.7) S-Thalidomide® teratogen

Figure 9.2

the spotlight when the drug Thalidomide® was offered for sale as a sedative. (See Figure 9.2 for the structures which, incidentally, are not terpenoid in origin.) The sedative properties are due solely to the *R*-isomer (9.6) and it was not until after Thalidomide® had been in use and its devastating side effects seen in real situations, it was discovered that the *S*-isomer (9.7) is a teratogen. Since this most unfortunate incident, the pharmaceutical industry has invested much research into developing enantioselective syntheses. The question of enantiomeric purity is less important for the fragrance industry, partly because of the lower toxicity of fragrance ingredients and partly because chirality does not always make a significant difference to the odour of a material, as will be seen in Chapter 10. Furthermore, many fragrance ingredients are manufactured from natural starting materials, such as limonene or pinene, which are enantiopure and so the products retain optical purity. With the much lower selling prices and profit margins of the fragrance industry, the economic balance between any increased performance of a homochiral product and the increased cost of producing it, relative to that of a racemate, is more often in favour of the racemate than is the case in the pharmaceutical industry. However, even in the fragrance industry, there is a growing interest in homochiral products.

9.2.4 Reproducibility

As with purity, the requirement for reproducibility may seem obvious but on looking in more depth, one will find many instances where it is, or has been, an issue. In principle, if a production process is run in the same way every time, the outcome should always be the same. Variations arise in two ways.

Firstly, variations in feedstock will give rise to variations in product. This is a particular issue when the feedstock is a natural product, such as turpentine. Variations in the weather, pests and in genetics of any plant source are some of the factors which can produce fluctuations in composition of oils produced from them. A good Purchasing Department will identify multiple suppliers of feedstocks whenever possible since reliance on a single supplier of either natural or synthetic raw materials, makes a company very vulnerable. One recent example was of a Japanese factory which was virtually the sole world supplier of a certain feedstock and whose factory was put out of commission by an earthquake. Companies which were dependant on this feedstock were subsequently unable to manufacture their products until repairs to the damaged plant in Japan were complete.

The other potential cause of variability of processes lies in their sensitivity to changes in conditions. It is obviously much better to run a process which can tolerate reasonable deviations in operating parameters since one which alters course when tiny changes take place will be susceptible to factors such as accuracy of weights/volumes, temperature or pressure control and reaction times. During development, it is therefore usual to determine the robustness of a process by running it many times to ensure that a consistent output is achieved. If a process is not robust, it would be very unwise to introduce it into a factory because of the risk of loss of material or much more seriously, the risk to the safety of the operators and others in the vicinity.

9.2.5 Capacity

When a process is introduced, the Marketing Department would have prepared an estimate of sales volume and the development chemists and chemical engineers would have planned to run the process in a vessel of suitable capacity. However, by definition, such estimates are only guesses and so the forward thinking development team will have, at least at the backs of their minds, a contingency plan for the eventuality that much more material is required than was anticipated. To take the fragrance industry for example, supposing that a new product has been developed and a marketing plan drawn up. This plan will be based on the growth of comparable materials in the past and will reflect the likely level of incorporation of the ingredient in a fragrance and the likelihood of fragrances containing it winning submissions against customers' briefs. So, for example, it might be estimated that 1 ton of the ingredient will be needed in its first year. However, it is always possible that an enterprising perfumer uses it at the level of an "overdose" in a fragrance and successfully competes for a very large volume briefly using this formula. The factory then might suddenly be confronted by an order of 10 tonnes of the ingredient. Failure to produce it will result in serious damage to the company's relationship with that customer. So, the development team and the factory staff must be prepared for such eventualities and know how to increase production capacity rapidly if required. Similarly, the development team and the Purchasing Department must be capable of finding increased supplies of feedstocks at short notice.

9.2.6 Cost

The cost of a process is the sum of six components. These are the raw materials cost, the overall yield of the process, the cost of waste disposal, the cost of the energy used to carry out the process, handling costs (including labour) and the capital cost of the necessary process plant.

Raw materials costs can be the subject of negotiation with suppliers but the prices asked by suppliers will depend on the intrinsic costs of the materials and these can be estimated before any development work commences. Obviously, before carrying out any practical work, it is worthwhile calculating the raw materials contribution to a product, using an estimate of 100% yield at every step. If the result is above the target cost of the product, then the route will never be a success. Reagent cost is also important. For example, if a base is required for a reaction, then the cost will be lower if sodium hydroxide can be used than it would if it was necessary to use *t*-butyl lithium. The more suppliers there are for a material, the stiffer will be the competition and consequently, the lower the price. Having a large number of suppliers also increases the security of supply. The general rule is therefore to use starting materials and reagents which are inexpensive and can be bought from a number of suppliers. Catalyst cost must be included in this section. Clearly, a catalyst is preferable to a stoichiometric reagent as the catalyst, in principle, can be recovered and re-used. However, handling losses and catalyst poisoning can result in the loss of active catalyst which can be recovered at the end of a reaction. Allowance must therefore be made in the costing for the loss of a catalyst. For example, if only 90% of the catalyst can be recovered in active form at the end of the reaction, then 10% of the cost of the catalyst charged to the reactor must be added to the total materials cost. Turnover number (TON) is the number of molecules of substrate converted by one molecule of catalyst before the catalyst loses its activity. TON is thus a measure of catalyst efficiency. A catalyst might be very expensive in terms of cost per kg from the supplier, but if it has a very high TON, then the contribution of the catalyst cost to the final price of the product will be small. A good example of this is the rhodium BINAP used to produce menthol as described in Chapter 4. Rhodium BINAP is an extremely expensive material but its TON is so high that its contribution to the total cost of the menthol produced is small and leads to a very competitive process.

In Chapter 7, when we looked at Stork's β-vetivone syntheses in comparison with that of Marshall and Johnson, we saw the effect on overall yield of multi-stage syntheses. Clearly, the fewer the total number of stages the better. The yield on each stage is made up of two components, conversion and selectivity. The yield of a reaction is the actual obtained weight of the desired product divided by the maximum theoretically possible weight of the desired product. The conversion is the weight of the starting material which was consumed in the reaction divided by the initial weight of the starting material. The selectivity of a reaction is the weight of the desired product divided by the total weight of all the products obtained. Thus the yield of a reaction is the selectivity times the conversion. The selectivity is the more important of the two. Low conversion is more acceptable than low selectivity since in the former case, the unreacted starting material can be recovered and recycled whereas, in the latter, the unwanted products are waste which must be disposed of. Chemists normally express selectivity, yield and conversion in terms of molar ratios. In this context, I have used weight instead since, for example in the case of inorganic salts generated, the limits imposed by the authorities on a production site will be in weight terms.

The formation of by-products increases the cost of the desired product by using up raw materials which otherwise could be converted to product. However, that loss is only part of the picture as the by-product represents waste which must be disposed of. The image of factories pumping such waste into local waterways or landfill sites is nowadays consigned to the history books. Waste must be disposed of in an acceptable manner and this inevitably involves cost. Similarly, the use of stoichiometric reagents generates waste and consequent disposal costs. For example, the use of sodium borohydride for a reduction will result in the formation of sodium borate on work-up. A much better alternative would be to use catalytic hydrogenation for the reduction of a ketone.

The cost of the energy used to carry out a process, must be considered in order to present a total picture. For example, there might be a choice between two reagents for a reaction and the less expensive of the two is less reactive and therefore requires a much higher temperature to effect conversion. All else being equal, there would be no point in avoiding the use of the more expensive reagent if the savings on using the cheaper alternative were less than the increase in the cost of heating the reaction to the higher temperature required by the cheaper reagent.

Every manual operation involved in a process will add labour costs to the total bill. Charging or discharging reactors, monitoring reaction progress, monitoring batch distillations, including solvent removal, washing regimes, and transferring the product to drums are some examples of such manual handling. Chemical plant operators are skilled people with a responsible job and so these costs will be significant. What a laboratory based chemist would regard as a one-step procedure, a chemical engineer will see as a number of unit operations, such as charging the reactor, removing the solvent, washing the product with water and so on. Because of the value of its products, the pharmaceutical industry might be able to stand multi-step processes with numerous unit operations per step, but most fine chemicals operations, such as fragrance ingredient synthesis, cannot. The cost of the average fragrance ingredient is such that process costs limit the majority of syntheses to three steps at most.

One solution to this issue is to use automation wherever possible and to use continuous (more properly referred to as steady-state) reactions rather than batch ones. Doing so, of course, is likely to increase the capital cost of the plant and reduce the versatility of the process and so the optimum balance must be found. The use of hazardous materials or high pressures will also increase the cost of the plant. In the former case, the materials of construction and the safety precautions will add to the cost. In the latter, the vessels must be stronger, hence usually thicker-walled and with additional safety features, again resulting in higher costs. The capital cost is usually depreciated over a set time, determined by the average lifetime of the type of plant in question. For example, if a plant costs £1,000,000 and is expected to last for 10 years, the annual depreciation would be £100,000. If the output is 10 ton per annum, then the capital depreciation will add £10 per kg to the cost of the product.

9.2.7 Sustainability

There are many definitions of sustainability. The most widely accepted is the one which states that "Sustainability constitutes meeting the needs of the present generation without reducing the ability of future generations to meet their needs." Sustainability is therefore a very big subject. All of the above factors are included since all of them must be right in order to have a sustainable process.

In addition to minimising waste, as part of ensuring a sustainable operation, the industrial chemist must think about their source of feedstocks. By using particular feedstocks, will they be diminishing the ability of future generations to provide for their needs? Petrochemicals are the obvious example. The world's reserves of crude oil are finite and their use does potentially compromise future generations. The percentage of crude oil used for chemical synthesis is very small compared to that used for energy production but, nonetheless, synthetic chemists are beginning to search for renewable feedstocks to replace oil. Non-food uses of crops and fermentation methods are two of the main avenues of research. The production of ethanol by fermentation of biomass is one obvious possibility though at present it does not compete either economically or environmentally with ethanol production from petroleum. The fragrance industry is fortunate in that there are natural, renewable feedstocks which can be used as an alternative to petrochemicals for many of its ingredients, as we shall see later.

The use of halogenated solvents is considered to be environmentally unacceptable nowadays. However, when seeking alternatives, the development chemist must keep the total picture in mind. For example, it might be shown to be possible to replace dichloromethane by liquid carbon dioxide in a given process. However, energy is required to provide the necessary pressure to liquefy carbon dioxide. This energy will probably come from burning fossil fuel and so the environmental damage from that might be greater than that caused by small losses of dichloromethane. Some calculations might give an indication of where the balance lies and, in any case, the chemist will probably look for other solvents which are free of both drawbacks. For instance, there is growing interest in the use of water and also ionic liquids as reaction solvents and, even better, in solvent-free systems.

At present, there are no really good, comprehensive methods for measuring sustainability and consequently, efforts are being made to find sound and practical ways of doing so. In the meantime, the development chemist uses his experience, instinct, creativity and lateral thinking in order to seek continual improvement in sustainability.

9.3 EXPERIMENTAL DESIGN

Finding the optimum conditions for a chemical reaction is far from being a trivial task. In every case, there are numerous parameters which will affect the outcome. Moreover, these factors may pull in different directions. Different factors may also interact with each other in a non-linear manner. Our understanding of chemistry is usually not exact enough to allow us to predict precisely the optimum position and so we must resort to statistics.

Simplex design, and the associated simplex procedure, is one of the many statistical methods for optimisation of reaction conditions. The concept is, as the name suggests, very straightforward and will be explained here in more detail since it is the easiest of the statistical techniques to explain and illustrate in a non-specialist text.

Consider a reaction in which only the reaction time and the reaction temperature affect the yield. We can produce a map of this reaction as shown in Figure 9.3 with the axes representing the time and the temperature. For every combination of time and temperature we can then measure a yield and plot something akin to a contour map. On a geographical map the axes represent distance North and distance East and the contours indicate height above sea level. In our reaction map, the axes represent temperature and time and the "contours" represent yield. The three-dimensional shape described by this map is known as the response surface. At first, of course, we do not know the shape of the response surface and have to guess where the highest point might be.

To start the simplex, we set up a number of experiments. This number is equal to the number of variables plus 1. In our example, we have two variables and so we select three experiments to try. These are represented by A (7 min at 28 °C), B (14 min at 36 °C) and C (10 min at 45 °C). We now run these experiments and measure the yield of each. The results of 65, 73 and 75%, respectively, tell us that we have probably not made a

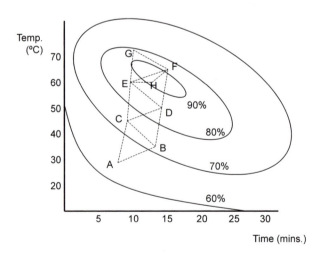

Figure 9.3

very good guess about the optimum conditions. What happens now in the simplex design is that we discard the lowest of the three results, A in this case. We then perform a simple mathematical calculation which reflects A through the plane between B and C, creating a new triangle BCD which is a reflection of the original triangle, ABC. The mathematics, therefore, tell us to carry out experiment D, that is a reaction time of 15 min at a temperature of 50 °C. The yield turns out to be 86%. In our current triangle therefore, the lowest point is now B and so we discard this and calculate, as before, our new point which is E (10.5 min at 60 °C). Triangle CDE is a reflection of BCD. The yield has now improved to 87% and the lowest point of our current triangle (CDE) is C. So, C is now rejected and a new triangle, DEF, calculated. The new result, F at 15.5 min and 66 °C, is 87%, which is the same as result E. This tells us that something new is happening, we are likely to be near the optimum and the slope is flattening out. The next point, G is at 11 min and 72 °C and the yield has fallen back to 81%. This tells us that the new triangle has flipped over a peak or a ridge and if we continue with the same mathematical process as before, we will simply flip back again over that high point. The technique now is to perform a different calculation which still reflects the triangle EFG but halves its size at the same time. This gives us point H with co-ordinates of 13 min and 61 °C and a resultant increase in yield to 93%. This procedure now continues until the working triangle shrinks in on the summit of the response surface.

For the purpose of illustration, the example given in Figure 9.3 is very simple, with only two variables and consequently a three-dimensional response surface. In a real situation there are likely to be many more variables. For example, consider a simple reaction in which the temperature is constant throughout each individual experiment and one reagent and the solvent are added to the flask first; then a second reagent added over time and stirring continued for a time after the addition is complete and before quenching the reaction. This reaction has five variables, namely, temperature, ratio of solvent to reagent 1, ratio of reagent 2 to reagent 1, time of addition and time of post-addition stirring. This gives a six-dimensional response surface which is impossible for us to visualise and so we need to rely even more on mathematics to predict our next experiment.

There are some things which the chemist must keep in mind when following the simplex procedure. For example, the procedure is very good at hill climbing but it will always climb the nearest hill. If there is more than one summit on the response surface and the first triangle is placed on the side of a hill which is not the highest, then it will climb the hill it is on and stop at that summit and will never see that a higher one exists.

Such a point is referred to as a local maximum, while the highest point possible is called the global maximum. Similarly, any surprisingly good result should be checked by repeating the experiment to ensure that the result was genuine and not due to experimental error since a false positive will anchor the simplex in that region because all surrounding results will be lower.

The simplex procedure is probably the best one for a chemist doing one experiment at a time and when each experiment is complex in terms of time, mode of operation, *etc.* If the experiments are easy to carry out (*e.g.* add two reagents and leave to stand at a given temperature for a given time) then it is probably better to do many reactions simultaneously, for example in a reactor block with numerous reaction wells, and explore a random distribution of conditions across the entire range of conditions using a factorial design. Mathematical analysis (nowadays using suitable software) of the results will then predict the optimum conditions. If the chemist has access to a synthesis robot, then even relatively complex experimental procedures can be easily carried out in a factorial design. Factorial designs usually give more information about the reaction and are less likely than a simplex to give a local maximum.

Statistical techniques can be employed to processes on production scale as well to those in the laboratory. However, on production scale, the changes to the reaction parameters will always be made in very small increments so as not to significantly jeopardise the yield and so each experiment will be repeated many times to ensure consistency before any conclusions are drawn. It is also essential to ensure that safety is never compromised as there are known cases where yield optimisation has transformed a safe process into a lethally dangerous one.

In a real-time situation there will be further complications in that the factor being optimised will probably be cost per kg of product rather than yield. Often environmental constraints and other such factors will also be included and one simple technique for doing so might be to assign a cash value to these and then optimise on cost per kg of product, as before.

Further details of the simplex design can be found in the paper by Hendrix.[9.4] Readers who wish to know more about statistical design of experiments in general could refer to the book by Box, Hunter and Hunter.[9.5] Nowadays, there are many commercially available software programmes for experimental designs. Most development chemists will choose to use one of these because of their convenience. Each software package will include a tutorial on the principles of the technique which it employs. There is also general guidance available on the Internet in the form of an electronic textbook.[9.6]

9.4 FRAGRANCE INGREDIENTS DERIVED FROM TERPENOIDS

In order to illustrate the above factors in operation, we will look at the history of the production of a group of monoterpenoids, related to citral, which are major ingredients of the fragrance industry. In order to understand the commercial background to the technical work, it will be beneficial first to look at some of the background to the fragrance ingredients business.

The terpenoids form the largest group of natural odorants and so it is only to be expected that they also form the largest group of modern fragrance ingredients. Thousands of different terpenoid structures, both natural and synthetic, will be found in perfumes from fine fragrance to household cleaners. Some of the more important ones are shown in Table 9.1.

The terpene hydrocarbons generally have weak odours and are used mainly as feedstocks. The higher molecular weight of the sesquiterpenoids results in their having lower vapour pressures than their monoterpenoid counterparts. Thus, sesquiterpenoids are present at a lower concentration in the air above a perfume than are monoterpenoids with the result that they must have a greater effect on the receptors in the nose in order to be detected. Hence, a lower percentage of sesquiterpenoids have useful odours than monoterpenoids. For the same reason, very few di- or higher terpenoids have odours. However, those sesqui- and higher terpenoids which do have odours are very tenacious because their lower volatility

Table 9.1 *Some of the more important terpenoid fragrance materials*

Material	Odour	Approx. usage ton per annum
Menthol	Mint, coolant	5000
α-Terpineol and acetate	Pine	3000
Dihydromyrcenol	Citrus, floral	2500
Borneol/*iso*borneol and acetate	Pine	2000
Carvone	Spearmint	1200
Acetylated cedarwood	Cedar	500
Amberlyn®/Ambrox®/Ambroxan®	Ambergris	10
Geraniol/nerol and esters	Rose	6000*
Citronellol and esters	Rose	6000*
Linalool	Floral, wood	4000*
Linalyl acetate	Fruit, floral	3000*
(Methyl)ionones	Violet	2500*
Hydroxycitronellal	Muguet	1000*
Total rose alcohol and citral related materials*		22500

means that they are lost more slowly from perfumes. Such materials form the basis of perfumes and serve also to fix the more volatile components.

These factors are reflected in Table 9.1 where it can be seen that most of the higher tonnage materials are monoterpenoids. Amberlyn® is used only in relatively low volume but it commands a very high price because of the intensity and persistence of its odour. The alcohols geraniol, nerol, citronellol and linalool are known as the rose alcohols because of their presence in roses and the fact that their odours are key parts of the rose complex. The ionone family are prepared from citral, which is the aldehyde corresponding to geraniol and nerol. Thus, it is clear from the table that this group of geraniol/citral related materials is of major importance to the fragrance industry. Their importance is even greater since citral is also the key synthetic intermediate for the preparation of vitamins A, E, K_1 and K_2. The consumption of citral for vitamins is comparable to that for fragrances.

It is probably worth mentioning that, strictly speaking, geraniol (9.8) is *E*-3,7-dimethylocta-2,6-dien-1-ol and nerol (9.9) is *Z*-3,7-dimethyl-octa-2,6-dien-1-ol and the corresponding aldehydes are geranial (9.10) and neral (9.11). However, the name geraniol is often used, particularly in the fragrance industry, to indicate a mixture of the two. Similarly, the name citral is used to indicate a mixture of geranial and neral with no specific isomer ratio being defined or implied (see Figure 9.4).

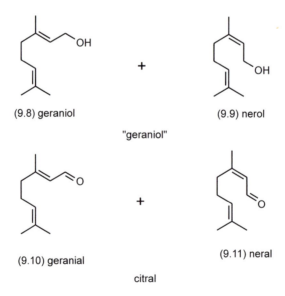

(9.8) geraniol   (9.9) nerol

"geraniol"

(9.10) geranial   (9.11) neral

citral

Figure 9.4

9.4.1   Interconversion of the Five Key Terpenoids

Figure 9.5 shows another reason for the importance of these five key terpenoids. They can be readily interconverted by isomerisation, hydrogenation and oxidation as appropriate. (The allylic isomerisation of geraniol to linalool and *vice versa* is effected by heating with vanadium pentoxide or a similar catalyst.) The ability to manufacture any one of these five terpenoids, therefore, opens up the potential to produce all of them and, hence, a wide range of other terpenes. Obviously, if one company produces geraniol/nerol initially and another linalool and both do so at the same cost per kg for their individual initial product, then they will not be able to compete with each other on both products. The first will have an advantage in geraniol/nerol and the second in linalool. Thus, the range of products which any terpene producer can market effectively will depend on a fine balance of its feedstock and process costs *vis-à-vis* those of its competitors.

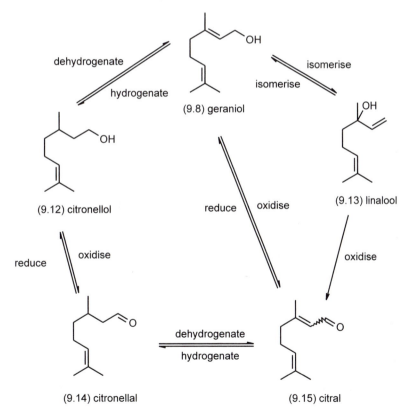

Figure 9.5

9.4.2 Commercial Aspects

The major producers of the rose alcohols and citral have become so for a variety of reasons. Companies which manufacture wood and paper products produce sulfate turpentine or similar by-products rich in pinenes. They, or their subsidiaries, may then produce terpenoid fragrance materials from pinenes as a way of generating income from their by-product. Pharmaceutical companies which manufacture vitamins use terpenoid intermediates and so will often diversify into the manufacture of aroma chemicals. Their basic feedstocks are likely to be of petrochemical origin. Similarly, manufacturers of synthetic rubber possess technology for the use of isoprene as a feedstock and so are also likely to diversify into terpene aroma chemical manufacture. Fragrance companies will develop a position in terpenoid chemistry because of the importance of terpenoids as ingredients.

The fragrance industry lies between the petrochemical and pharmaceutical industries in terms of scale of production and cost per kg of product. The production scale is closer to that of the pharmaceutical industry but the prices are closer to those of the bulk chemicals industry.

The largest volume fragrance ingredients are produced in quantities of 5000–6000 ton worldwide per annum and some ingredients, mostly those with extremely powerful odours which limit their use in a fragrance, are required in only kilogram amounts. These figures are dwarfed by products such as nylon 66 and nylon 6 (polycaprolactam), each produced at the level of about 4 million tonnes annually.

The cost restraints on the industry are, ultimately, imposed by the consumer. In many products, such as soap and laundry powder, the fragrance may be a significant contributor to the overall cost of the finished goods. If the price of one product is not acceptable, the consumer will select a competitive brand. The manufacturer of these products, therefore, puts considerable pressure on the fragrance supplier to come up with the most effective fragrance at the lowest possible cost. The highest volume fragrance ingredients cost only a few pounds per kilogram. This is not far above the cost of many basic petrochemical building blocks which are, typically, in the £0.5–1 per kg price range.

All these factors mean that chemists working in the fragrance industry have to work hard and think creatively and opportunistically in order to provide materials at an acceptable cost and without the advantages of scale that the bulk chemicals industry enjoys.

9.4.3 Evolution of Petrochemical Routes to Citral

At this point, we will refer back to Chapter 3 and two of the earliest syntheses of citral. These were shown as Figures 3.6 and 3.7 and, for convenience, are reproduced here as Figures 9.6 and 9.7, respectively. Let us look first at the Barbier–Bouveault–Tiemann synthesis shown in Figure 9.6.

This synthesis had its place as an academic synthesis for proof of structure. However, it is not suitable for large-scale production of citral. It starts with a dibromide (9.16) and also uses ethyl iodoacetate as a reagent in the third step. Bromine and iodine are expensive elements and their high atomic weights means that significant weight is carried through part of the process only to be lost later on. Acetylacetone, which is used in the first step, is relatively expensive and almost half of its

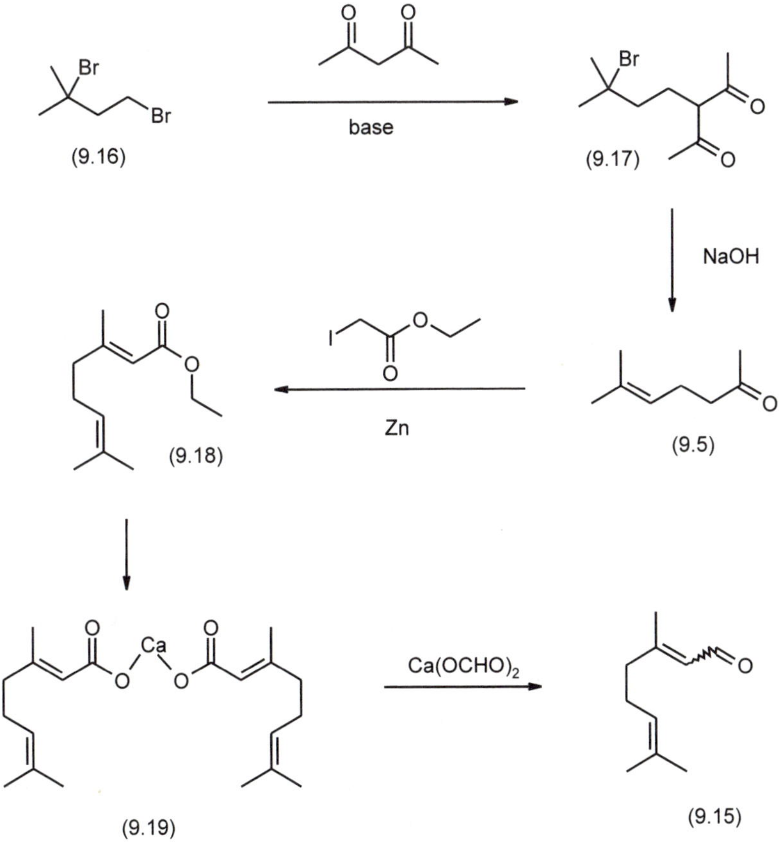

Figure 9.6

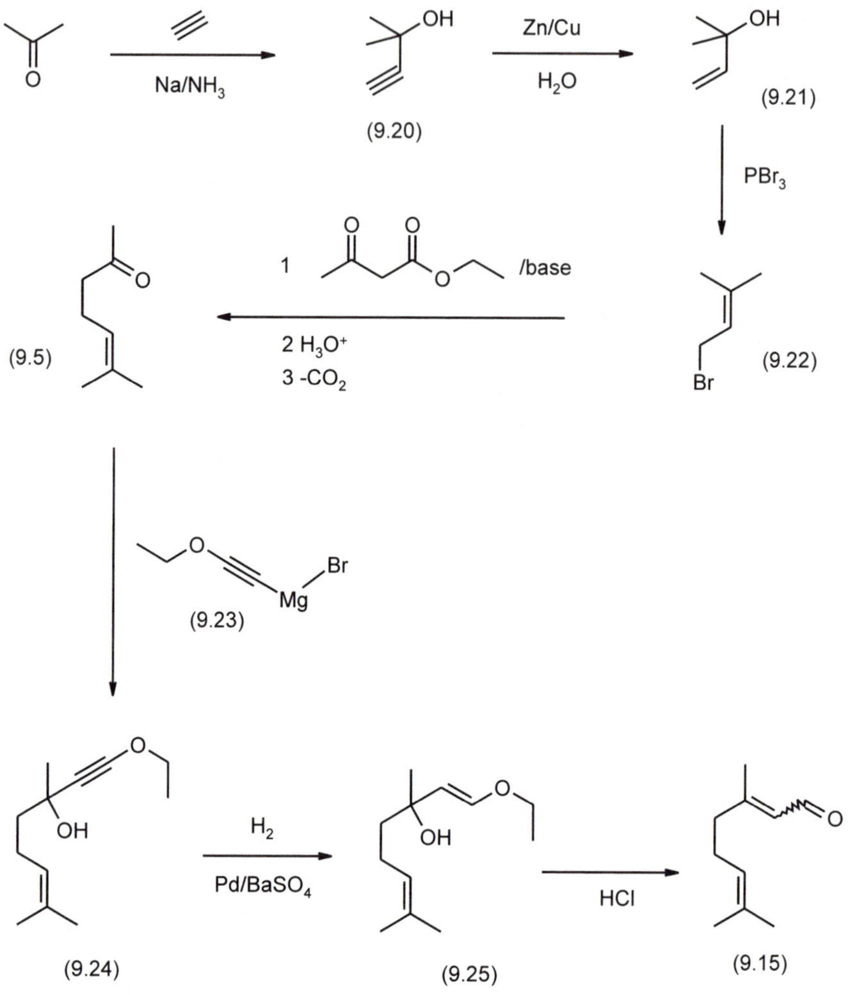

Figure 9.7

weight is discarded as waste in the second stage. The pyrolysis in the last step proceeds in poor yield with the production of a great deal of tar and other by-products including carbon dioxide from the calcium formate, another loss of weight from the system. The overall atom efficiency is very poor and the volume of waste which would be generated would be totally unacceptable in a modern manufacturing process.

The synthesis of Arens and van Dorp also has a poor atom efficiency, though not quite as bad as that of Barbier, Bouveault and Tiemann. The use of sodium in liquid ammonia would require refrigeration and the ability to distil and recover ammonia and this would lead to high engineering costs. Safety issues around large-scale handling of sodium metal would also increase engineering costs. The acetylenic Grignard reagent (9.23) is not readily available and would have to be prepared; thus adding to the total number of process steps which is already high at eight (the decarboxylation at step 3 would follow spontaneously after the hydrolysis in step 2 and so the two would constitute only a single step). However, despite all of these weaknesses, the synthesis of Arens and van Dorp was a major advance in thinking concerning the synthesis of citral and it is the basis of the best modern processes because of the basic strategy which they employed. The key features of their synthesis are firstly the construction of an isoprene unit from a C3 unit and a C2 unit, and then addition of a C3 unit to give methylheptenone (9.5) followed by the addition of a C2 unit to complete the monoterpenoid skeleton.

The discovery of the Carroll reaction in 1940 allowed significant improvements to be made to Arens and van Dorp's synthesis. The Carroll reaction involves heating the acetoacetate ester of an allyl alcohol. The ester can be formed *in situ* from the alcohol and ethyl acetoacetate. The reaction is an electrocyclic one which results in the elimination of carbon dioxide and the addition of an acetone fragment to the terminal carbon of the double bond of the alcohol. The mechanism is shown in Figure 9.8.

When this is applied to Arens and van Dorp's synthesis, and with a few other additional modifications, we obtain the route shown in Figure 9.9. This looks like a much better prospect for large-scale synthesis and the changes from the route of Figure 9.7 are very characteristic of the intuitive thinking of a good process development chemist. Firstly, the use of liquid ammonia has been dispensed with. This is possible because of the Favorskii – Babayan synthesis in which potassium hydroxide is employed as the base to remove the acidic hydrogen atom of acetylene. The zinc/copper couple used by Arens and van Dorp to reduce methylbutynol (9.20) to methylbutenol (9.21) has been replaced by catalytic hydrogenation, thus increasing atom efficiency and removing the generation of heavy metal waste. Now comes the Carroll reaction and removal of the need for bromination with the accompanying generation of phosphorus

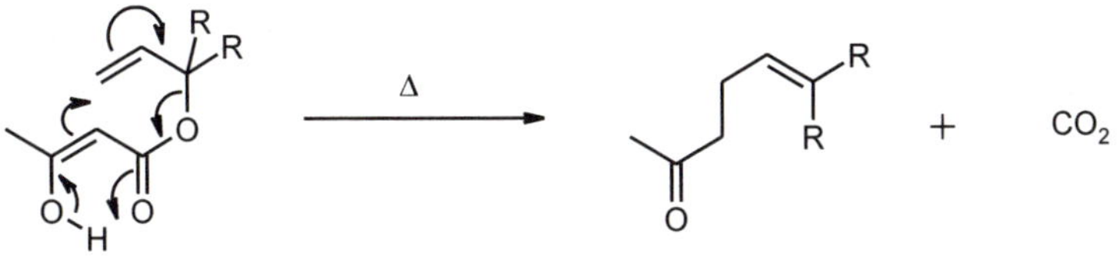

Figure 9.8

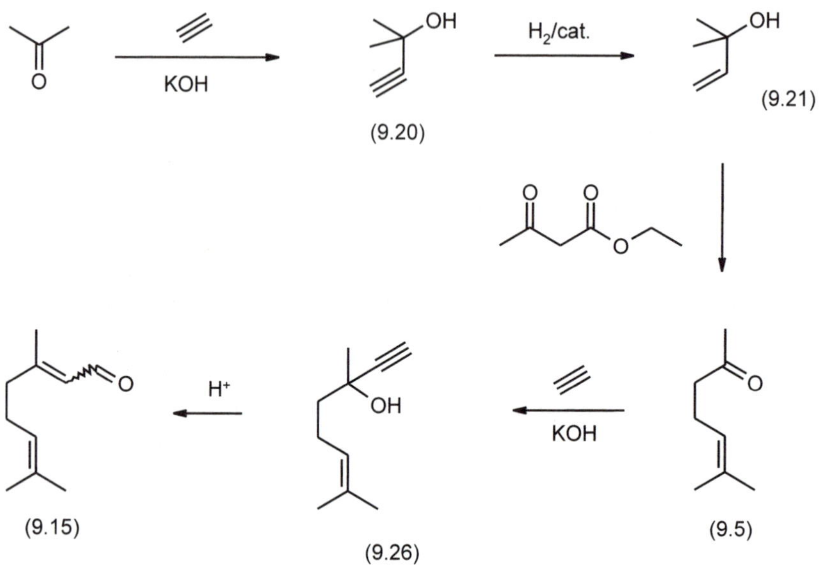

(9.20)

(9.21)

(9.15)   (9.26)   (9.5)

Figure 9.9

and bromine waste. The acetylenic Grignard reagent (9.23) has been replaced by acetylene which is much cheaper, and again the Favorskii – Babayan conditions are employed. This gives dehydrolinalool (9.26) instead of the more complex acetylenic material (9.24). There is now no need to hydrogenate the product since treatment of dehydrolinalool (9.26) with acid enables the Meyer–Schuster rearrangement to give citral (9.15) directly. This process can, and has been, run successfully on production scale. However, its atom efficiency is still not as good as we would like because of the loss of carbon dioxide and either loss of, or need to recycle, ethanol. The next improvement therefore was to replace the Carroll reaction by a Claisen rearrangement. This is shown in Figure 9.10 for the steps from methylbutenol (9.21) to methylheptenone (9.5).

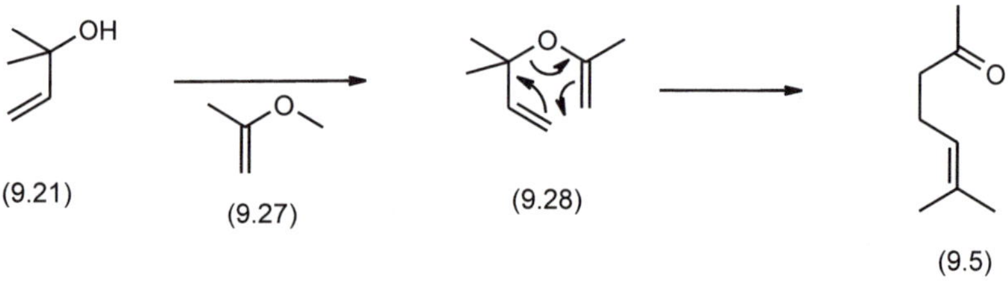

(9.21)   (9.27)   (9.28)   (9.5)

Figure 9.10

On heating methylbutenol (9.21) with 2-methoxypropene (9.27) (which is the methyl enol ether of acetone), a *trans*-etherification occurs with the elimination of methanol and the formation of the allyl vinyl ether (9.28). This ether is perfectly set up for a Claisen rearrangement, which it undergoes, to produce methylheptenone (9.5).

The evolutionary changes described above clearly demonstrate the desirability of electrocyclic reactions in chemical manufacture. These reactions possess the advantages of high selectivity and high atom efficiency. Having established this principle in the synthesis of citral, it was then not surprising that other electrocyclic reactions were investigated. One example is shown in Figure 9.11.

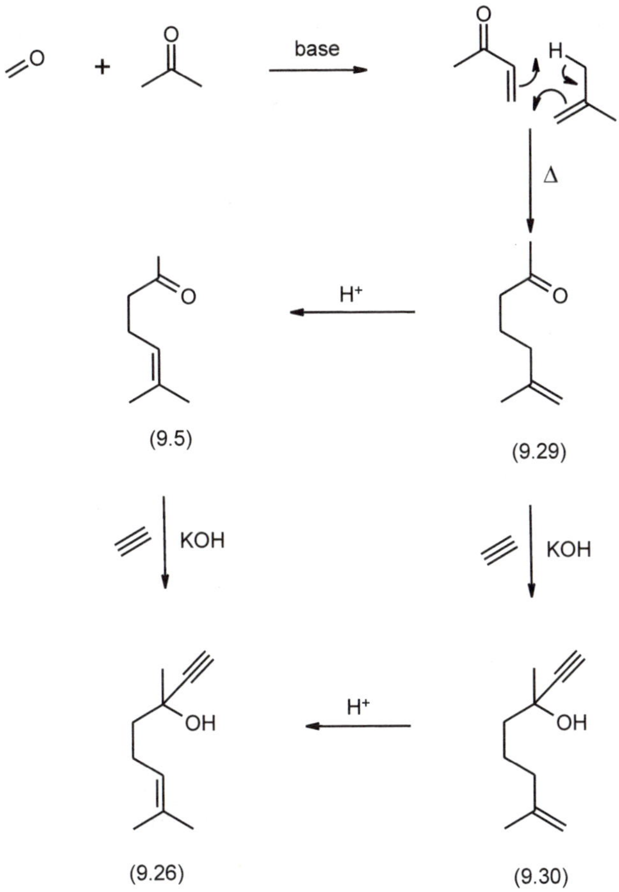

Figure 9.11

The synthesis in Figure 9.11 starts with an aldol condensation between formaldehyde and acetone to give methyl vinyl ketone. This is then heated with isobutylene and an ene reaction takes place giving ketone (9.29) which is a double bond isomer of methylheptenone (9.5). The use of an acid catalyst will isomerise (9.29) to methylheptenone, if required. The Favorskii–Babayan reaction will convert (9.5) to dehydrolinalool (9.26) and similarly (9.29) to (9.30) which is an isomer of dehydrolinalool and which can be isomerised to it. If one wishes to prepare linalool by this method then either route to dehydrolinalool will be suitable. In principle, either route could be used for the preparation of ionones and vitamins but there is one slight advantage in using the isomer (9.30) rather than dehydrolinalool (9.26). The isomeric ψ-ionone prepared from (9.30) is easier to cyclise than normal ψ-ionone because of the increased polarisation in the 1,1-disubstituted double bond of (9.30)-derived ψ-ionone relative to the 1,1,2-trisubstituted double bond of (9.26)-derived ψ-ionone making it more susceptible to attack by acids.

Electrocyclic reactions feature prominently in the most recent and most elegant of commercial citral syntheses. This scheme was developed and patented by BASF and is shown in Figure 9.12.

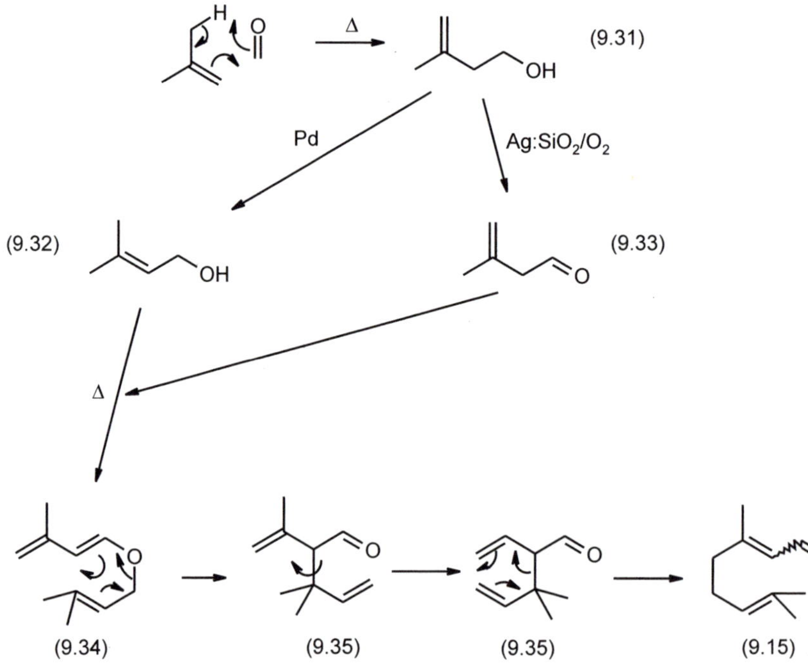

Figure 9.12

The synthesis starts with the addition of formaldehyde to isobutylene. This reaction is catalysed by acids but is essentially an electrocyclic addition. It falls within the definitions of both the ene and the Prins reactions and is sometimes referred to as one of these and sometimes as the other. The mechanism is shown in the figure and the product is isoprenol (9.31). Treatment of isoprenol with a palladium catalyst causes isomerisation of the double bond into the more thermodynamically stable position, giving prenol (9.32). Passing a mixture of isoprenol and air over a silver catalyst leads to oxidation of the alcohol function and formation of isoprenal (9.33), and giving water as a by-product. Now comes the most elegant part of the scheme. A mixture of prenol and isoprenal is heated to produce citral in one process step! On mixing an alcohol with an aldehyde, the hemi-acetal forms and water can be eliminated to produce a vinyl ether. In this case the vinyl ether is (9.34) which is, of course, an allyl vinyl ether and therefore perfectly set up to undergo the Claisen rearrangement as shown. This gives the unsaturated aldehyde (9.35). This aldehyde is also a 1,5-hexadiene and rotation about the central carbon–carbon bond of the diene system, brings it into perfect alignment for the Cope rearrangement to proceed to give citral (9.15). Thus, starting from two readily available and inexpensive starting materials, using air as the only reagent, in four process steps (two of which are run in parallel) we can obtain citral with the only by-product being water (two moles per mole of citral). This synthesis demonstrates superbly the elegance, cleanliness and cost-efficiency which are the goals of a modern chemical process development.

9.4.4 Isoprene to Citral

Addition of hydrogen chloride to isoprene gives predominantly prenyl chloride (4-chloro-2-methylbut-2-ene) (9.36). This is as would be expected from carbocation chemistry, as the initial protonation will occur at C-1 of isoprene, the more electronegative end of the more polarisable bond. This gives the carbocation (9.37). The bulky chloride anion then adds to the less hindered end of the allylic cation system to give (9.36). Some chloride ions do add at C-2 to give the isomeric halide (9.38). However, this is of little practical consequence since steric factors dictate that nucleophilic attack will occur at the less hindered end of the allylic halide function. Thus, whilst (9.36) undergoes straightforward nucleophilic displacement *via* an S_N2 reaction to give (9.39), the isomeric (9.38) undergoes an allylic S_N2' displacement and the product is also (9.39). This phenomenon is shown in Figure 9.13. The effect exists also in higher homologues such as myrcene and the importance of that will be seen shortly.

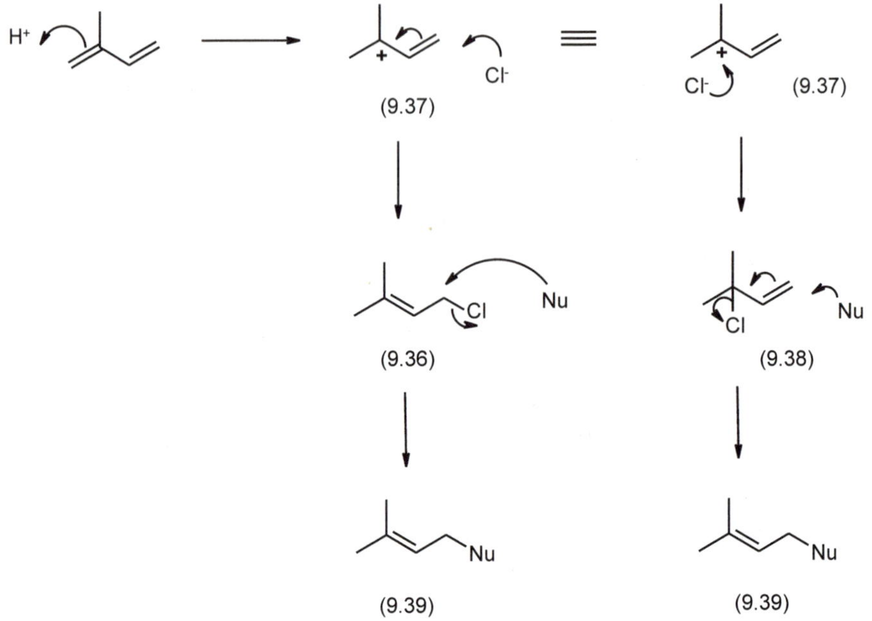

Figure 9.13

The synthesis of methylheptenone shown in Figure 9.14 uses isoprene as the feedstock and the initial addition of hydrogen chloride as discussed above. Reaction of prenyl chloride with acetone in the presence of a base, gives methylheptenone which can then be used in the synthesis of linalool, citral, *etc*.

This synthesis has the disadvantage of introducing chlorine into the molecule. Chlorinated organic materials are not desirable and the removal of chlorine from the product must be assured. It is very difficult to remove trace organochlorine components down to the detection threshold as is required and this therefore adds to the process costs. However, one advantage of this route lies in its potential contribution to sustainability. Isoprene is normally produced from petrochemicals

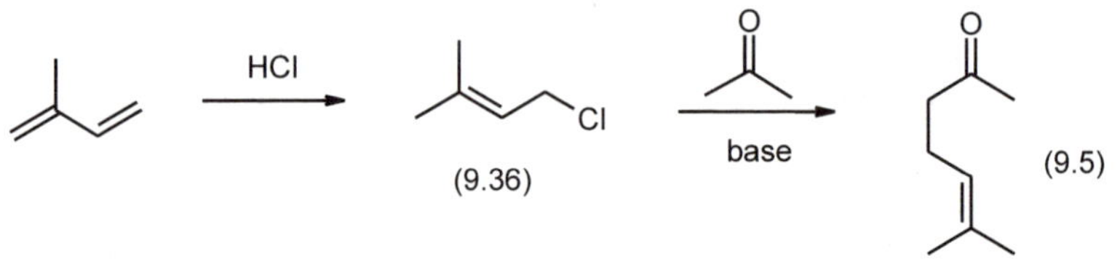

Figure 9.14

nowadays, but it is also available from natural sources, for example by pyrolytic cleavage of limonene, and so it would be easy to switch from petrochemical to sustainable sources.

9.4.5 Citral, Geraniol *etc.* from Turpentine

For thousands of years, turpentine has been obtained from conifers by a process known as tapping. A cut is made in the bark which prompts the tree to exude an oil which can be drained off into vessels attached to the tree. Turpentine thus obtained is referred to as gum turpentine. Nowadays, another form of turpentine is available in large quantities as a by-product of paper manufacture. When softwood (pine, fir, spruce) is converted into pulp in the Kraft paper process, the water insoluble liquids which were present in it are freed and can be removed by physical separation from the process water. This material is known as crude sulfate turpentine (CST). Fractional distillation of CST gives a number of products as shown in Table 9.2. (Dipentene is the name given to racemic limonene.) The residue from the distillation is known as tall oil and contains diterpenes such as abietic acid.

Thus, pure α- and β-pinenes can be obtained by fractional distillation of turpentine. The two can be interconverted by catalytic isomerisation but this leads to an equilibrium mixture. The equilibrium could be driven in one direction by continuous removal of the lower boiling component through distillation. However, α-pinene is the lower boiling of the two and is already the more abundant. To increase the yield of β-pinene, it is necessary to fractionate – isomerise α to β - fractionate - isomerise α to β - and so on. This is obviously a costly process in terms of time and energy. As a result of these factors concerning availability, β-pinene is about twice the price of α-pinene and this affects the economics of the processes to prepare other terpenoids from them.

One of the earliest commercial routes into this key group of terpenes involved pyrolysis of β-pinene (9.40) and is shown in Figure 9.15. When

Table 9.2 *Composition of distillate from sulfate turpentine*

Product	Percentage
Lights	1 – 2
α-Pinene	60 – 70
β-Pinene	20 – 25
Dipentene	3 – 10
Pine oil	3 – 7
Estragole, anethole, caryophyllenes	1 – 2

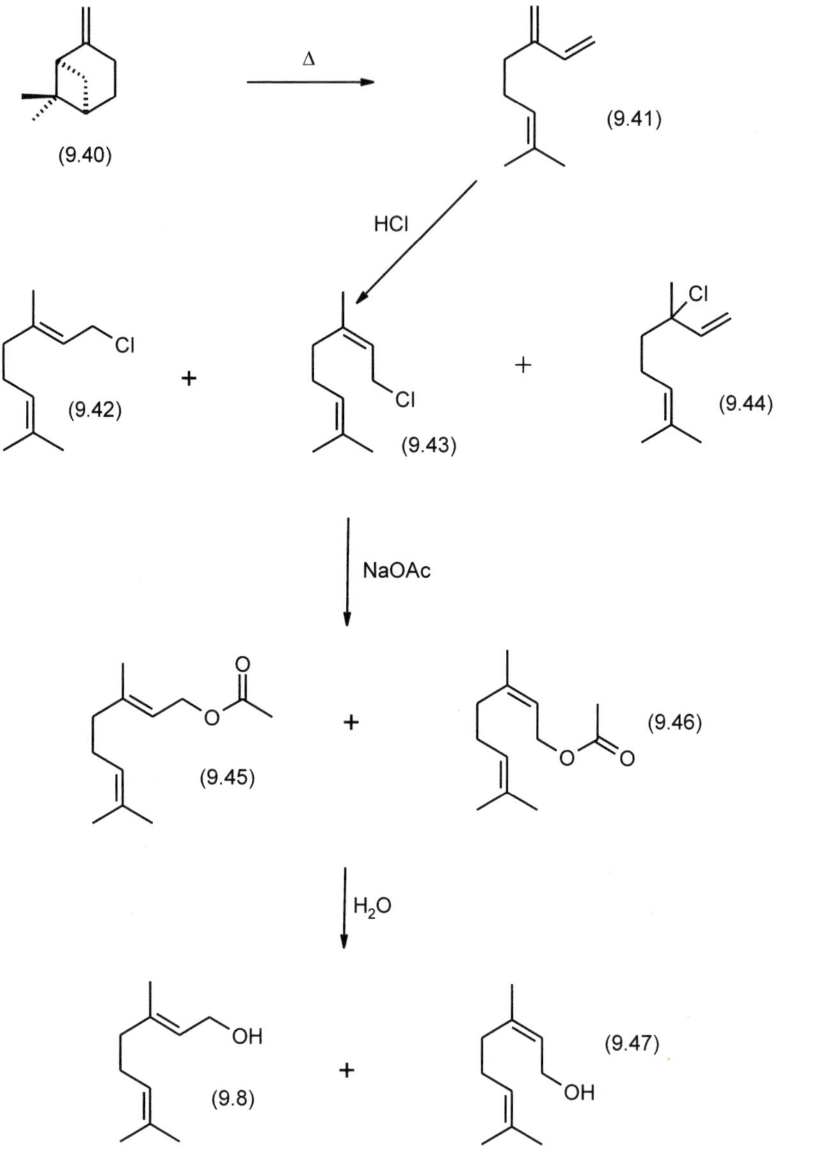

Figure 9.15

β-pinene is heated to 500 °C, the cyclobutane ring breaks *via* a *retro*-2 + 2 cycloaddition. This ring opening is regioselective and produces the triene, myrcene (9.41). Addition of hydrogen chloride to myrcene gives a mixture of geranyl, neryl and linalyl chlorides, (9.42), (9.43) and (9.44), respectively. These can be hydrolysed directly, but the reaction with acetate anion is more efficient than that with hydroxide and so the acetate esters are usually the initial products in commercial syntheses. As is the case with prenyl chloride which was discussed above, geranyl and neryl chlorides react *via* an S$_N$2 reaction. The steric crowding around the chlorine bearing carbon of linalyl chloride makes the S$_N$2 reaction difficult and so an S$_N$2' reaction occurs instead. Thus, the product of the acetolysis of the mixture of the three chlorides is a mixture of only two products, geranyl acetate (9.45) and neryl acetate (9.46). Hydrolysis of the acetates then produces geraniol (9.8) and nerol (9.47). Geraniol is the major product and various grades of "geraniol" are commercially available containing different levels of nerol and other materials. These are produced by fractional distillation. The distillation is complicated by the presence of traces of impurities and isomers formed from the opening of the cyclobutane ring in the "wrong" direction.

The two main disadvantages of this route are the cost of β-pinene, its lower availability relative to that of α-pinene, and the presence of trace amounts of chlorinated materials which must be removed from the product.

Hydrogenation of the less expensive major component of turpentine, α-pinene (9.48), gives pinane (9.49) which can be oxidised by air under radical conditions to give the hydroperoxide (9.50) which is then reduced by hydrogenation to 2-pinanol (9.51). Pyrolysis of this alcohol gives linalool (9.13). This sequence of reactions is shown in Figure 9.16.

The disadvantage of this process lies in a side reaction. Linalool (9.13) is not stable under the pyrolysis conditions and some of it undergoes an intramolecular ene reaction to give a mixture of isomeric alcohols known as plinols (9.52). These have boiling points which are close to that of

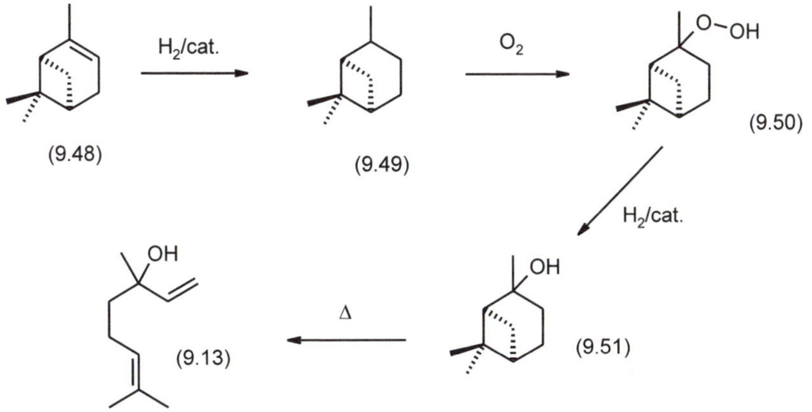

Figure 9.16

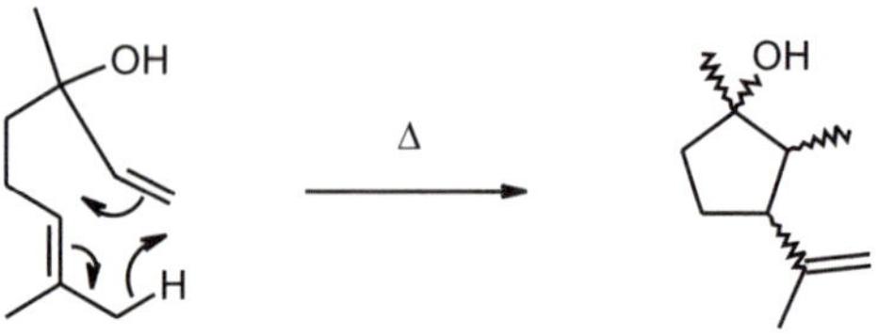

(9.13)   (9.52)

Figure 9.17

linalool making separation by distillation difficult. The pyrolysis is therefore run at below total conversion in order to minimise plinol formation. The mechanism of the ene reaction and the structure of the plinols are shown in Figure 9.17. In order to separate the plinols from the desired products, it is usual to isomerise the linalool (9.13) to geraniol (9.8) and then fractionally distil since the boiling point of the plinols is sufficiently different from that of geraniol to allow separation.

Readers who are interested to know how other terpenoids are produced commercially could consult the books by Pybus and Sell and Muller and Lamparsky, details of which are given in the bibliography under the heading "Chemistry and the Perfume Industry".

9.6 PRODUCT TREES

As discussed above, there are various reasons why a company will enter a given sector of the fine chemicals market. Once it has a position in a sector, two of the main driving forces for extension of its product portfolio will be the available intermediates and available technology. In order to achieve a strong competitive position for a given product, a company's production costs must be the lowest or close to the lowest in the marketplace. Such a position will be easier to achieve if it has access to low cost feedstocks and appropriate technology. For example, if it is processing large volumes of intermediate A and is strong in a technology which would enable it to produce product B from this, then it will at least consider the possibility of entering the market with product B. By considering some fictitious examples, we can see some of the factors involved in the growth of product trees.

For example, a company which is manufacturing citral through a technology like that of the BASF process shown in Figure 9.12, will have ready access to significant volumes of prenol (9.32) at a lower cost than a non-producer. They might then consider manufacturing some prenol derivatives such as are shown in Figure 9.18. Preparation of the acetate (9.53) and benzoate (9.54) would be easy chemically and these are useful fragrance ingredients. If the company possessed the expertise to purify materials to perfumery quality, it might make good sense to enter production of these esters. The synthesis of chrysanthemic acid (9.55) (an intermediate in pyrethrin production) would be more difficult and the company would probably only consider adding these to its portfolio if it had experience in the manufacture of agrichemicals or intermediates for these.

In the next example, we will consider a company which manufactures geraniol (9.8) from β-pinene (9.40) *via* myrcene (9.41). This company will have pure β-pinene and myrcene in hand and will also have experience in producing perfumery grade materials. Natural extensions of their product portfolio would therefore include well-known fragrance ingredients such as nopyl acetate (9.56), prepared by a Prins reaction on β-pinene, and the hydroxy aldehyde (9.57), prepared by a Diels–Alder reaction with acrolein and hydration of the side chain double bond. Nowadays, since suitable chiral catalysts are becoming commercially available, they might also be tempted to develop a process like that of Noyori (see Chapter 4) in order to gain a foothold in the lucrative menthol market. These possible schemes are shown in Figure 9.19.

A company making geraniol (9.8) and linalool (9.13) from α-pinene (9.48) *via* pinane (9.49) might also be expected to produce dihydromyr-cenol (9.58), α-terpineol (9.59) and camphor (9.60) since they have the feedstocks and almost certainly the technology to do so. This tree is

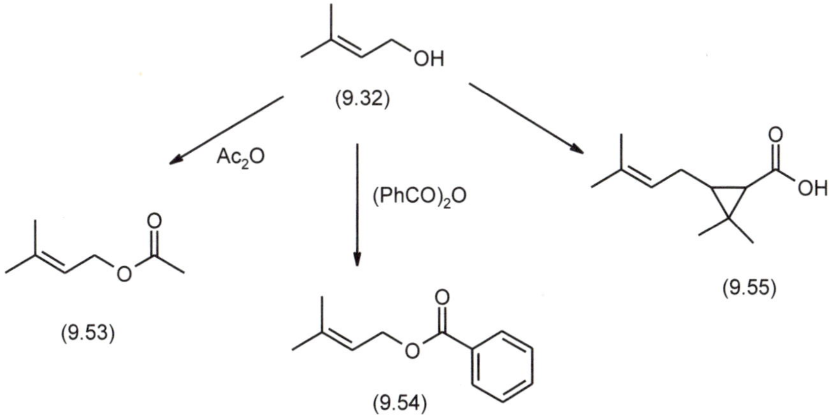

Figure 9.18

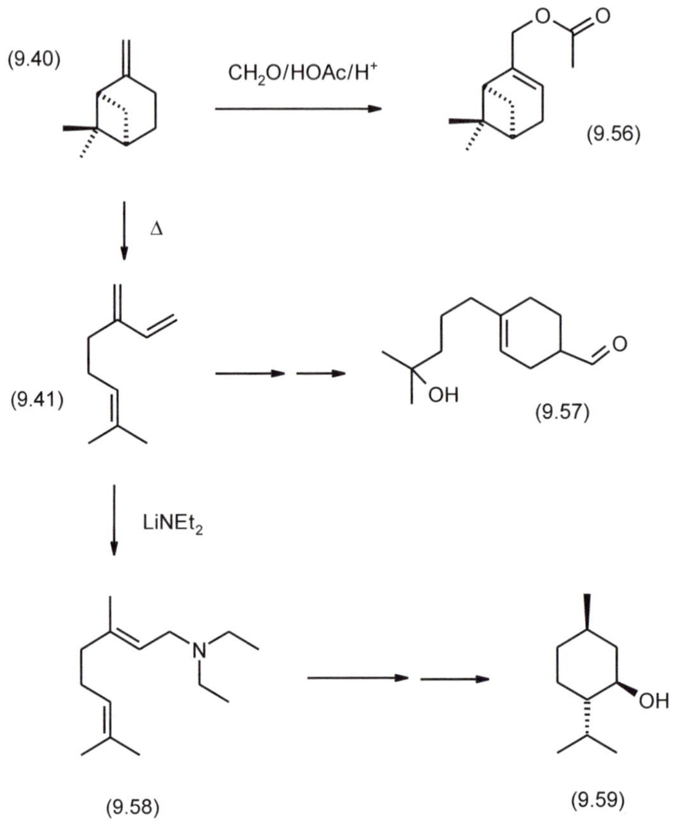

Figure 9.19

shown in Figure 9.20. An attractive opportunity for this company might be to develop epoxidation technology. This would enable them to produce α-pinene oxide (9.61) and hence campholenic aldehyde (9.62) and the sandalwood ingredients which can be manufactured from it. Epoxidation currently requires the use of peroxy compounds and there are explosion hazards associated with processes involving these. Safe introduction of such technology is referred to as a barrier technology because the cost of doing so represents a barrier to other companies doing the same thing. In other words, possession of such a technology provides a measure of security against competition.

Figure 9.21 shows a more extensive product tree and this would be the representative of a major chemical company with interests in both the perfumery ingredients and pharmaceutical markets. Perhaps this company was initially interested in the synthesis of vitamins as part of their activity in the pharmaceutical industry. The initial targets were

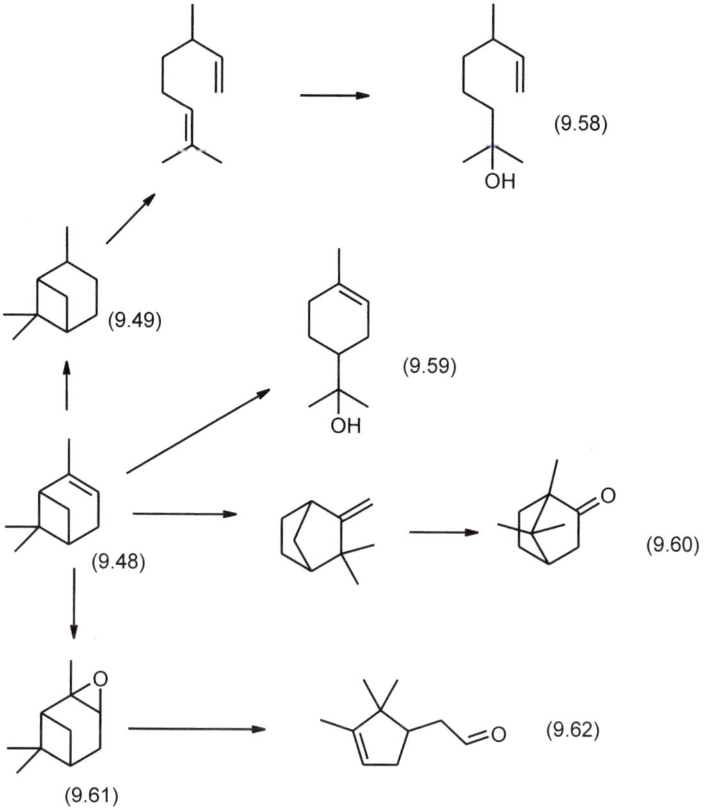

Figure 9.20

therefore vitamins A (9.63), E (9.64), K₁ (9.65) and K₂ (9.66). Their existing technology favoured methylheptenone (9.5) as a starting material. The most obvious route then would be to go *via* linalool (9.13), citral (9.15) and α-ionone (9.67) to vitamin A (9.63). They find that they can sell linalool into the fragrance industry and so begin to build up an expertise in manufacturing to odour as well as pharmaceutical specifications. It is then an easy thing to broaden their portfolio to include the linalyl esters and then dihydrolinalool (9.68) and tetra-hydrolinalool (9.69) and their esters. The reason the hydrogenated forms of linalool are useful in perfumery relates to their stability in aggressive consumer products in which the double bonds of linalool would undergo rearrangement or oxidation. Another material useful for its stability in such products is geranyl nitrile (9.70) which has an odour similar to that of citral but which shows much greater stability on products than does the aldehyde. This nitrile is easily accessible from methylheptenone (9.5)

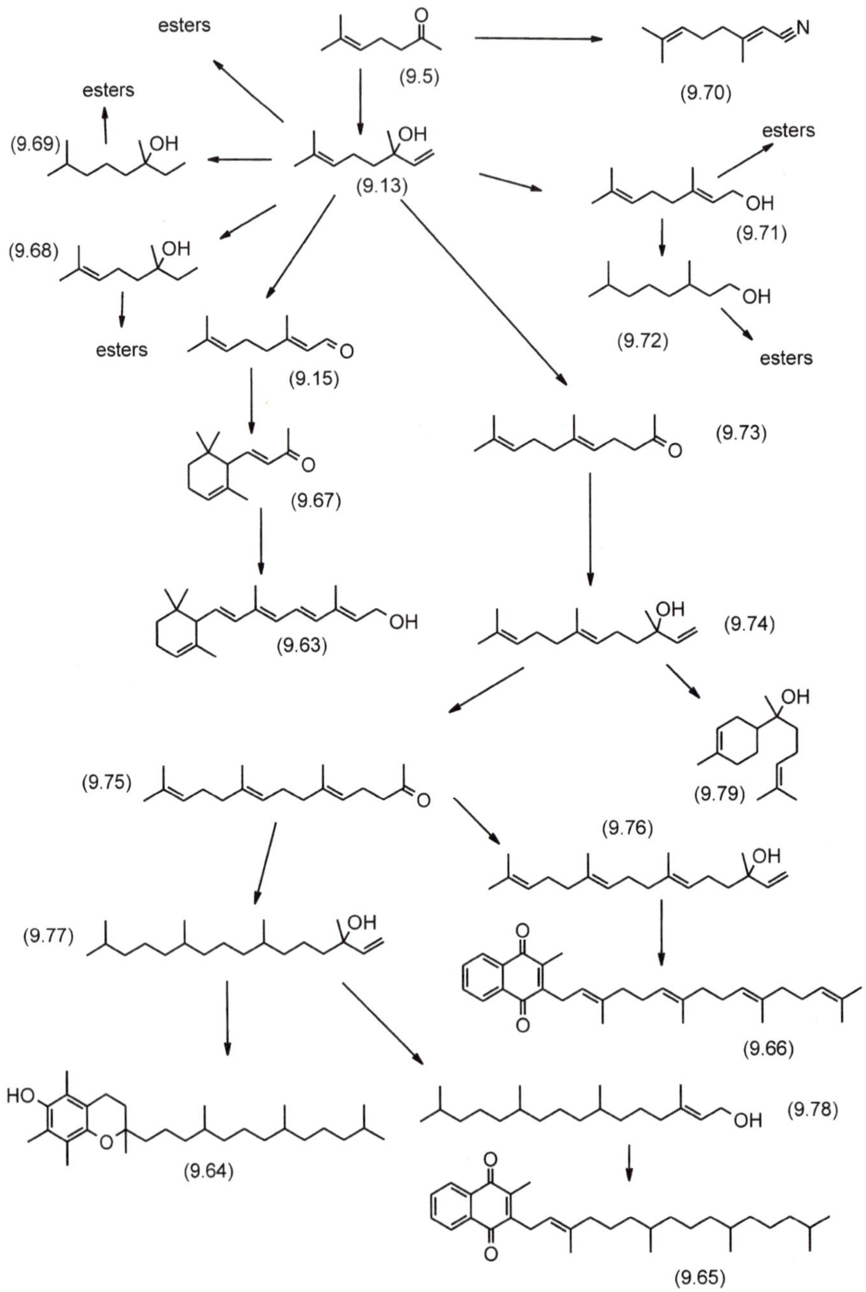

esters

esters

(9.69)

(9.68)

esters

(9.5)

(9.13)

(9.15)

(9.67)

(9.63)

(9.70)

esters

(9.71)

(9.72)

esters

(9.73)

(9.74)

(9.79)

(9.75)

(9.76)

(9.77)

(9.66)

(9.64)

(9.78)

(9.65)

Figure 9.21

through the Döbner reaction with cyanoacetic acid and so it is easily added to the portfolio. Success in linalool chemistry would then lead the company to consider geraniol also and so geraniol (9.71) and tetrahydrogeraniol (9.72) and esters of these might be added to the palette. However, with the geraniol family, market competition from companies starting from turpentine, will be much stiffer. To reach the other vitamins, the route chosen is to carry out a Carroll type synthesis form linalool to geranylacetone (9.73), acetylene addition and reduction to give nerolidol (9.74) and then a repeat of these two steps to give farnesylacetone (9.75) and geranyllinalool (9.76). Vitamin K_2 (9.66) is accessible from geranyllinalool (9.76) and farnesylacetone (9.75) can be converted to isophytol (9.77) and hence phytol (9.78). Vitamin E (9.64) is accessible from isophytol (9.77) and vitamin K_1 (9.65) from phytol (9.78). Having this chain in hand then provides other opportunities such as the production of α-bisabolol (9.79), an anti-inflammatory and anti-bacterial agent used in cosmetics, from nerolidol (9.74). So the build-up of an extensive product portfolio can continue by addition of new products and technologies.

Most fine chemicals companies produce printed copies of their product trees along the lines of Figure 9.21. Such information is invaluable to the development chemist as it not only tells him what a potential supplier does make, but it also gives a very good indication of what they might be able to add to their portfolio. This means that, should a required feedstock not be commercially available at the time, the development chemist will have a very good idea of which company to approach, to see if they might be able to introduce it. There have been some attempts to produce databases containing such information but there seems to be no real good alternative to the memory of an experienced chemist who can hold, not just structures, but reaction pathways (both real and extrapolated) in his head.

9.7 CONCLUSION

Human society will always have a need for the production of chemicals since these are essential for the maintenance and improvement of our quality of life. Modern chemical synthesis, despite its bad press, is generally relatively clean, efficient and inexpensive. However, there is always room for improvement. For example, some reactions still require stoichiometric reagents and some require relatively high temperatures and pressures. The use of volatile organic solvents leads to recycle costs and loss of solvents can produce environmental effects. We also still need to improve the selectivity of catalysts and particularly their enantio-selectivity.

There is also a considerable effort into biochemical approaches to synthesis at the moment. Nature's catalysts, the enzymes, are very specific in terms of individual reactions but the overall selectivity of any process involving intact organisms, whether bacteria, yeasts, fungi or higher plants, is always very poor compared to a chemical process and the dirtiest processes (by a very wide margin) I have ever come into contact with were biotechnological processes. There are two major drawbacks with fermentation or extraction from plant material. Firstly, the vast amount of biomass produced relative to the desired product creates an enormous effluent problem. Secondly, the energy required to separate the desired product from the biomass is often orders of magnitude higher than the total energy consumed in a chemical synthesis. Both of these problems constitute serious issues as regards the environment and the sustainability of the operation.

Current biotechnological research programmes are attempting to address these issues. One approach is to use enzymes *in vitro* as catalysts in processes which then resemble conventional chemical technology more closely. This is fine for enzymes which catalyse reactions such as hydrolysis and esterification. However, for redox processes in which an enzyme requires a co-factor, the co-factor must either be supplied in stoichiometric amounts (and at great cost for co-factors such as NAD) or recycled. One interesting concept is to combine chemical and biochemical methods, for example, using simple chemical reagents to regenerate enzyme co-factors or even to set up more complex cycles using both chemical and enzymic catalysts in a one-pot, multi-step process. Advances in biochemistry, molecular biology and genomics are making it possible to introduce more fundamental changes. For example, there are many active programmes in "directed evolution" where the objective is to "breed" new enzymes with enhanced functionality and improved robustness. It is also possible nowadays to alter the metabolic pathways in cells by switching off expression of a native enzyme and introducing new DNA so that an alternative enzyme is produced instead. This is known as metabolic engineering and is the subject of an excellent recent review by Burkart.[9.7]

At the same time, chemical research is far from dormant. Even a cursory inspection of the current chemical literature will reveal that there is considerable activity being devoted to "green" chemistry. For example, the use of ionic liquids as solvents and the development of solvent free systems are not only reducing the need for the traditional solvents of organic chemistry and hence reducing recycle costs and the environmental effects of solvent loss, but also creating new opportunities for controlling the outcome of a reaction through new solvent effects. Chemists are also becoming more adept at learning from nature's secrets as far as catalysis is concerned and consequently devising "designer catalysts" with much higher selectivity than ever before.

The areas of research where chemistry and biology meet and overlap are very exciting and will no doubt be a breeding ground for even more Nobel Prize winning works in future.

REFERENCES

9.1. B.M. Trost, *Science*, 2001, 254.

9.2. R.A. Sheldon, *Chem. Ind.*, 1992, 903.

9.3. R.A. Sheldon, *Chem. Ind.*, 1997, 12.

9.4. C. Hendrix, *Chemtech*, 1980, 488.

9.5. Statistics for Experimenters, An introduction to design, data analysis and model building, G.E.P. Box, W.G. Hunter and J.S. Hunter, John Wiley, London, 1978, ISBN 0-471-09315-7.

9.6. www.statsoft.com/textbook/stathome/html

9.7. M.D. Burkart, *Org. Biomol. Chem.*, 2003, **1**, 1.

© Christie's Images/CORBIS

CHAPTER 10

Discovery and Design of Novel Molecules

Taking inspiration from nature, the chemist can produce whole new families of useful materials.

Perfume and incense bring joy to the heart

Proverbs 27, 9

Perfume has given pleasure to mankind since the first humans smelt those created by nature. The antiquity of perfumery as a profession is demonstrated in many ancient texts. For example, the book of Exodus tells us how Moses used perfumers to manufacture incense for the first Jewish tabernacle and, in his "Histories" which were written in the fifth century BC, Herodatus reports his interviews with Arabian perfumers on the secrets of their trade. (Though, as we discovered in Chapter 8, he did not get much truth from them.) Perfumery, like almost all other aspects of our everyday lives, has changed enormously as a result of the advances made in synthetic organic chemistry starting from the nineteenth century. Once the preserve of only the wealthy and powerful, perfume is now enjoyed by everyone. What is more, nowadays we see, or rather smell, perfume being used in a huge variety of applications which could not have been dreamt before the advent of organic synthesis. In order to design effective novel molecules for perfumery, the chemist must understand their application and this requires knowledge of physical chemistry, biochemistry, manufacturing technology and the art of perfumery in addition to his skills in organic synthesis.

In this chapter, perfumery is used as an example of discovery chemistry but the basic underlying principles are the same as those in discovery of novel molecules for applications including pharmaceuticals, flavour ingredients, adhesives, lubricants and so on. The first part of the chapter outlines these basic principles and serves as an introduction for any area of discovery chemistry. The second part of the chapter uses fragrance ingredients as an example of the discovery process.

KEY POINTS

> The discovery of novel molecules is an important part of our efforts to improve our material standard of living.
>
> Novel molecules can be discovered by random screening, by copying nature, by statistical design (using structure/activity relationships, SARs) or by understanding the basic mechanism of the required activity.
>
> Attempting to correlate SARs with mechanisms, or *vice versa*, can be useful but is fraught with intellectual pitfalls.
>
> Scientific theories are never established beyond doubt and are always subject to testing.
>
> Careful consideration must be given to the quality of data used in building SARs.
>
> Successful design of materials for commercial application requires an awareness of all aspects of the application in question from production to consumption and all the requirements which will be made of the material during that process.
>
> The biological process by which we perceive odours is not fully understood. However, recent advances in our understanding indicate that the combinatorial nature of the mechanism of odour perception presents serious problems for the design of novel fragrance ingredients.
>
> Fragrance ingredients must meet certain criteria in terms of safety, performance, availability and odour.

10.1 WHY SEARCH FOR NOVEL MOLECULES?

There is no doubt that our material standard of living has risen dramatically over the last century. Much of this improvement is either the direct result of, or would be impossible without, synthetic organic chemistry. For example, modern drugs to combat a disease, electronic equipment such as telephones and computers, credit cards, cars, sports equipment, non-iron shirts, the list is endless, are all possible only because of synthetic organic chemistry. Our desire to improve our material well being has been a part of human life since it first started (we date periods of history by the technology and materials in use, hence terms such as stone age, bronze age, iron age) and is unlikely ever to cease. We will always seek to improve yet further on the materials available to us in order to provide complementary improvements in our standard of life. When I was a student, one computer would occupy a whole room. The lap-top on which I am writing this book, fits in a briefcase and is thousands of times more powerful than the biggest computer used 30 years ago. We are now seeking to build logic processors on the molecular scale and thus bring about another dramatic change in computational power. The computers of today are only possible because of new molecules such as those in the high performance adhesives used in their construction and the liquid crystals used in their display screens. Future computers will almost certainly depend even more heavily on new molecules and new chemical technology. The developing world seeks to enjoy a standard of living like that of the West and we seek continual improvement over what we enjoy. Nowadays, we are also concerned about the sustainability of all areas of human activity. In order to ensure sustainability without lowering our standard of living, we will need to find new molecules which will be part of a new technology aimed at improving both living standard and sustainability. It has been estimated that the cost of bringing a new molecule into the market as a pharmaceutical is currently about $600,000,000. Only about 10% of those that enter development actually reach the market and the number of new molecules screened in order to find one candidate for development runs into hundreds of thousands. In other fine chemicals industries, such as fragrance, the total costs are lower and the success rate is higher but the R&D costs are still very significant in comparison to the value of the appropriate total market. The design of novel molecules is therefore a subject of intensive study, not only of specific molecules and applications but also into the methods used in this design.

10.2 MOLECULE DISCOVERY THROUGH RANDOM SCREENING

The simplest approach to finding new chemicals for any given application is simply to screen known molecules and see which, if any works. The pharmaceutical industry has used this approach for years and has invested large amounts into automation of screening methods. This is necessary as the hit-rate on such random screening is very low indeed. The candidates for screening could simply be molecules that are available in libraries of research samples and every major pharmaceutical has vast candidate libraries. The candidates could also be materials which are known in one application and are therefore available for screening in another. An example from the history of fragrance is that of the nitromusks. These pleasantly odoured nitrobenzene derivatives were discovered by Baur while he was working on explosives related to TNT. They played an important role in perfumery in the late nineteenth and early twentieth century but are now obsolete, as safer alternatives have been developed. Another source of candidates for screening are chemicals which could be easily prepared from known feedstocks using available technology. This last approach is useful in applications where price and speed to market are important since the choice of candidate for synthesis can easily be directed towards those which will prove to be less expensive and for which development would be relatively straightforward.

10.3   NATURE AS A SOURCE OF NOVEL MOLECULES

Nature serves as a wonderful inspiration for discovery chemists in all application areas. Folk medicine, such as European herbal remedies and Indian Ayurvedic medicine, has supplied many good leads for modern pharmaceutical agents. Similarly, the natural oils and resins used in traditional perfumery have been the source of inspiration for many modern perfumery ingredients. The basic approach used is to seek out a natural material with a traditional use, analyse it to identify the active chemicals which it contains and then reproduce them in the laboratory and eventually the factory.

Having identified a natural lead and developed a synthetic route to it, it is then a small step for the chemist to prepare simple analogues of the natural material and investigate their properties. As a result, the chemist then begins to build up a library of chemical structures the properties of which are known as far as the application in question is concerned. This database then enables the pursuit of one of the approaches to rational design of novel molecules, that of design through statistics.

An example of the process from natural chemicals to high performance synthetic ingredients is given in chapter 3 of the book by Pybus and Sell.[10.1] In this example, the development of modern jasmine ingredients based on the key odour components of natural jasmine extracts is outlined.

10.4   DESIGN THROUGH STATISTICS

The molecular structure of any chemical will determine all of its physical, chemical and biological properties. Discovery chemists will therefore seek correlations between molecular structure and a desired property and the search for novel materials is therefore often guided by structure/activity relationships (SARs). These can also be referred to as structure/property relationships (SPRs) and when carried out with data inputs in a numerical form and using statistical methods for their analysis, the adjective quantitative is added giving quantitative structure/activity relationships (QSARs) or quantitative structure/property relationships (QSPRs).

Chemists looking at a set of structures, some of which possess a desired activity and others of which do not, will almost instinctively derive simple SARs on sight. For instance, Figure 10.1 shows some of the sandalwood materials described in Chapter 6. The chemist's eye will quickly spot the fact that they are all monohydric alcohols with a molecular weight in the region of 200–240 Da and a bulky hydrophobic centre 5 or 6 carbon atoms away from the carbon carrying the alcohol function, giving the effect of a molecular ball and chain with the alcohol function at the end of the chain. This is a simple SAR and quite an effective one. Materials meeting these criteria are quite likely to possess a sandalwood odour and materials which do not fit the pattern are unlikely to have such odours. However, the exact quality of the odour and its intensity relative to the rest of the family are much more difficult to predict.

With a large data set of structures and properties, it is possible to use multivariate statistical methods such as principal components analysis to try to identify patterns. The applications of statistics to chemistry is known as chemometrics and an internet search for this keyword will lead to a variety of useful sources of information on the field. Some QSAR methods are based on the molecular orbitals of the compounds in the

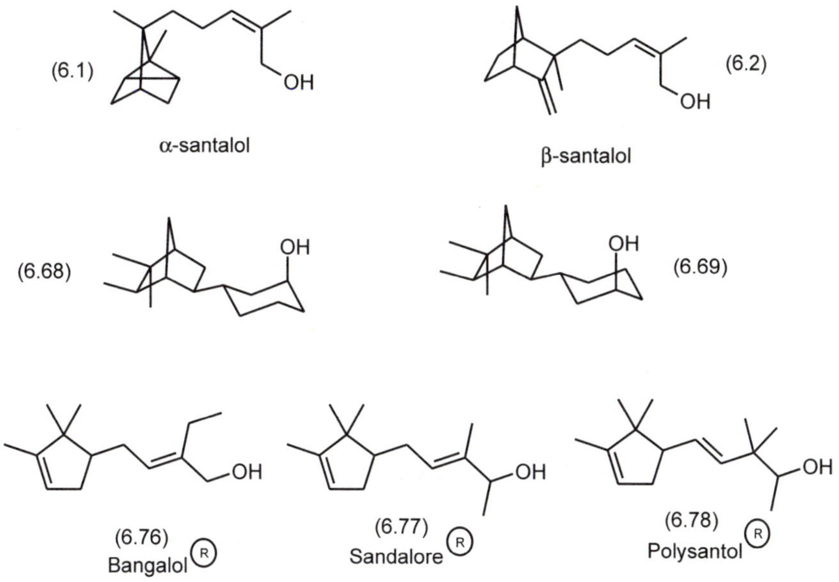

Figure 10.1

data set. For example, using computerised molecular modelling, it is possible to calculate the electrostatic and steric fields around a molecule and then seek a pattern in these. Similarly, the electron topological approach looks at the frontier orbitals of the molecules in the study. Both of these techniques relate to biological properties of molecules and are based on the concept that, as a substrate molecule approaches its biological target, usually a protein, the target will initially sense the electrons in the outermost orbitals of the drug, odorant or whatever. An adequate description of any of these techniques is far beyond the scope of this book and the reader who wishes to know more is directed to specialist textbooks such as that by David Livingstone.[10.2] As far as smell is concerned, the excellent review by Karen Rossiter.[10.3] gives a very good account of the statistical, and other (Q)SAR, techniques which have been applied to structure/odour correlations.

QSARs are second nature to the discovery chemist and contribute much to the search for efficacious novel molecules. However, there are a number of points which must be borne in mind when using QSARs.

The quality and consistency of the data which is fed into a QSAR is obviously very important. If the data is acquired from a number of different sources, then measurement techniques may vary from one to another with resultant variation in the results. For example, if one data source is measuring length in metres and another uses imperial yards, then simply taking their figures without any attempt to convert one unit to the other, would produce a fairly meaningless QSAR. This example is designed to make the point by using a very obvious error which would be most unlikely to go undetected. However, similar, but much less obvious, errors can and do make their way into QSARs with the inevitable result. The accuracy and precision of data can also vary from one source to another and are often affected by sample purity. A small amount of a very potent ingredient can produce a false positive result from an inactive compound. Conversely, the activity of an active ingredient might go undetected because of the presence of a trace of a potent inhibitor of that activity. One way in which such false results can become apparent is that they may appear as outliers in a statistical QSAR. An outlier is a result which is far away from the pattern of the other results. Genuine outliers do exist and study of them can often give useful information about the mechanism of the activity in question. However, the first thing to do when one encounters an outlier is to check the input data on which it is based.

Molecules are not rigid. Rotation is possible around all single bonds in acyclic structural fragments and rings can flex to a greater or lesser extent depending on their exact nature. There are many software packages available nowadays which can compute the relative free energies of different conformations of a molecule and predict a minimum energy configuration for it. Such computerised molecular modelling is very useful in that it gives an indication of the likely shape of a molecule. However, there are two important points to remember. First, at the absolute zero of temperature, the minimum energy configuration will be adopted but at ambient temperature the molecule is likely to have sufficient energy to occupy a number of higher energy states as well. Furthermore, if it is interacting with a protein, the energy stored in the tertiary structure of the protein may well be sufficient to force the ligand into quite a high energy conformation. Secondly, the energy minimisation software works on gas phase effects and a molecule in solution phase may experience very different forces, for instance from the formation of hydrogen bonds with the solvent. In summary, conformational flexibility is the bane of the molecular modeller's life.

By their very nature, QSARs must always be interpolative. Given a training set of compounds, all of the parameters upon which the QSAR model is built will centre around the properties of the training set. Predictions for new compounds will always direct the researcher back into the training set and molecules with radically different structures will not be predicted as likely to be effective. In other words, QSARs will find the best member of a known set of chemicals but fundamental leaps into new types of structures will come from random screening or serendipity rather than QSARs. To illustrate this point we can look at the area of musks. Figure 10.2 shows the structures of 10 materials which possess musk odour. The top row shows three naturally occurring musks, *viz.* muscone (10.1), hexadecanolide (10.2) and civetone (10.3). The next row shows three nitromusks, *viz.* musk xylene (10.4), musk ketone (10.5) and musk ambrette (10.6). The third row comprises the polycyclic musks Tonalid® (10.7), Galaxolide® (10.8) and Traseolide® (10.9). A QSAR on macrocyclic musks would result in the design of new macrocyclic materials since that is the only experience available in the dataset. The discovery of the nitromusk family was the result of serendipity as they were made during work on potential explosives in the TNT (trinitrotoluene) series. Similarly, a QSAR on nitromusks would predict more nitromusk structures. Neither of these QSARs would lead us to the polycyclic musk family. The range of predictions would increase if a new QSAR were to be developed using compounds from all three families. However, even that combined QSAR would be most unlikely to predict a structure such as that of the terpenoid-derived musk Helvetolide® (10.10) since its structure is so distinct from the others. For example, it is

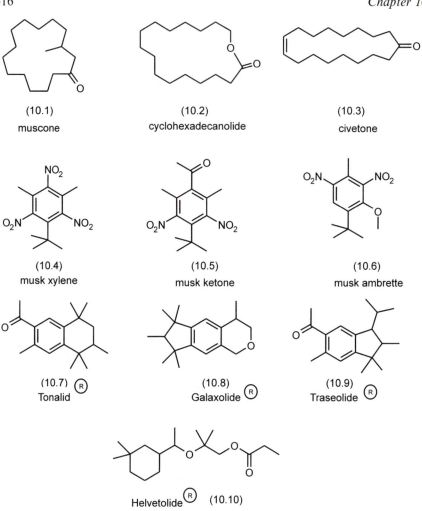

Figure 10.2

by far the most conformationally mobile, the others all having fairly rigid
molecular structures. A QSAR model built solely on rigid molecules is
unlikely to indicate that a flexible molecule could possess the same
biological properties.

Having acquired a large dataset of active and inactive materials and
QSARs derived from it, it becomes very tempting for the chemist to
postulate mechanisms of action based on the QSAR model. For
example, consider the case of the musk family of odorants represented
by structures (10.1) – (10.9) in Figure 10.2. One SAR model derived from

the structures of the macrocyclic, polycyclic and nitromusk materials, suggests that musks should have a molecular shape resembling a flat oval disc of certain dimensions, which is essentially hydrophobic and contains an oxygen atom at one "corner".[10.4] It is easy to see that the first nine musks of the figure fit this model and, indeed, one can imagine the more recently discovered Helvetolide® folding round into a conformation which would allow it to adopt a shape to fit this mould. It is therefore easy to suggest that the receptor contains a tray or cleft of the same size and shape and that this binding pocket is lined by hydrophobic fragments of the protein structure except for a hydrogen bond donor at the corner where the complementary acceptor (the oxygen atom) can bind. QSAR models are valid and useful tools even if they do not give mechanistic information. If we try to extract mechanistic information from them, there are a number of intellectual traps too easy to stray into and the chemist must always be alert to these.

Science advances by the formulation of theories. The scientific method comprises observation of the facts, hypothesis of a theory which accommodates all of the facts and testing the theory by using it to predict the results which are unknown at the time. No theory can ever be proved, it can only be disproved and so is valid only as long as there are no known exceptions. The platitude "The exception that proves the rule" is misunderstood by most of those who use it. They forget that here, the word "proves" is being used in its original sense of "tests" rather than in the modern sense of "establishes beyond doubt". Any exception means that the rule is either inadequate or invalid and must be revised or rejected. QSAR models are based on statistics and outliers tend to be ignored or played down. However, in terms of understanding a biological mechanism, they can be the most useful of all the experimental data. Why should one particular molecule, the structure of which falls outside the pattern of the model, still behave in the same way as those which fit the model perfectly? Tackling this question could give more information about the real mechanism than speculating about the significance of the core features of the model. Of course, if one's aim is purely to synthesise more candidates for testing, then the outlier is not such a good lead to follow as it is less likely to give active candidates in the short term.

Another intellectual trap is that of failing to distinguish between cause and effect. Very often, someone finds a correlation between two parameters and assumes a causal relationship without asking if this correlation could simply be between two effects of a common cause. For example, finding a correlation between the infrared spectra of a set of materials and a given biological activity does not imply that molecular vibrations are the cause of the activity. The biological activity and the spectra are separate effects of a common cause, *i.e.*, the molecular structure.

It is also important to consider how many discrete steps are involved in the process being modelled. In most cases, when the activity involves living organisms, there will be a number of steps leading up to the final point at which the drug or fragrance molecule exerts its final effect. For example, there will be various transport mechanisms involved in getting the substance from the point of administration to the site of action. In consequence, we cannot assume that any correlation between agent and effect is due to only one of these. Often it is assumed that only the interaction between the agent and its ultimate biological target is of significance and that structure/activity correlations will give mechanistic information about this specific event. This is not necessarily the case.

10.5   DESIGN THROUGH UNDERSTANDING

Chemists are rational creatures and so have an inbuilt tendency to try to design new molecules based on an understanding of the chemistry involved in achieving the desired effect. This is, in principle, the best way of approaching molecular design. The difficulty is that the total pattern of physical, chemical and biological interactions is usually unclear and exceedingly complex. A good discovery chemist will have as good a grasp as possible of, not only these physical, chemical and biological processes, but also of the legislative and commercial realities of the industry in which he is operating. These are different for each application and so I will now concentrate on perfumery and the design of novel fragrance ingredients. This is the chemical application area which I understand best and the problems encountered in it do serve as a good illustration of the general principles involved in discovery chemistry.

10.6   PERFUMERY

As stated earlier, the discovery chemist must understand the application for which he is designing novel ingredients and so it is necessary at this point to give a brief overview of the technology of the perfume industry. Readers who wish to know more than what follows here, could consult one or more of the texts recommended in the bibliography. For an overview of the industry, the books by Pybus and Sell,[10.1] Curtis and Williams[10.5] and Muller and Lamparsky[10.6] are recommended. For a deeper understanding of the art of perfumery, the book by Calkin and Jellinek[10.7] is a very good starting point.

Perfumery is an ancient art. References to it are available in the writings of all ancient civilisations. The Egyptians were skilled perfumers as is evident from their hieroglyphic writings and, in the Jewish Torah, we read that the incense for the tabernacle which Moses built in the desert of Sinai, was the work of a perfumer.[10.8] Herodatus included a whole chapter on perfumery in his "Histories".[10.9] In the days of the ancient Egyptian and Greek civilisations, the distinction between perfumery and pharmacy was very blurred and even today, many plants which are used in perfumery are also used for herbal medicine. People went to great lengths to find the raw materials for their perfumes, as we read in Herodatus, and a very good illustration of this is found in the New Testament where Mary Magdalene used a perfume containing Spikenard to anoint Jesus.[10.10] Spikenard is an extract rich in terpenoids which is extracted from the plant *Nardostachys jatamansi*. This plant is known to have medicinal properties such as anti-spasmodic and stimulant activity and is also known by the Indian name of jatamansi which is derived from Sanskrit and is mentioned in Ayurveda where these properties were first described. It grows only at high elevation in the Himalayas and so is an expensive ingredient today just as it was 2000 years ago when Mary's bottle of it was described as being worth a year's wages.[10.10] Marco Polo's trek to China and Columbus' voyage to attempt to find a new route to the East Indies were both motivated, in part, by the trade in spices as Europeans sought to make the fragrances of the far Eastern cultures more accessible.

Perfumery, like so many aspects of our lives was changed dramatically by the emergence of synthetic organic chemistry. The classic fragrance Jicky, launched in 1889, was one of the very first to use the products of synthesis. It contained coumarin, heliotropin and vanillin all of which were prepared in factories using the Reimer–Tiemann reaction as part of the process. The Reimer in question founded a perfumery company in Holzminden in Germany and it still bears his name today. These three chemicals are, of course also available from natural sources; coumarin from tonka beans, heliotropin from heliotrope flowers and vanillin from vanilla beans. However, Reimer's chemistry made them available at a lower price, a more consistent quality and a more secure supply. Synthetic ingredients which are the same as chemicals present in a natural source are known as nature identical materials. The first highly successful use of non-nature identical materials in perfumery was that of the inclusion of aliphatic aldehydes in Chanel 5, launched in 1921. This most famous of fragrances uses many natural ingredients such as rose and jasmine but it was the aldehydes which gave it its unique first impact. The perfumer who created it, Ernest Beaux, said, "One has to rely on chemists to find new aroma chemicals creating new, original notes. In perfumery, the future lies primarily in the hands of chemists." The use of novel ingredients to create new fashion trends in fine perfumery is only part of the story. The modern perfumery industry creates fragrances for all sorts of applications such as soaps, detergents, hard surface cleaners, lavatory cleaners and so on. These applications put serious demands on the ingredients, demands which natural materials such as rose and jasmine would not be able to meet.

Humans are capable of perceiving a seemingly endless range of smells and of judging very small differences between them. However, most people do not have a very good vocabulary for describing what they experience. So, one of the first things the beginner has to do is to develop a fragrance vocabulary. Fragrance companies usually have an agreed odour classification and everyone who works in one company will learn the language of that company. This is an important point in terms of structure/odour correlation as we will see shortly. We do not have fixed reference points for odour type and so all of our odour descriptions are associative, *i.e.* we describe the smell of one ingredient by comparing it to that of something which it resembles. Again, this is a serious issue for structure/odour correlation.

Common categories used for odour classification include citrus, herbal, fruit, floral, spice, wood and amber. Each is then sub-divided. Under 'citrus', we will find orange, lemon, lime, mandarin, grapefruit and bergamot. 'Herbal' will include basil, rosemary, thyme, sage and so on. 'Fruit' includes category apple, pear and banana, and usually there are two significant sub-categories of red fruit (*e.g.* raspberry, blackberry and blackcurrant) and tropical (*e.g.* mango and passionfruit). Floral is a large category and includes, for example, rose, geranium, lilac, honeysuckle and violet. Spices include ginger, clove and pepper, all of which (like the fruits) are used in perfumes as well as flavours. The sub-divisions of wood and amber have been discussed in earlier chapters.

Perfumers divide these various notes into top, middle and bottom or, more in keeping with the industry image, into head, heart and base. These classifications are based on a combination of volatility and impact of ingredients. The head notes are those which are most obvious on opening a bottle of perfume and the citrus notes, for example, fall into this category. The heart are the medium volatility notes which are responsible for the basic character of the perfume and the base notes are the least volatile notes which help to hold the others back and blend them together. The base notes become more obvious after the rest of the perfume has evaporated. A well balanced fragrance will typically have about 25% head, 50% heart and 25% base. Apart from the fruits, all the examples listed above contain many terpenoid components in their essential oils. Components which are responsible for a significant part of the characteristic smell of a plant, are known as character impact compounds. Figure 10.3 shows the structures of terpenoid character impact compounds for some of the odour types mentioned above. Chapters 6 and 7 contain many examples for wood and Chapter 8 for amber.

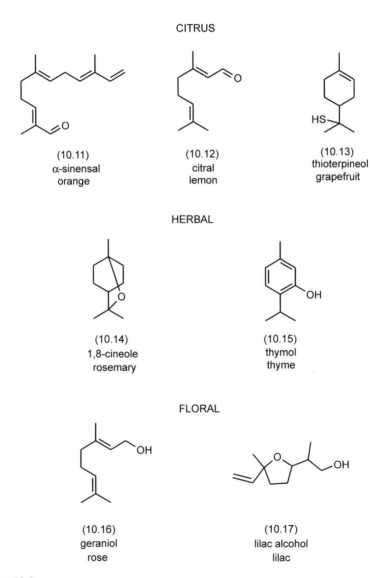

CITRUS

(10.11)
α-sinensal
orange

(10.12)
citral
lemon

(10.13)
thioterpineol
grapefruit

HERBAL

(10.14)
1,8-cineole
rosemary

(10.15)
thymol
thyme

FLORAL

(10.16)
geraniol
rose

(10.17)
lilac alcohol
lilac

Figure 10.3

10.7 REQUIREMENTS OF FRAGRANCE INGREDIENTS

In order to design a novel fragrance ingredient, it is important to know what will be required of it since all factors must be taken into consideration in the design. We will look at these factors under the five headings of safety, odour, performance, additional benefits and availability. In some instances the requirements will pull in opposite directions. For example, a requirement for stability in hypochlorite bleach means that a molecule must be resistant to a pH of 14 and to oxidants of the strength of hypochlorite. Such a molecule is likely to prove resistant to the enzymes of the bacteria present in a sewage treatment plant and thus it is less likely to satisfy criteria for rapid biodegradation. In other cases, there is no "correct" result for a given parameter. For example, in the case of odour character, a chemist might have had the intention of designing a new sandalwood ingredient but instead produced a superb muguet (lily of the valley) material. This would not be seen as a "wrong" result but rather as a different opportunity. Incidentally, this last example is neither random nor far-fetched. The structural requirements for a molecule to possess a sandalwood odour are similar to those for a muguet scent. Having a superb odour, however, is not a sufficient justification on which to develop a novel material. No matter how wonderful its odour, it will not be successful if it cannot be produced at a price which the customer is prepared to pay and it most certainly will never see production if it exhibits undesirable toxicological properties. Thus, each decision to develop a novel fragrance ingredient is based on a consideration of all of the criteria in balance with each other, rather than on any single one of them.

10.7.1 Safety

Safety is the single most important criterion for any novel product. If it cannot be produced safely, used safely and disposed of safely after use, it will never be developed. Safety in manufacture was discussed in Chapter 9 and so will not be covered here, except to say that a good discovery chemist will be aware of the issues involved and will endeavour to design molecules which he knows or believes would be amenable to manufacture through safe processes.

10.7.1.1 Safety – In Use. Product safety is of paramount importance to the chemicals industry. No company wishes to harm its customers and so will test products to ensure that there is no risk in the use of the product. There is also a legal requirement in every country to ensure safety for use of all chemicals which are produced in anything other than small quantities used in research.

Fragrance ingredients are based on the natural chemical components of essential oils and plant extracts. Some of these natural chemicals do have undesirable properties. For example, bergamot is a citrus fruit, the oil of which is one of the major components of the classic cologne family of fragrances. However, bergamot contains a furanocoumarin derivative known as bergapten, which is phototoxic, *i.e.* produces a skin rash when placed on the skin and then irradiated with the ultraviolet component of sunlight. Therefore nowadays, bergamot oil is always treated to remove the bergapten before being used in perfumery. However, this is an exceptional case and the vast majority of natural fragrance chemicals have very low toxicity towards humans. Since synthetic fragrance ingredients are based on these naturals, they too are unlikely to display toxic properties. Nonetheless, all fragrance companies go to considerable lengths to ensure that the risk of harm from use of their products is as low as possible.

The major concern is that of skin sensitisation. When a material comes in contact with the skin, it may react with a protein present in the skin to form a modified protein and this could become an antigen as far as the immune system is concerned. If this happens, the immune system will produce antibodies to combat the antigen. If the immune system over-reacts, an allergic reaction will result and a rash will break out on the skin. Thus, skin sensitisers do not produce a reaction on the first contact with the skin, but only after repeated exposure. Fortunately, we now know the basic features of molecular structure which are associated with skin sensitisation and can design molecules which will not behave in this way. This is a good example of the importance of safety SARs to the discovery chemist. Of course, predictions are not the ultimate test and therefore many companies, especially in the pharmaceutical industry, invest considerable amount in the development of rapid *in vitro* screens to detect any adverse toxicological effects.

In order to introduce any new chemical onto the market, a company must present a dossier of safety tests to the government of each country in which it intends to produce, use, sell or even transport the material. Government appointed experts will then decide whether or not the material is safe for the stated purpose. Penalties for not following the statutory procedures are so severe that compliance with the legal system is total. Most companies, particularly the large multinational ones, are strongly ethical in their safety policies and actually exceed the standards laid down in the law. Similarly, they will operate to the same high standard in every country even if the legislation is less stringent in some.

10.7.1.2 Safety – In the Environment After Use. In Chapter 9, we considered the environmental issues relating to the production process. Obviously, the same consideration must be given to the environmental fate of the desired product. In the case of fragrance materials, their fate in sewage treatment plants is the most important factor to consider since the largest volumes of fragrance are used in products such as laundry powder, soaps, shampoos and other cleansing products and they find their way into waste water treatment plants after use. Therefore, the discovery chemist should be aware of the biodegradability test methods described in the previous chapter and should use SARs based on results from these in the design of novel molecules.

10.7.2 Odour

Odour has three properties, *viz.* character, intensity and tenacity. The character is the property by which we recognise the material and is usually named by association to a natural source with a similar odour, for example, rose, sandalwood, lemon and so on. The intensity is the perceived strength of the smell and is sometimes measured against a standard of defined intensity or in terms of the odour threshold, *i.e.* the lowest concentration at which it can be detected. The tenacity, also known as persistence, is the length of time for which the smell lasts after it has been applied to a surface or medium.

All three are subjective and can only be measured in sensory terms using a human subject as the measuring instrument. Because of the subjectivity of the measurements, the best practice is always to use a panel of subjects and to give the result as a mean with a standard deviation. For example, when measuring the odour threshold of a material, we may find that some people can only detect it at 10 ppm whereas others can reliably detect it at a concentration of 1 ppm. If the spread of results follows the normal Gaussian distribution, then in this case, the mean threshold of detection might be 5 ppm. The reliability of any one subject's assessments will be measured by giving them repeated triangle tests. In a triangle test, the subject is presented with three samples and asked to identify the odd one out. So, when measuring threshold, two of the samples will contain the odorant at a given concentration and the third will be a blank sample devoid of odorant. Of course, there could equally be one sample with odorant and two without. Only if the subject can correctly identify the odd sample every time, is it considered that they can detect the difference. So, for example, for a subject who could reliably determine the odd one out when the odorant is present at 5 ppm but not when it is present at 4 ppm, the material would be considered to have a threshold of 5 ppm. Some workers report threshold figures which have been measured rigorously in this way. At the other extreme, some workers report assessments using a single subject and without using triangle tests. In between these extremes lies a continuum of intermediate degrees of rigour. This presents a serious pitfall for anyone attempting to carry out SAR work using literature data because the data input for an SAR must be consistent for the SAR model to be valid. If the various pieces of data have been measured in different ways, then the results of the SAR are questionable. For further reading on the subject of odour measurement, I would recommend the chapter by Neuner-Jehle and Etzweiler in the book by Müller and Lamparsky.[10.6]

Sample purity is always important when measuring properties for SAR work. In the field of olfaction, the organoleptic purity (*i.e.* odour purity) is even more important than chemical purity, especially when measuring odour character. A small trace of an impurity with a very intense odour is likely to lead to incorrect results in the odour measurement and thus render the data, and hence any resultant SAR, invalid. For example, 1 ppm of thiomethanol will distort the odour of geraniol far more than would 10% of citronellol. Therefore any SAR on odour will be suspect if the data originates from a source which is unfamiliar with measurement of organoleptic purity.

10.7.2.1 Odour Character. Most odour correlation work has been done on character since it is, superficially, the easiest to measure. However, this is a misconception based on the simplistic assumption that, since a subject presented with, for example, samples of rose oil and eucalyptus oil will normally correctly identify the botanical origins of the samples, it can be taken that all odour character descriptions are equally facile. However, this is not the case. For example, one serious issue is that the description of an odour is associative since we have no hard reference points.

Odour character is subjective at three levels and this introduces problems in trying to find a coherent set of data. The first level of subjectivity is one of semantics. Everyone knows what a rose is and most people use the term rose in a similar way when describe an odour. However, what is meant by amber? To the layman, it probably relates to fossilised resin and has nothing to do with odour. To those with some experience of perfumery, it will relate to an odour type but the exact meaning of the word will vary from person to person, unless they are working to an agreed standard for the amber odour. Therefore the issue of semantics can be overcome by training to a standard language with an agreed set of standard odours. Of course, there will only be data consistency between people trained in the same way and comparison between the results from different groups of workers is prone to errors arising from differences in terminology. The second level of subjectivity is that of hedonics. The sense of smell is closely linked to memory and emotions because of the physical, and hence neurological, proximity of the olfactory bulb to the hypothalamus in the brain. The olfactory bulb is a part of the brain which connects with the olfactory receptors and where the initial signal processing is carried out. The hypothalamus is another part of the lower brain and is associated with memory and emotion. Therefore an odour, even before the higher brain is conscious of its presence, can trigger memories and emotions and these can distort the evaluation of it. For instance, some people will associate the odour of eugenol with apple pies since cloves are often used as spices in them. Other people will associate the smell of eugenol with visits to a dental surgery because of its use in dentistry. The contrasting positive and negative emotional responses could therefore affect the cerebral cortex as it begins to identify the odour. In my experience, odours which trigger warnings through memory or emotion usually provoke a more intense reaction than those with no such alarm mechanism. This makes sense in evolutionary terms since a large part of the reason why we have a sense of smell is to provide warnings of danger whether through fire, poison, presence of predators or whatever. Thus if an odorant triggers various patterns in the brain, one which is associated with past trauma in one individual is likely to override the other aspects in that subject's judgement. This aspect of subjectivity can be dealt with if each evaluator is aware of such triggers in their own response to smell. The third level of subjectivity is much more difficult to handle in odour evaluation panels. It is easily demonstrated that each of us live in our own sensory universe. This can be done using a simple triangle of samples as shown in Figure 10.4.[10.11] In this experiment, a group of people are exposed to the odours of $(+)$-Z-α-santalol (10.18) and cyclopentadecanolide (10.19) and told that these represent the characteristic odours of sandalwood and musk respectively. They are then given a sample of Bangalol® (10.20) and then each of them is asked whether, in their opinion, its odour is closer to that of musk or sandalwood. Overall, about 40% of people perceive Bangalol® as musk whereas 60% perceive it as sandalwood. This phenomenon is not unique to Bangalol® or to these two odour types. In other words, we each perceive odours in our own individual way. The discovery chemist must therefore work with the majority verdict on each ingredient and accept that there will be an element of variability in any results he produces from a structure–odour correlation.

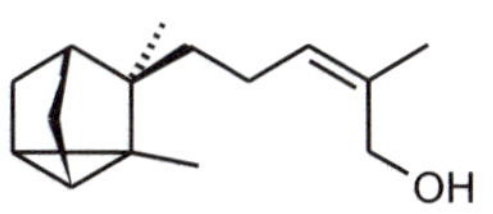

(10.18)

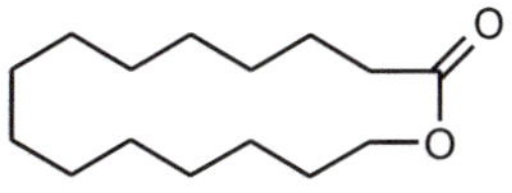

(10.19)

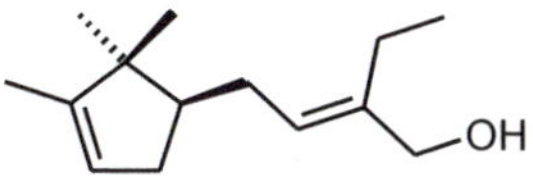

(10.20)

Figure 10.4

Language is an important factor in describing odours. Most people have a relatively poor odour vocabulary and the degree of precision of odour descriptions increases dramatically when evaluators are trained in odour language. Of course, different organisations use different language conventions and so a comparison of data from them is suspect because different results would have been described using different terms of reference. The nature of the language used will influence the basic odour perception. For example, Morrot, Brochet and Dubourdieu[10.12] have shown that the language used by oenologists to describe the aroma of red wines is different from that used to describe the aroma of white wines and that, when a tasteless red dye is used to trick an expert into using the "wrong" mental frame of reference, there is a dramatic effect on the perceived sensation. As before, using data gathered from different sources will probably result in an inconsistent data set and therefore a questionable SAR.

Related to the issue of language, is the problem of odour classification. Different groups use different classification systems and so data gathered from different laboratories is likely to be inconsistent. There is a more fundamental problem also in that we usually classify odours based on their biological source. For example, the classifications of fruit, flower and wood are almost universal and are obviously based on their natural origin. The sub-divisions of each are also based on botany. Thus, fruit divides into apple, orange, banana, mango, *etc*; flower into rose, jasmine, gardenia, honeysuckle, *etc*; and wood into pine, cypress, cedar, sandalwood and so on. Our conscious brain classifies them in this way because of associations with plants and the botanical relationship between plants.

However, this does not necessarily have any relevance whatsoever to the process of olfaction. One example is that of apples and pears. The characteristic odour of each does depend on having an ester or equivalent osmophoric group present but the requirements regarding the steric environment of the group are very different for the two.[10.13] Lumping the two together under the general description of "fruit" may be very tempting because of the morphological similarities between apples and pears and the respective trees, but in terms of SAR it introduces a distortion into interpretation of the data.

10.7.2.2 Odour Intensity. Odour intensity can be measured either as perceived intensity against a standard or as a threshold of detection. The measurement of the latter is used above as an example of odour measurement. Measurement against standards introduces a huge potential for variability between subjects and inconsistency in any one subject. For example to ask someone to measure the intensity of a woody material using a rose odorant as a standard, or even another woody material as a standard, will rely on that subject's relative sensitivity to the two odours in question and this might not be the same for another person. Therefore it might seem better to measure threshold as an indication of intensity. However, that also is fraught with issues. The relationship between perceived intensity and concentration is governed by Stevens' Law and Figure 10.5 shows a Stevens' Law plot for two odorants. The log of the concentration is plotted on the *x*-axis and the log of the perceived intensity is plotted on the *y*-axis. In each case there is a straight line relationship between the two logarithmic values. Of course, we must also remember that the slope and intercept

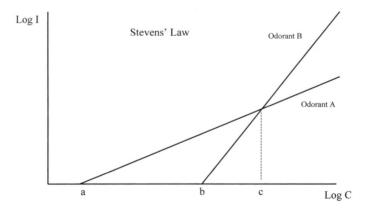

Figure 10.5

for each plot will vary from one individual to another and normally we are working with averages across a group of people. The threshold concentration of odorant A is at concentration *a* and for odorant B, at concentration *b*. If we compare the intensities of the two odorants at a concentration below *c*, then odorant A will be described as the stronger of the two. On the other hand, if they are compared at a concentration above *c*, B will be considered stronger. Thus odorant A has the lower threshold, but whether or not it is more intense than odorant B will depend on the concentration at which the intensity is measured and so threshold is not a sufficient criterion for evaluation of intensity. There are many good SAR models for odour character but none for intensity and this is possibly partly due to the difficulties involved in measuring the latter.

10.7.2.3 Odour Tenacity. The tenacity, or persistence, of an odour is simply the time for which it remains detectable after application to a surface or to a medium. The simplest test for tenacity is to dip a strip of absorbent paper (known as a smelling strip or perfumer's blotter) into the liquid and leave it to stand on a desk. The time taken for the odour to disappear is a function of both physical–chemical and sensory properties. The rate at which material evaporates is elated to its vapour pressure (hence boiling point) and so a low boiling material is likely to be lost faster than a high boiling one and therefore have lower tenacity. However, if the material is too high boiling, then it will never evaporate and thus never reach the nose and therefore will be odourless. Physical interactions (such as hydrogen bonding and van der Waals forces) with the paper surface will slow down the rate of evaporation and therefore increase the tenacity. Clearly, these forces will depend on the nature of the surface as well as the nature of the odorant and so the tenacity of any odorant will vary according to the surface from which it is evaporating. If that surface is the skin, then a whole new set of parameters enters into the equation since enzymes and bacteria present on the skin may well degrade the odorant and therefore reduce its tenacity. Until the material is lost from the paper, there will be an equilibrium concentration of it in the air around the paper (known as the headspace) and the ability to perceive this is related to the Stevens' Law plot of the molecule in question. Thus, perception of a material with a low odour threshold will require only small amounts in the headspace. Conversely, a material with a high odour threshold will require a higher headspace concentration and if its volatility is too low, it will not be perceived. Thus, intensity is superimposed on the physical–chemical parameters in determining tenacity.

10.7.2.4 Mechanism of Olfaction. Medicinal chemists seek to understand the molecular mechanisms of disease states in order to help in the intelligent design of potential new pharmaceutical agents. Similarly, fragrance chemists have sought to understand the mechanism of olfaction in order to help in the design of novel odorants. Until fairly recently, the dream was that the olfactory code could be cracked by identifying the specific receptor requirements for individual odours and then designing perfect substrates for the olfactory receptors. However, recent developments in our understanding of the mechanism of olfaction have revealed that, for the foreseeable future, this is no more than a pipe dream.

In comparison with fragrance design, the medicinal chemist has an easy task. The medicinal chemist tries to design substrates for a single target protein and this protein will be one with a very tight set of criteria regarding the structural features of its substrate. This latter feature results from the fact that hormone receptors, enzymes and so on, have all evolved to recognise one substrate and one substrate only, otherwise the chemistry taking place in organisms would run out of control. The sense of smell has evolved to allow organisms to detect as wide a range of chemical signals as possible, even molecules which the organism has never encountered before. Indeed, many thousands of molecules which never existed in the universe before a synthetic chemist produced them in the laboratory can be easily detected by the human nose. This is because our olfactory receptors are very broadly tuned and each receptor type responds to a range of odorants. We have genes for over 1000 different types of olfactory receptors and latest estimates are that at least 350 of these are expressed in humans. Thus each odorant molecule is detected by a receptor array and interacts with numerous receptors in that array. The pattern of receptor interaction is the key and so, to understand the role of a single odorant, we must know the structures of every receptor with which it interacts and how all of the resultant signals are processed at each stage in the various parts of the brain. We are a very long way from being able to do this and the cost of carrying out all the research would require an investment that would never be recovered by the fragrance industry. A brief review of the subject was published by the author.[10.14]

The olfactory receptors belong to the 7-transmembrane–G-protein coupled family of receptors. A schematic structure is shown in Figure 10.6. They are coupled to G-proteins in the same way as is rhodopsin and generate the nerve signal through a similar second messenger system to that used by rhodopsin (see Chapter 8). They differ from rhodopsin in that they lack the additional tether back into the cell

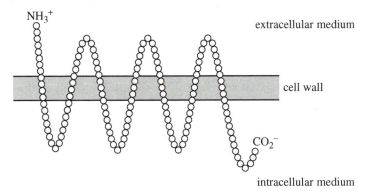

NH$_3^+$

extracellular medium

cell wall

CO$_2^-$

intracellular medium

Figure 10.6

wall and the need for a co-factor. The odorant is detected on the receptor surface on the outside of the cell and this induces a change on the intracellular part of the receptor protein which, in turn, sets the G-protein/second messenger system into operation.

The olfactory receptors are located in the cells of the olfactory epithelium. This is a yellowish patch of tissue, several square centimetres in area which is found on the roof of the nasal cavity, approximately on a level with the eyes. It is about 100–200 µm thick and the cells terminate in hairs, or cilia, which are 20–200 µm long and are bathed in an aqueous liquid known as the olfactory mucus. The mucus is about 35 µm thick and flows backwards across the epithelium at about 1–6 cm min^{-1}.

In order to reach the olfactory epithelium, a molecule must be volatile at ambient temperature and pressure. Thus fragrance molecules have a molecular weight under 300 Da and generally are hydrophobic in nature. This raises an interesting question, *viz.* since odorants are hydrophobic, how can they cross the aqueous mucus to reach the receptor cells? The answer lies, at least in part, in another family of proteins, the odour binding proteins (OBPs). These proteins are members of the family of lipocalins. Liopocalins are proteins which serve to transport small molecules around the body and the OBPs are a sub-set which have developed to transport small hydrophobic molecules across the mucus to the receptor cells. Their exact role is not fully understood. They may serve simply in transport but, as they are proteins, they might also play a part in recognition either through selective transport or through the receptor cells discriminating between substrate-bound and free OBPs. Such a mechanism is known to operate with some other lipocalins. Each OBP is really a complex of two cup-shaped proteins with a "hinge" formed by the amino acid backbone of one passing through that of the other. Once they have bound a substrate, the cups come together at the rims and so form a spherical hydrophilic unit which can easily travel through the mucus. This process is shown schematically in Figure 10.7 However, our sense of smell is even more complex than this since there is another part of the nervous system involved in detecting odorants. The trigeminal nerves which are found all around the nasal cavity are involved in the detection of irritants but also respond to about 70% of odorants. Vanillin is detected only by the olfactory system but the odour of nicotine is a composite of signals from both the olfactory and trigeminal systems. Indeed, odour discrimination between the enantiomers of the latter actually seems to be achieved through the trigeminal rather than the olfactory system.

The olfactory cells run through the cribriform plate at the base of the skull and feed directly into the olfactory bulb. Neurons from the epithelium converge on areas of the olfactory bulb known as glomeruli. Evidence suggests that all of the signals from one type of receptor converge on a common glomerulus. There is considerable processing of these signals in the olfactory bulb and the output travels upwards *via* the thalamus to the cerebral cortex. It is only in the cortex, the highest part of the brain, that the phenomenon of odour exists and nerve signals lower down cannot be correlated with odour. The final odour impression is a synthesis of signals from the olfactory epithelium, the trigeminal system, the taste receptors on the tongue and all of this is subject to the influence of other inputs such as vision (as mentioned earlier). In order to understand the significance of any individual odorant/receptor interaction, it would be necessary to understand the detail of the entire processing system. Hence the conclusion that knowing what takes place at the olfactory receptors will not help in the design of bespoke odorants.

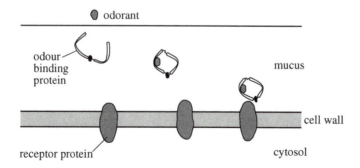

Figure 10.7

10.7.3 Performance in Formulae

A fragrance is a harmonious blend of individual perfume notes, just as a chord is a harmonious blend of individual musical notes. In view of the complexity of odour perception as seen above, it is not surprising that the way in which one perfume ingredient will blend with others is not easily predictable. The only way to find out how well a novel ingredient works in perfume formulae is by experience and so an essential part of the screening of novel ingredients is their use by perfumers in test formulae.

10.7.4 Performance in Product

The word perfume usually conjures up images of elegance, haute couture and precious liquids in exquisite glass bottles. This is an important part of the business. However, nowadays there is another side to it. We now expect, often unconsciously, to find pleasant fragrances in a plethora of consumer goods from cosmetics through soaps to detergents, household cleaners and even bleaches. The use of essential oils in such products is restricted by their chemical stability. For example, linalyl acetate, a component of lavender oil, would be hydrolysed in dishwash powder, laundry powder and even soap because of the high pH of these products (13–14, 10–11 and 9–10 respectively) and it is also susceptible to the oxidants present in the first two. Anti-perspirants contain aluminium chloride as the active ingredient and their pH is consequently rather low (about 3.5). Many natural odorants are unstable to this degree of acidity and so more stable analogues must be sought if we are to have a pleasant smelling product.

Table 10.1 shows the percentage of fragrance ingredients which are stable at various levels of pH and also some common consumer products which would possess such a level. It is clear that, as the pH moves away from neutral, the perfumer's palette becomes increasingly restricted. The fall off is particularly severe on the alkaline side; this is partly because these products usually also contain bleaches such as perborate or hypochlorite. What makes matters worse for the perfumer is that the loss

Table 10.1

pH	Typical product	Stable ingredients (%)
1	Acid lavatory cleaner	25
3	Fabric conditioner	65
7	Perfume	100
10	Laundry powder	45
13	Dishwash powder	5

of notes is not uniform across the palette but often leaves gaps. So, for example, asking a perfumer to create a lily of the valley fragrance for an anti-perspirant is rather like asking an artist to paint a landscape without using green. A major target for discovery chemists in the fragrance industry is therefore to plug such palette gaps.

10.7.5 Additional Benefits

Nowadays, we expect perfume ingredients not only to work in difficult product media, but also to perform roles other than purely that of providing an odour. Many perfumery ingredients are repulsive to insects; citronella oil is perhaps the best known example. Inclusion of such ingredients into a fragrance will therefore impart some degree of insect repellency to it. One of the reasons for personal use of fragrance in the past was to cover up undesirable body odours. Some ingredients do this more effectively than others and some will also cover up malodours associated with cooking smells, pet smells, tobacco smoke and so on. There are various mechanisms by which such deodorant effects arise. These include chemical, anti-bacterial and sensory effects. By understanding these mechanisms we can design ever more efficacious deo-materials. Anti-bacterial effects of perfume ingredients (remember Chapter 1, plants often produce essential oils as a chemical defence against bacteria) can also be used for hygiene effects in perfumed products and to prevent spoilage of the products.

10.7.6 Availability

The issues concerning availability in terms of both supply and cost of ingredients have already been covered in Chapter 9 and so need not be repeated here. Suffice it to say that the discovery chemist should be aware of the problems which may be encountered by the development chemist and would be well advised to bear these in mind when designing novel structures. Indeed, awareness of available feedstocks and technologies often serves as a source of ideas for the discovery chemist and the matrices of targets generated in this way are ideally suited for exploration using synthesis robots.

10.8 FROM NATURAL TO HIGH PERFORMANCE, EXAMPLES OF DISCOVERY OF TERPENOID ODORANTS

In this last section, we shall look at some typical terpenoid fragrance ingredients in order to illustrate the flow from naturally derived materials to the high performance ingredients of the twenty-first century. The terpenoid hydrocarbons generally have weaker odours and are used mainly as feedstocks although they do play a part as top notes some in herbal and floral odours and are important in the overall character of lime. As far as perfume materials are concerned, the most important members of the terpenoid family are the oxygenated monoterpenoids. The higher molecular weight of the sesquiterpenoids results in their having lower vapour pressures than their monoterpenoid counterparts. Thus, sesquiterpenes are present at a lower concentration in the air above a perfume than are monoterpenoids with the result that they must have a greater effect on the receptors in the nose in order to be detected. Hence, a lower percentage of sesquiterpenes have useful odours than mono-terpenoids. For the same reason, very few di- or higher terpenoids have odours. However, those sesqui- and higher terpenoids which do have odours, are very tenacious because their lower volatility means that they are lost more slowly from perfumes. Such materials form the base of perfumes and serve also to fix the more volatile components.

The following sections each give some examples of fragrance ingredients illustrating the progress from natural, through simple natural analogues to high performance designed ingredients. The reader who is interested in finding more such examples and demonstrations of their use in well-known fragrances, could begin with the reviews by Frater, Bajgrowicz and Kraft;[10.15] Sell;[10.16] Kraft, Bajgrowicz, Denis and Frater;[10.17] and Gautschi, Bajgrowicz and Kraft.[10.18]

10.8.1 Examples of Natural Fragrance Ingredients

Figure 10.8 shows three monoterpenoid alcohols; geraniol (10.16), citronellol (10.21) and linalool (10.22); all of which have been mentioned earlier as being key fragrance ingredients which can be obtained from natural sources. Geraniol is one of the character impact components of rose but it is only present in rose oils to the extent of about 30%. Much

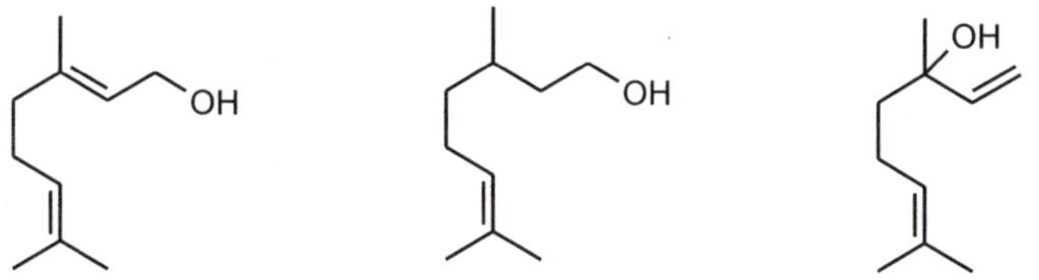

(10.16)   (10.21)   (10.22)

Figure 10.8

richer sources are Palmarosa Oil (distilled from the tropical grass *Cymbopogon martini*) which contains between 60 and 90% geraniol, according to the exact strain and place where it was grown, and the oils of some *Monarda* species which contain 85–95% geraniol. (Monarda is a herb which resembles the mint family.) Rose oil would never be a commercial source of geraniol because of the cost of the oil; however, palmarosa and monarda are much less expensive and so can be used to produce high quality geraniol. Citronellol (10.21) is present in rose at levels between 30 and 50%, however, as for geraniol, the high price of rose oil precludes its use as a source of citronellol. The less expensive oils in which it is found, contain much lower levels. For example it is present at only 10–15% in oils such as those of *Eucalyptus citriodora* and citronella (*Cymbopogon nardus*). There is therefore no commercially attractive source of citronellol. The richest source of linalool (10.22) is the leaf oil of the Ho tree (*Cinnamomum camphora*) which contains 90–97% of it. This oil can be used as a source of high quality natural linalool.

10.8.2 Examples of Analogues of Natural Fragrance Ingredients

The double bonds present in the three alcohols described above, render them susceptible to oxidation by air (autoxidation) and bleaching agents such as perborate and hypochlorite. It is a common practice in fragrance discovery to hydrogenate unsaturated odorants to see whether the odour quality can be maintained whilst increasing stability on storage and in products. Thus tetrahydrogeraniol (10.23) and tetrahydrolinalool (10.24) are obvious targets. Indeed both have become established fragrance ingredients and are of particular use in harsh product media. Having discovered that tetrahydrolinalool (10.24) is a useful fragrance ingredient, the obvious production method is to hydrogenate, not linalool (10.22), but dehydrolinalool (10.25) which is an intermediate in the manufacture of linalool, as was seen in Chapter 9. In fact, this hydrogenation is easier than the hydrogenation to linalool as it is not necessary to achieve selective reduction of the triple bond (see Figure 10.9).

Production intermediates often serve as sources of inspiration for discovery chemists. For example, we saw in Chapter 9 how pinane (10.27) (prepared by hydrogenation of α-pinene (10.26)), is used as an intermediate in the preparation of the rose alcohols by oxidation to pinanol and subsequent pyrolytic cleavage of the cyclobutane ring. If the pyrolysis is carried out without introduction of the alcohol function, the product is dihydromyrcene, or citronellene (10.28). Hydration of the trisubstituted double bond of dihydromyrcene (10.28) gives the alcohol

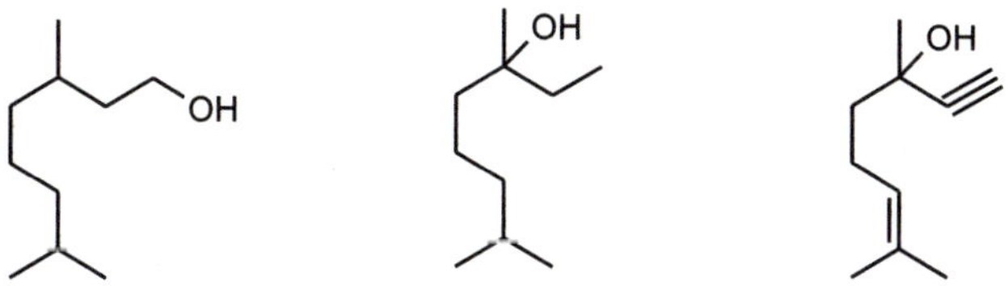

(10.23) (10.24) (10.25)

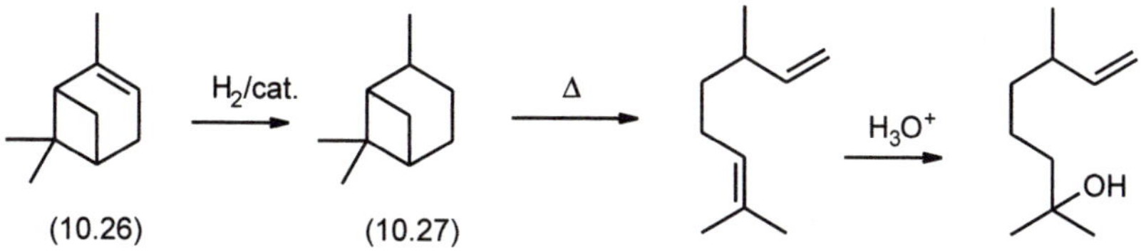

(10.26) H₂/cat. (10.27) Δ (10.28) H₃O⁺ (10.29)

Figure 10.9

dihydromyrcenol (10.29), first prepared in 1975. Dihydromyrcenol, because of its low price and stability relative to geraniol and linalool, would be expected to be a good ingredient for such applications as soap and laundry powder. However, it is also very useful in fine fragrances. In 1982, it was used in the men's fragrance "Drakkar Noir" to add citrus freshness to the basic fougere type perfume. Its ability to impart freshness was taken even further when, in 1988, higher levels of it were used in "Cool Water" and a new fashion in masculine freshness was created.

10.8.3 Examples of Designed Fragrance Ingredients

In the preceding section, we saw how the chemical stability of geraniol (10.16) and citronellol (10.21) is improved by hydrogenation. There is also a need to improve the substantivity of rose notes. This can be done by increasing the molecular weight and consequently reducing the vapour pressure meaning that the rate of evaporation is reduced and the material is lost more slowly. One tactic used in all branches of discovery chemistry is to use groups of similar size and polarity to replace each other in structures. If the role of the group is to occupy space, rather than to form hydrogen bonds to the macromolecular target, the groups are known as isosteres. The isobutenyl fragment occurs frequently in terpenoid chemistry and a suitable isostere for it is the phenyl group. The sizes of the two are similar and both are fairly hydrophobic but do contain π-electrons which can form non-bonded interactions or even weak hydrogen bonds. Substitution of the isobutenyl group of citronellol by a phenyl group, gives the alcohol (10.30) which contains 12 carbons and therefore is less volatile than citronellol. It is known under the trade names of Phenoxanol® or Mefrosol®. Mefrosol® can be prepared by Prins reaction between isoprenol and benzaldehyde followed by hydrogenation, or by hetero-Diels–Alder reaction between benzaldehyde (10.31) and isoprene (10.32), again followed by hydrogenation.[10.19] In both cases, the double bonds in the intermediate pyran rings are hydrogenated and the ring itself is cleaved through hydrogenolysis of the benzylic ether bond. The intermediate, Rosyrane® (10.33), also has a rosy odour though its odour is more like that of rose oxide (10.34), another component of natural rose oils, than like that of geraniol or citronellol. The Diels–Alder synthesis of Rosyrane® and Mefrosol® and the relationship of these molecules to their natural counterparts are shown in Figure 10.10. By way of example, it can be noted that the rose character of (10.30) was used to build a natural rose note as part of the total floral complex in fragrances such as "Bulgari eau de parfum" in 1994 and "Birmane" in 1999.

Lily of the valley, known in the fragrance industry by its French name, muguet, is an elusive plant as far as chemistry is concerned. No essential oil can be prepared from it and there is ongoing speculation about the nature of the components responsible for its characteristic odour. Dihydrofarnesol (10.35) has been identified in the flowers and does possess an odour which is part of the overall profile.[10.20] However, the muguet note has been characterised in recent years by three synthetic ingredients all of which have odours strongly suggestive of that of the flowers. These three key materials, hydroxycitronellal (10.36), Lyral® (10.37) and Lilial® (10.38) are shown in Figure 10.11.

Hydroxycitronellal is generally considered to have the odour closest in character to that of the natural product and has been used in such fragrances as "Diorissimo" which is a classic muguet type of perfume. Hydroxycitronellal is clearly terpenoid in origin, Lyral® is prepared from a terpene, myrcene, and the structure of Lilial® is very reminiscent of that of the p-menthane family of terpenoids. The syntheses of hydroxycitronellal and Lyral® are both described in Chapter 3. All three of these materials contain aldehyde functions which limits their stability in oxidising media and in products where the pH is not neutral. There has therefore been a considerable amount of work on trying to find replacements for them and some examples are shown in Figure 10.11.

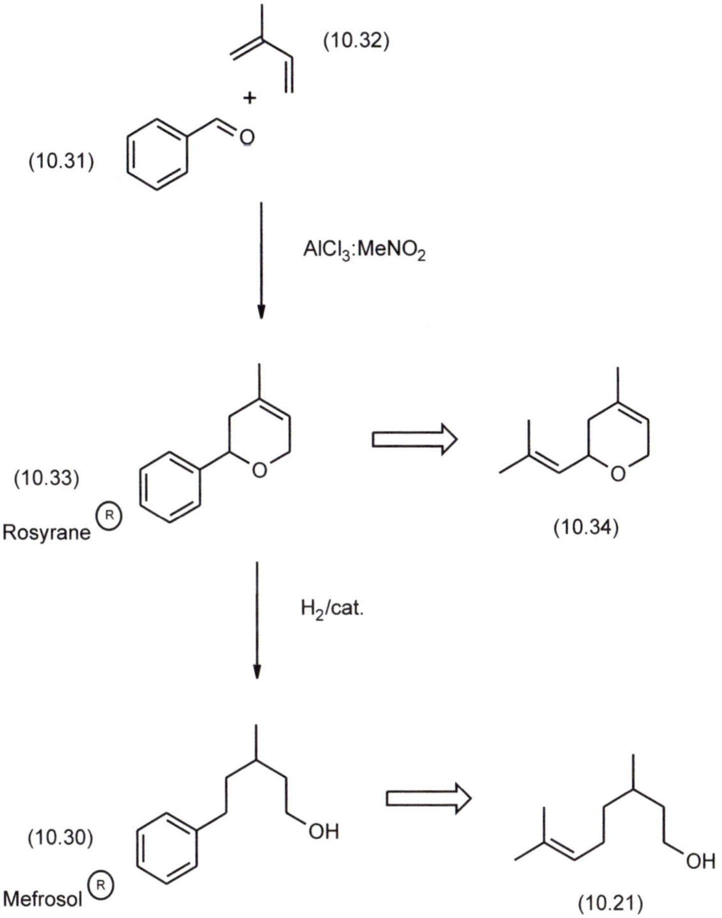

Figure 10.10

Mayol® (10.39) has an overall shape rather reminiscent of that of Lilial® but has a terpenoid structure similar to that of hydroxycitronellal (10.36). It not surprising, therefore, that it has a very fine muguet character which has been used to good effect, for example, in "Dolce Vita". Also present in "Dolce Vita" is Majantol® (10.40) where a phenyl group has been used in place of the aldehyde end of hydroxycitronellal or the unsaturated terpenoid tail of dihydrofarnesol. Osmophores (or olfactophores) are the fragrance industry's equivalent of pharmacophores. An osmophore is therefore an idealised spatial distribution of electronic charge and steric bulk which describes the average shape and surface electronic properties required for a molecule to possess a given odour. Using this approach, chemists from the Swiss fragrance house,

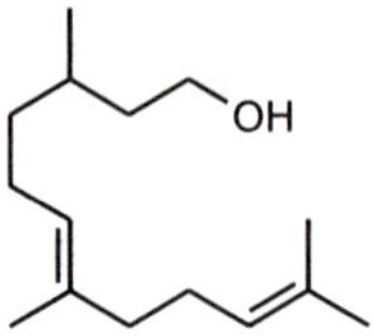

(10.35)
dihydrofarnesol

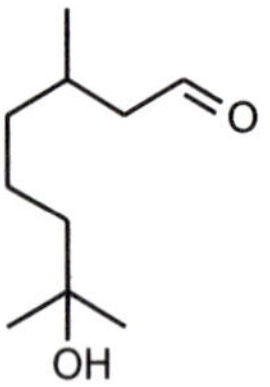

(10.36)
hydroxycitronellol

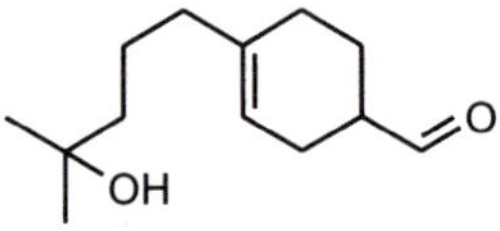

(10.37)
Lyral ®

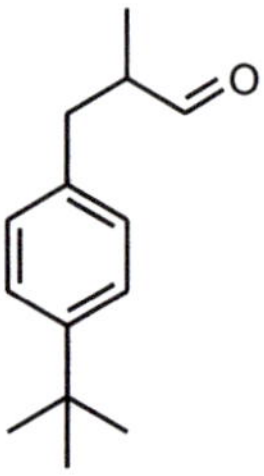

(10.38)
Lilial ®

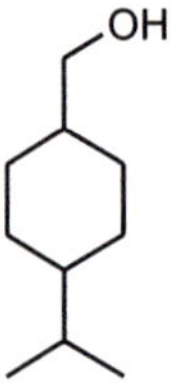

(10.39)
Mayol ®

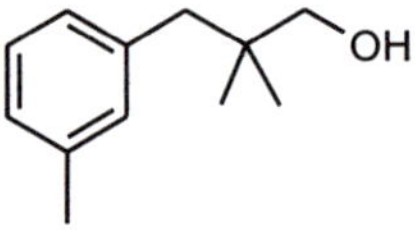

(10.40)
Majantol ®

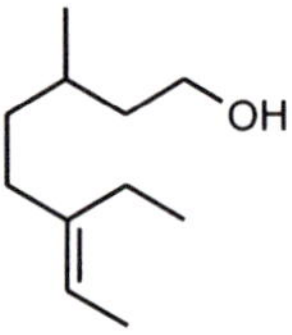

(10.41)
Meo Parf ®

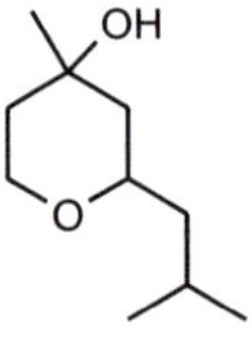

(10.42)
Florosa ®

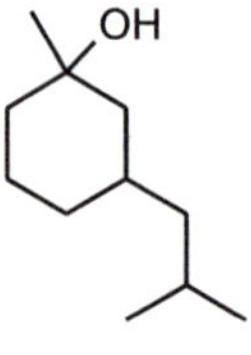

(10.43)
Rossitol ®

Figure 10.11

Givaudan, were able to design Meo Parf® (10.41) as a new muguet material.[10.17] The tetrahydropyranyl alcohol (10.42) is known under the tradenames of Florol® and Florosa®. It has been used to add muguet notes to many fine fragrances such as "XS for Her" and, more recently, "J'Adore" and "Fragil". It was during some work using molecular modelling to try to understand the differences in intensity of the four stereoisomers of Florosa®, that an even more effective muguet ingredient Rossitol® (10.43) was designed.[10.21] Rossitol has since been used in fragrances such as "Miracle".

The sesquiterpenoid ketone nootkatone (10.44) is an important part of the odour of grapefruit. The reader will recall the structure of α-vetivone (10.45), discussed in Chapter 7 and, because of the similarity of their structures, it is not surprising that vetiver oil does indeed possess a grapefruit-like character for which α-vetivone is responsible. The two double bonds and the stereochemical detail make nootkatone a difficult synthetic target and so it is not unusual to produce simplified analogues such as Decatone® (10.46). Rhubofix® (10.47) is even further away from the natural material as it has a tricyclic ring system and an epoxide rather than a ketone as the osmophoric group. (An osmophoric group is the term generally used to indicate a potential hydrogen bond accepting group in an odorant.) Rhubofuran® (10.48) uses a phenyl substituted furan to mimic the decalone ring system and the ethereal oxygen of the furan ring serves as the osmophoric group. Rhubofuran® produced by Quest, has an extremely low odour threshold and can therefore be used effectively at very low concentrations. Its chemical stability results in good performance even in the most aggressive product bases. As their names suggest, Rhubofuran® and Rhubofix® also have considerable rhubarb character in their odours. Grapefruit and rhubarb odours do seem to be related and small structural changes can tip an odorant from one of them to the other (Figure 10.12).

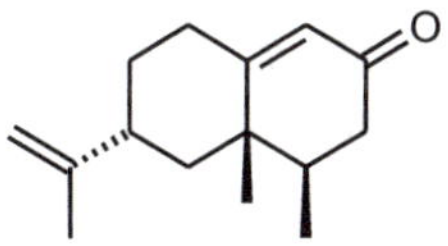

(10.44)
nootkatone

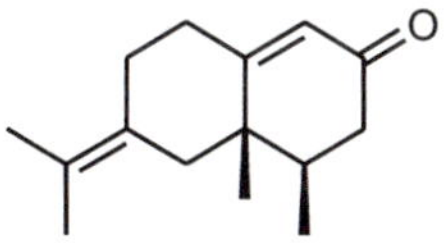

(10.45)
α-vetivone

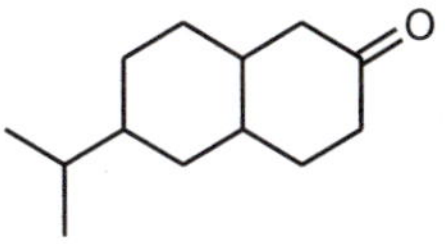

(10.46)
Decatone ®

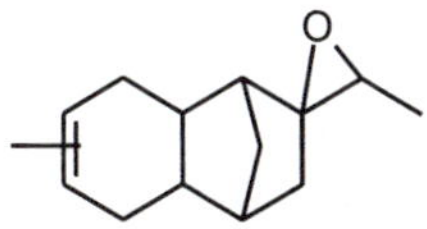

(10.47)
Rhubofix ®

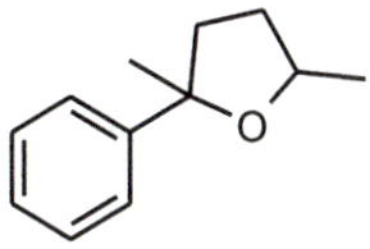

(10.48)
Rhubofuran ®

Figure 10.12

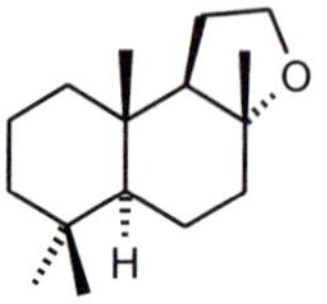

(10.49) ®
Amberlyn

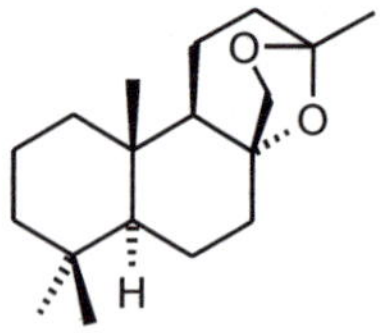

(10.50)
Jeger's ketal

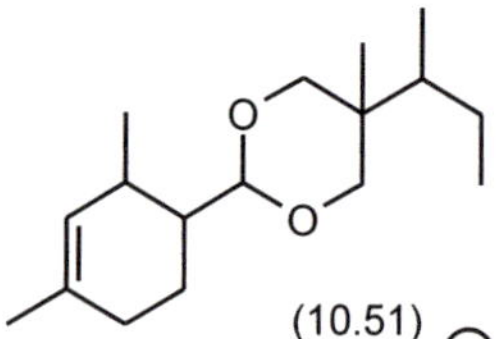

(10.51) ®
Karanal

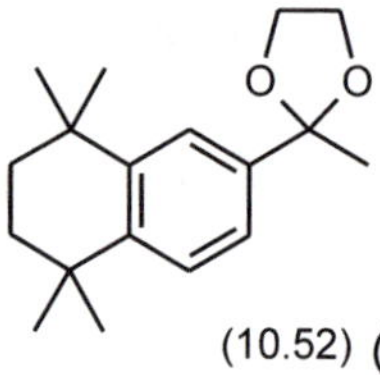

(10.52) ®
Okoumal

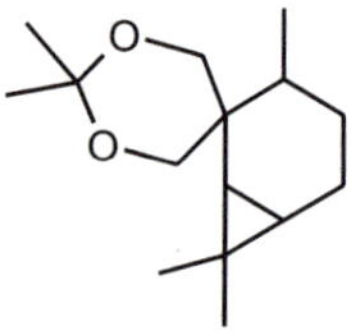

(10.54) ®
Spirambrene

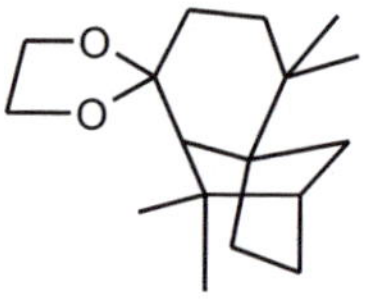

(10.53) ®
Ysamber K

Figure 10.13

In Chapter 8, we came across the naphthofuran (10.49) and Jeger's ketal (10.50) as important ambergris materials. Both of these are very expensive and the supply of manool, the starting material for the latter, is severely restricted. Not surprisingly, there has been a great deal of effort put into the search for alternatives for these two valuable ingredients. Figure 10.13 shows four ketals which have been developed as ambergris odorants which can be used in place of the more expensive Jeger's ketal. In each case the molecule contains a ketal group, serving as an osmophore in the same way that the ketal function of Jeger's ketal does. Each molecule also contains a complex hydrocarbon skeleton which provides a steric bulk resembling that found in Jeger's ketal. The relationship between the structure of a molecule and the ambergris odour is complex and many hundreds of materials with similar structures to these four do not possess any detectable ambergris odour. Many SARs have been proposed for the ambergris odour and the reader who wishes to know more is referred to the review by Karen Rossiter.[10.3]

(Incidentally, this is the person after whom Karanal® (10.51) was named.) The exact reason why these four materials do replicate the odour of Jeger's ketal whilst close analogues do not, remains a mystery at present.

Let us look at the four Jeger's ketal analogues in alphabetical order. Karanal® (10.51) is the most surprising of the four structures since its ketal function is located in the middle of the molecule whereas those of all the other materials shown in Figure 10.13 are at the edge. Only some of the 16 stereoisomers of Karanal® possess the ambergris odour, the others being odourless. The intense ambergris odour was, indeed a surprise, Karanal® having been designed for performance in terms of substantivity and stability in alkaline media rather than with the intent of producing an ambergris odour. Its performance in products more than lived up to expectations and the fine odour was a bonus for the discovery team. This odour character accounts for its use in fine fragrances such as "Diesel + + masc" and "Boss". The structure of Okoumal® (10.52) is actually based on those of tetralin musks such as Tonalid® (10.7) and Traseolide® (10.9) but it can be seen that superimposing it onto the structure of Jeger's ketal would give a reasonable fit in terms of the relative positions of the ketal function and the hydrophobic bulk. It has been used to add ambergris notes to "Escape for men" and "Hugo". A more recent addition to the ambergris palette is Ysamber K® (10.53). Like Spirambrene® (10.54) it uses a readily available terpenoid building block for the construction of the hydrophobic backbone. Ysamber K® (10.53) is the ethylene ketal of isolongifolanone, the preparation of which from longifolene is described in Chapter 7. Spirambrene® (10.54) is produced from carene by hydroformylation, Tollens reaction and acetalisation with acetone. It has been used both in ladies' fragrances such as "Eden" and in men's such as "Kenzo pour Homme".

10.9 CONCLUSION

Discovery chemistry is a very exciting area to work. The discovery chemist can design a new molecule, something which, to the best of our knowledge, has probably never existed before in the history of the universe. He can then synthesise it in the laboratory and study its properties to see if it can perform some useful function. In the fragrance industry the excitement is heightened by the fact that the chemist has instant feedback on the biological activity, he simply smells the product. The relatively short timescales of the fragrance industry (compared to those of the pharmaceutical and agrichemical businesses) give added reward in that the chemist will see the fruits of his research on the supermarket shelves and in department stores within a few years of his first making the material.

Albert Einstein said that, "The most beautiful thing we can experience is the mysterious. It is the source of all true art and science." The mechanism of odour perception and the nature of interaction of small molecules with biological macromolecules both still hold many mysteries. Perfumery is a blend of art and science, principally chemistry. I agree with Einstein and conclude that perfumery and the design of novel ingredients for it are beautiful and rewarding occupations.

REFERENCES

10.1.   D.H. Pybus and C.S. Sell (eds.), *The Chemistry of Fragrances*, RSC, Cambridge, 1999.

10.2.   D. Livingstone, *Data Analysis*, Oxford Science Publishers, Oxford, 1995.

10.3.   Karen J. Rossiter, Structure-Odour Relationships, *Chem. Rev.*, 1996, **96** (8), 3201–3240.

10.4.   J.E. Amoore, *Molecular Basis of Odour*, Charles C. Thomas, Illinois, 1970.

10.5.   T. Curtis and D.G. Williams, *Introduction to Perfumery*, Ellis Horwood, 2nd edn., 2001.

10.6.   P.M. Müller and D. Lamparsky (eds.), *Perfumes, Art, Science and Technology*, Elsevier, Amsterdam, 1991.

10.7.   R.R. Calkin and J.S. Jellinek, *Perfumery, Practice and Principles*, John Wiley, London, 1994.

10.8.   Exodus 29, 35.

10.9.   Herodatus, *The Histories*, translated by Aubrey de Selincourt, Penguin Classics, Harmondsworth, England, 1968.

10.10. John 12, 5.

10.11. C.S. Sell, *Perfumer and Flavorist*, 2000, **25** (1), 67.

10.12. G. Morrot, F. Brochet and D. Dubourdieu, *Brain and Language*, Academic Press, London, 2001, ISBN0093-934X/01.

10.13. C.S. Sell, *Seifen, Oele, Fette, Wachse*, 1986, **112** (8), 267.

10.14. C.S. Sell, *Perfumer Flavorist*, 2001, **26** (1), 2.

10.15. G. Frater, J. Bajgrowicz and P. Kraft, *Tetrahedron*, 1998, **54** (27), 7633–7703.

10.16. C.S. Sell, Chapters 3 and 4 in *The Chemistry of Fragrances*, D.H. Pybus and C.S. Sell (eds.), RSC, Cambridge, 1999.

10.17. P. Kraft, J.A. Bajgrowicz, C. Denis and G. Frater, *Angew. Chem. Int. Edn.*, 2000, *39/17*, 2981–3010.

10.18. M. Gautschi, J.A. Bajgrowicz and P. Kraft, *Chimia*, 2001, **55** (5), 379.

10.19. N.L.J.M. Broekhof, J.J. Hofma, H. Renes, and C.S. Sell, *Perfumer and Flavorist*, 1992, **17**, 11.

10.20. R. Pelzer, U. Harder, A. Krempel, H. Sommer and P. Hoever in *Recent Developments in Flavour and Fragrance Chemistry*, R. Hoppe and K. Mori (eds.), VCH, Weinheim, 1993, pp. 29–67.

10.21. K.J. Rossiter, *Chimia*, 2001, **55** (5), 388

Bibliography

This bibliography is intended to give an introduction to the wider literature for those who wish to know more on specific aspects of the subject. Some of the references given below are already to be found as cited sources in the text. However, here, they are classified by subject and so should be easier to find. Similarly, some of the books below cover more than one topic and so are entered under more than one heading.

REFERENCE TEXTS ON TERPENOIDS

J.L. Simonsen, *The Terpenes*, 3 Vols, Cambridge University Press, Cambridge, 1949.

A.R. Pinder, *The Chemistry of the Terpenes*, Chapman & Hall, London, 1960.

A.A. Newman, *Chemistry of Terpenes and Terpenoids*, Academic Press, New York, 1972.

T.K. Devon and A.I. Scott (eds.), *Handbook of Naturally Occurring Compounds*, Vol. 2, *The Terpenes*, Academic Press, New York, 1972.

S. Dev, A.P.S. Narula and J.S. Jadav, *Handbook of Terpenoids*, 2 Vols, CRC Press, Boca Raton, 1982.

Terpenoids and Steroids, Specialist Periodical Reports now incorporated into *Natural Product Reports*, Royal Society of Chemistry, Cambridge.

Tse-Lok Ho, *Carbocycle Construction in Terpene Synthesis*, VCH, Weinheim, New York, 1988.

USEFUL CHAPTERS ON TERPENOIDS

I.L. Finar, *Organic Chemistry*, Vol. 2, Longmans, London, 1968.

E.T. Theimer (ed.), *Fragrance Chemistry*, Academic Press, New York, 1982.

P.-J. Tesseire, *Chemistry of Fragrant Substances*, VCH, Weinheim, New York and Cambridge, 1993.

J. Mann, R.S. Davidson, J.B. Hobbs, D.V. Banthorpe and J.B. Harbourne, *Natural Products: Their Chemistry and Biological Significance*, Longman, New York, 1994.

R.A. Hill, Terpenoids, in *The Chemistry of Natural Products*, R.H. Thompson (ed.), Blackie Academic and Professional, Chapman & Hall, London, 1993, pp. 106–139.

BIOLOGICAL ACTIVITY OF TERPENOIDS

H.G. Cutler, Natural product flavour compounds as Potential anti-microbials, insecticides and medicinals, *Agro-Food-Ind. Hi-Tech.*, 1995, **6**, 19.

H.G. Cutler, R.A. Hill, B.G. Ward, B.H. Rowitha and A. Stewart, *Antimicrobial, Insecticidal and Medicinal Properties of Natural Product Flavours and Fragrances in Biotechnology for Improved Foods and Flavors*, ACS Symposium Series 637, G.R. Takeoka, R.Teranishi, P.J. Williams and A. Kobayashi (eds.), American Chemical Society, Washington, 1996, pp. 51–66.

BIOSYNTHESIS

J. Mann, R.S. Davidson, J.B. Hobbs, D.V. Banthorpe and J.B. Harbourne, *Natural Products: Their Chemistry and Biological Significance*, Longman, London, 1994.

J.D. Bu'Lock, The Biosynthesis of Natural Products, McGraw-Hill, New York, 1965.

K.B.G. Torssell, *Natural Product Chemistry*, John Wiley, New York, 1983.

J.D. Rawn, *Biochemistry*, Harper & Row, New York, 1983.

R. Croteau, *Biosynthesis and Catabolism of Monoterpenoids, Chem. Rev.*, 1987, **87**, 929.

C.K. Matthews and K.E. van Holde, *Biochemistry*, Benjamin/Cummings, Redwood City, 1990, ISBN 0-8053-5015-2.

ESSENTIAL OILS

E.J. Parry, *The Chemistry of Essential Oils and Artificial Perfumes*, Scott, Greenwood and Son, 1921.

E. Gunther and D. van Nostrand, *The Essential* Oils, 6 Vols., 1948.

E. Gildemeister and Fr. Hoffmann, Die Aetherischen Oele, 11 Vols., Akademie-Verlag, Berlin, 1956.

S. Arctander, *Perfume and Flavour Materials of Natural Origin*, S. Arctander, New Jersey, 1960.

B.D. Mookherjee and C.J. Mussinian, *Essential Oils*, Allured, Illinois, 1981.

B. M. Lawrence, A review of the World Production of Essential Oils, *Perfumer & Flavorist*, 1985, 10.

R.K.M. Hay and P.G. Waterman, *Volatile Oil Crops*, Longman, New York, 1993.

D.G. Williams, *The Chemistry of Essential Oils*, Micelle Press, Dorset, 1996.

STEREOCHEMISTRY

E.L. Eliel and S.H. Wilen, *Stereochemistry of Carbon Compounds*, John Wiley, New York, 1994.

CHEMICAL PROCESS DESIGN AND OPTIMISATION

R. Turton, R.C. Bailie, W.B. Whiting and J.A. Shaeiwitz, *Analysis, Synthesis and Design of Chemical Processes*, 2nd edn., Pearson Professional Education, Harlow, 2002.

G.B. Tatterson, *Process Scaleup and Design*, Gary Tatterson, North Carolina, 2002.

J. Oakland, *Statistical Process Control*, 5th edn., Butterworth-Heinemann, Oxford, 2002.

F.X. McConville, *The Pilot Plant Red Book*, FXM Engineering and Design, Worcester, Massachusetts, 2002.

BIOCHEMISTRY AND BIOLOGY OF PERCEPTION (SIGHT, TASTE AND SMELL)

D. Lancet and U. Pace, The molecular basis of odour recognition, *TIBS* **12**, 1987, 63.

C.K. Matthews and K.E. van Holde, *Biochemistry*, Benjamin/Cummings, Redwood City, 1990, ISBN 0-8053-5015-2.

H. Breer, K. Raming and J. Krieger, Signal recognition and transduction in olfactory neurons, *Biochem. Biophys. Acta*, 1994, **1224**, 277–287.

H. Kandori, The chemistry of vision, turning light into sight, *Chem. Ind.*, 1995, 735.

R. Axel, The molecular logic of smell, *Sci. Am.*, 1995, October, 130–137.

B. Malnic, J. Hirono, T. Sato and L.B. Buck, Combinatorial receptor codes for odours, *Cell*, 1999, **96**, 713.

C. Blakemore and S. Jennett (eds.), *The Oxford Companion to the Body*, Oxford University Press, Oxford, 2001, ISBN 0 19 852403 X.

C. S. Sell, *Perfumer & Flavorist*, 2001, **26** (1), 2.

STRUCTURE/ACTIVITY RELATIONSHIPS AND THE DESIGN OF FRAGRANCE INGREDIENTS

P. Laffort, *Relationships between Molecular Structure and Olfactory Activity*, in *Odours and Deodorization in the Environment*, G. Martin and P. Laffort (eds.), VCH, New York, 1994.

K.J. Rossiter, Structure-odour relationships, *Chem. Rev.*, 1996, **96** (8), 3201–3240.

C.S. Sell, *On the Right Scent, Chem. Br.*, 1997 March, 39–42.

C.S. Sell, *Perfumer & Flavorist*, 2000, **25** (1), 67.

CHEMISTRY AND THE PERFUME INDUSTRY

H.V. Daeniker, *Flavours and Fragrances Worldwide*, SRI, Menlow Park, 1987.

S. van Toller and G. H. Dodd, *Perfumery, The Psychology and Biology of Fragrance*, Chapman & Hall, London, 1988.

P.M. Muller and D. Lamparsky (eds.), *Perfumes, Art, Science and Technology,* Elsevier, Amsterdam, 1991, ISBN 1-85166-573-0.

R.R. Calkin and J.S. Jellinek, *Perfumery, Practice and Principles*, John Wiley, New York, 1994, ISBN 0-471-58934-9.

R.L. Doty, *Handbook of Olfaction and Gustation,* Marcel Dekker, New York, 1995.

D.G. Williams, *The Chemistry of Essential Oils*, Micelle Press, Dorset, 1996, ISBN 1-870228-12-X.

D.H. Pybus and C.S. Sell (eds.), *The Chemistry of Fragrances*, RSC, Cambridge, 1999.

T. Curtis and D.G. Williams, *Introduction to Perfumery*, 2nd edn., Micelle Press, Dorset, 2001.

FRAGRANCE INGREDIENTS

G. Frater, J. Bajgrowicz and P. Kraft, Fragrance Chemistry, *Tetrahedron*, 1998, **54** (27), 7633–7703.

K. Bauer, D. Garbe and H. Surburg, *Common Fragrance and Flavour Materials*, Wiley-VCH, New York, 4th edn., 2001.

P. Kraft, J. Bajgrowicz, C. Denis and G. Frater, *Angew. Chem. Int. Edn.*, 2001, *39/17*, 2981–3010.

M. Grautschi, J.A. Bajgrowicz and P. Kraft, *Chimia*, 2001, **55** (5), 379.

Problems

Human beings enjoy solving problems, hence the popularity of cross-word puzzles, "Whodunnit?" novels and the like. Terpenoid chemistry is full of interesting problems and solving them helps to improve our understanding of chemistry and ability to solve further problems. This chapter of the book contains a selection of problems designed to stimulate and entertain as well as to educate. Some of them rely on the understanding gained in the text to solve new problems whereas others are almost integral parts of the whole, in that they explain in more detail certain aspects which were deliberately glossed over in the main text. The sequence of the problems is related approximately to the order of material in the 10 main chapters of the book and the relevant chapters are indicated appropriately. The problems are of varying degrees of difficulty, the most difficult probably being some of those related to Chapters 6 and 7. Solutions to the problems are given in the next chapter.

CHAPTERS 1 AND 2

Problem 1

Figure P1 shows the structures of six natural products. Which of these are terpenoids? Justify your choice by outlining the isoprene units in their carbon skeleta.

Problem 2

Figure P2 shows the structures of six natural products. Which of these are terpenoids? Justify your choice by outlining the isoprene units in their carbon skeleta.

Problem 3

Figure P3 shows the structures of six natural products. Which of these are terpenoids? Justify your choice by outlining the isoprene units in their carbon skeleta.

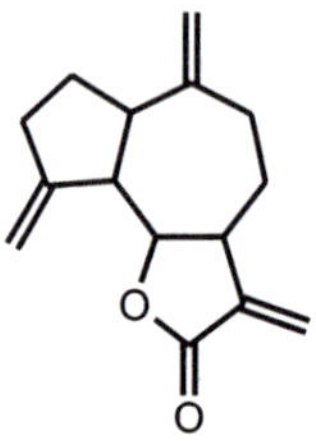

dehydro costus lactone

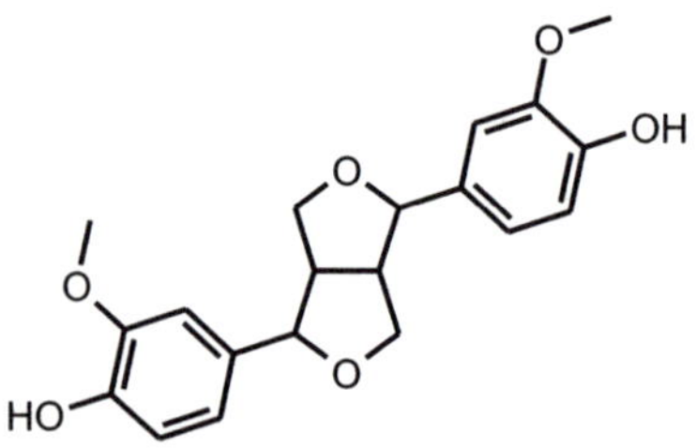

pinoresinol

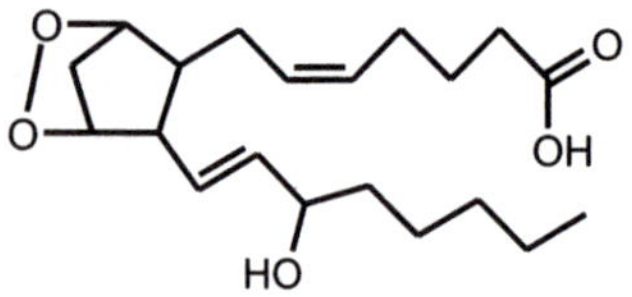

prostaglandin PG-H2

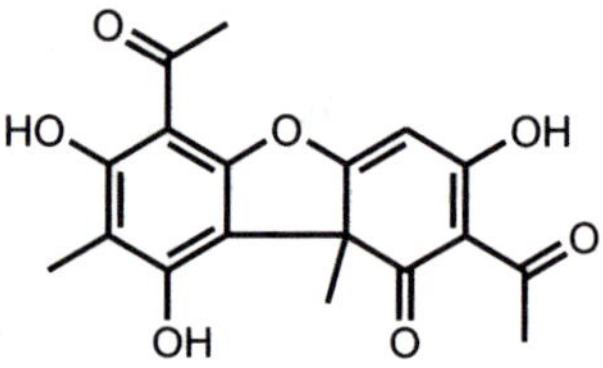

usnic acid

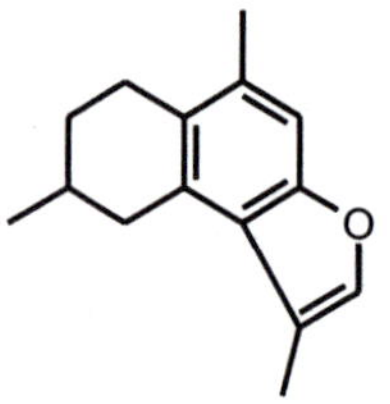

dihydropyrocurzerenone

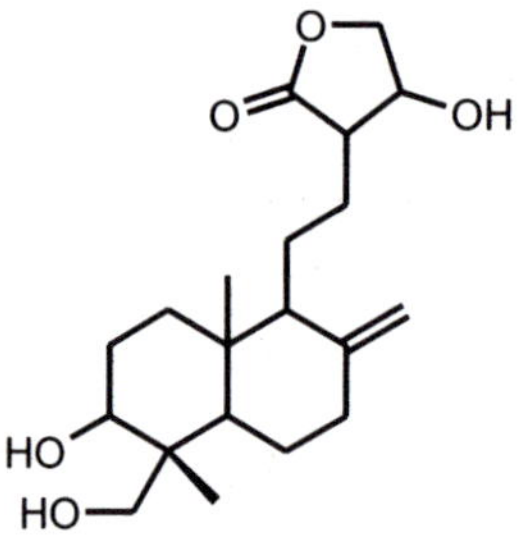

andrographolide

Figure P1

Problem 4

Figure P4 shows the structures of six natural products. Which of these are terpenoids? Justify your choice by outlining the isoprene units in their carbon skeleta.

CHAPTER 3

Problem 5

Compound A was isolated from rose oil. Elemental analysis showed it to have the empirical formula $C_{10}H_{18}O$. It does not form any derivatives

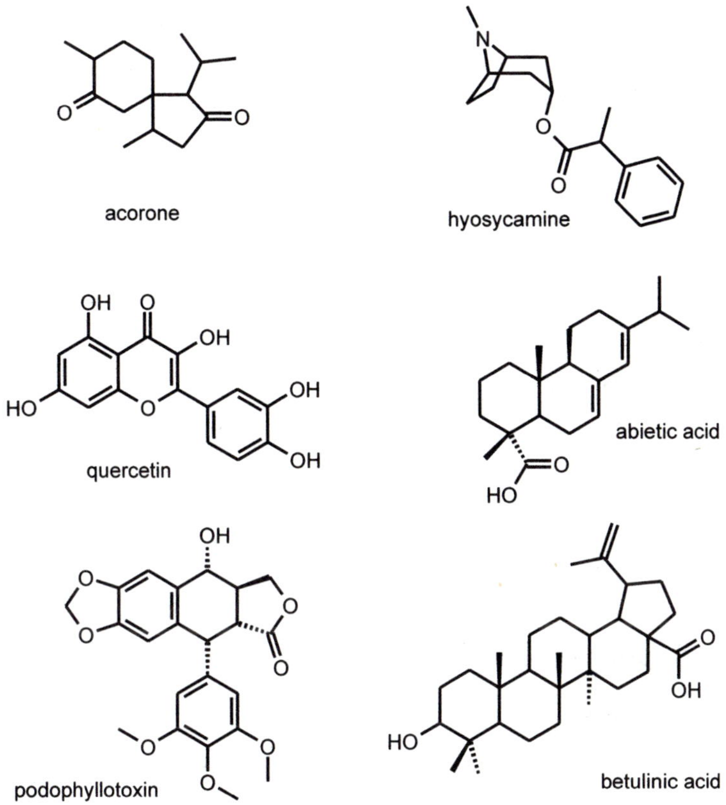

acorone

hyosycamine

quercetin

abietic acid

podophyllotoxin

betulinic acid

Figure P2

with hydroxylamine or 2,4-dinitrophenylhydrazine. When treated with sodium, it was observed to release hydrogen, although slowly. Compound A is not readily oxidised under Oppenauer conditions but, when treated with an acid, it isomerised to compound B which was readily oxidised to an aldehyde (C) by activated manganese dioxide. Treatment of compound A with potassium permanganate gave laevulinic acid ($CH_3COCH_2CH_2CO_2H$), acetone and carbon dioxide. Hydrogenation of compound A gave compound D which had the empirical formula $C_{10}H_{22}O$. Treatment of compound D with acid gave a mixture of two isomeric olefins, E and F. Treatment of compound E with permanganate gave acetic acid and a ketone, which was shown by comparison with a standard, to be 2-methylheptan-6-one. What are the structures of compounds A, B, C, D, E and F? Show how you arrived at these conclusions and explain how your knowledge of the isoprene rule might help in determining the structures and give you confidence in your answers.

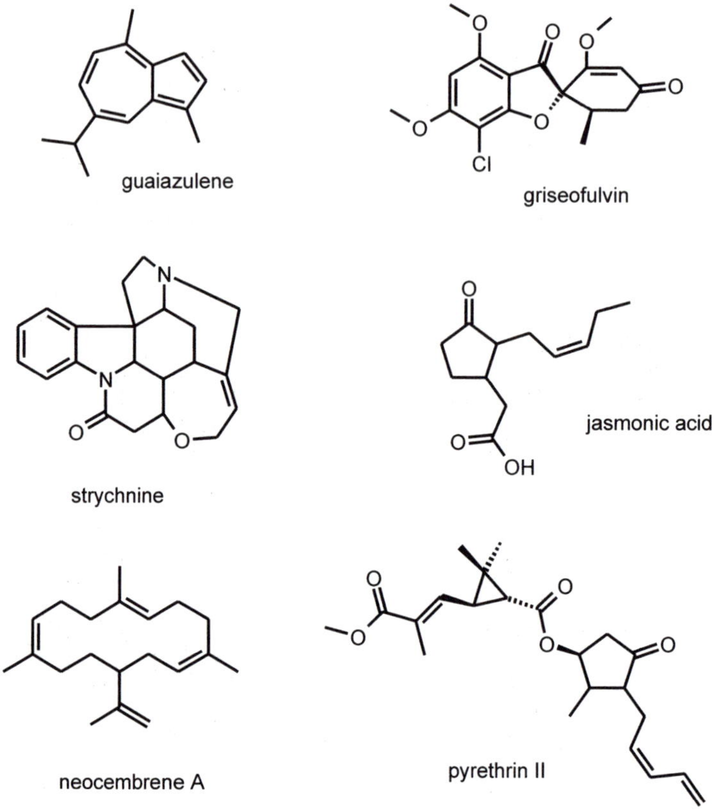

guaiazulene

griseofulvin

strychnine

jasmonic acid

neocembrene A

pyrethrin II

Figure P3

CHAPTER 4

Problem 6

Why must care be taken to ensure that acids are absent from the system when distilling carvone?

CHAPTER 5

Problem 7

The important perfumery ingredient dihydromyrcenol (P7.1) can be prepared by hydration of dihydromyrcene (P7.2), but the isomeric alcohol (P7.3) is formed as a by-product as shown in Figure P7. Write a mechanism to account for the formation of (P7.3) and suggest ways of producing only dihydromyrcenol.

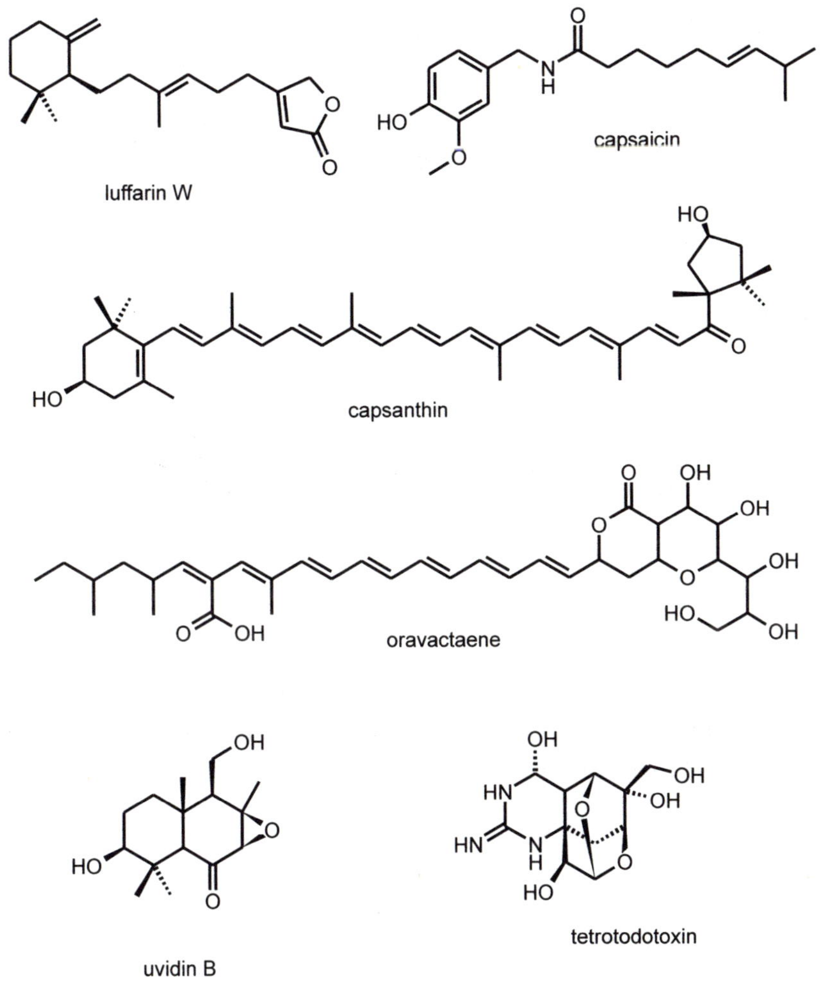

luffarin W

capsaicin

capsanthin

oravactaene

uvidin B

tetrotodotoxin

Figure P4

Problem 8

Treatment of a mixture of dimethylbutadiene and methacrolein with stannic chloride at room temperature gives the bicycloheptanone (P8.1) as shown in Figure P8. Write a mechanism to account for this observation.

CHAPTER 6

Problem 9

Ozonolysis of α-santalol gives the acid (P9.1). Treatment of this with HCl gives isomeric acid (P9.2) which can also be obtained from β-santalol as

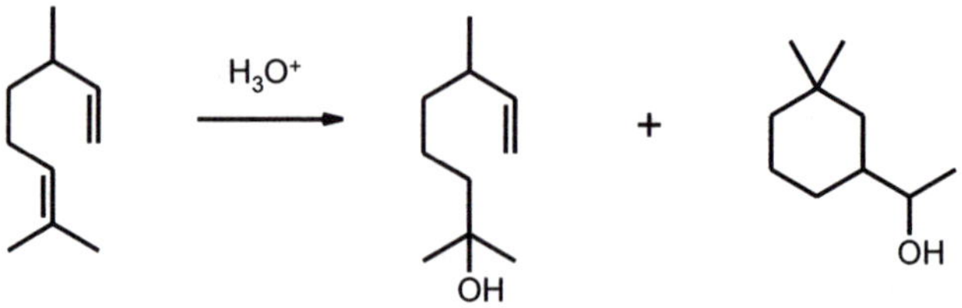

(P7.2)   (P7.1)   (P7.3)

Figure P7

(P8.1)

Figure P8

described in Chapter 6. Write a mechanism for this reaction and explain the driving force behind it (Figure P9).

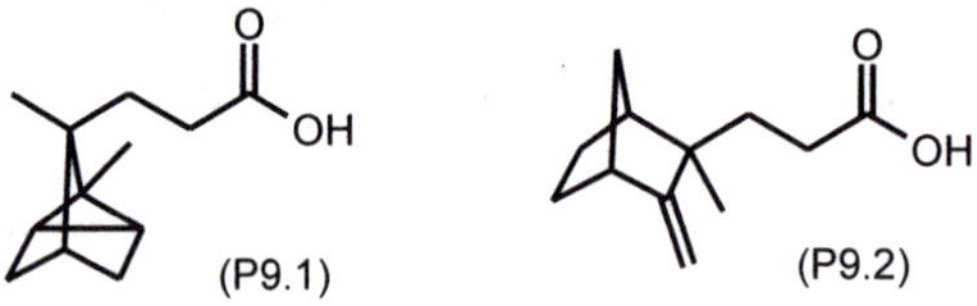

(P9.1)   (P9.2)

Figure P9

Problem 10

The bisabolyl cation (P10.1) is the key intermediate in the biosynthesis of the bergamotane, α- and β-santalane families of terpenes. The biogenetic reactions involved in converting (P10.1) to the final products, are all enzyme mediated, but their chemistry follows the basic principles of carbocation mechanisms. Draw mechanisms to account for the formation of α-bergamotene (P10.2), α-santalene (P10.3) and β-santalene (P10.4) from the cation (P10.1) and propose a structure for the intermediate (X) (Figure P10).

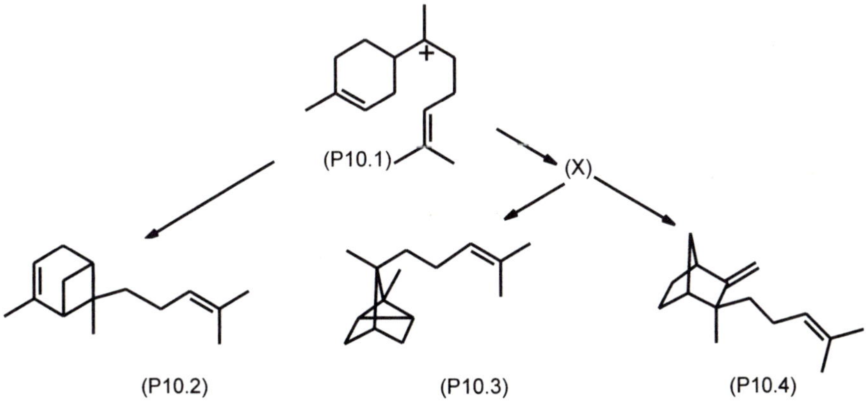

Figure P10

Problem 11

Which driving forces are present in each cation rearrangement during the biogenesis of the thujopsane skeleton from the bisabolane skeleton? (Figure P11).

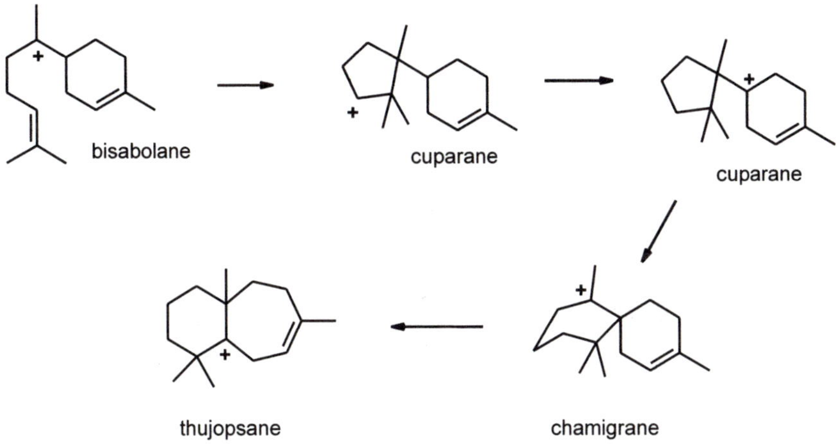

Figure P11

Problem 12

Which driving forces are present in each cation rearrangement during the biogenesis of the khusane skeleton from the acorane skeleton? (Figure P12).

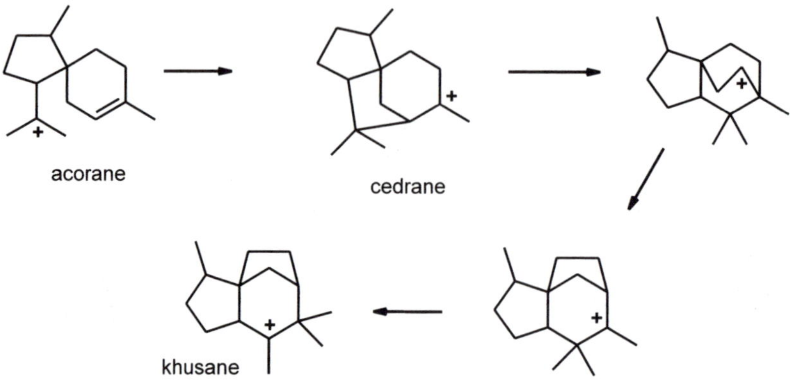

acorane

cedrane

khusane

Figure P12

Problem 13

In Chapter 6, we saw how bornylguaiacols are formed when camphene (P13.1) and guaiacol are reacted together in the presence of an acidic catalyst. The three key carbocations which add to the guaiacol are shown in Figure P13. How is each of them formed from camphene?

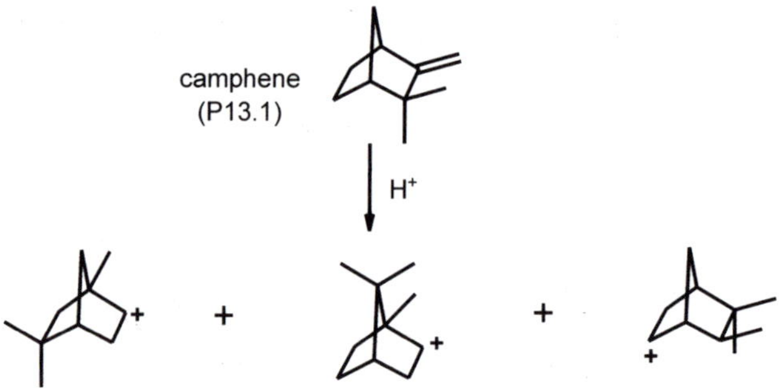

camphene
(P13.1)

H⁺

Figure P13

Problem 14

In Chapter 6, we learnt that campholenic aldehyde is an important precursor for a family of synthetic sandalwood derivatives. Write a mechanism to account for the formation of campholenic aldehyde when α-pinene oxide is treated with zinc bromide (Figure P14).

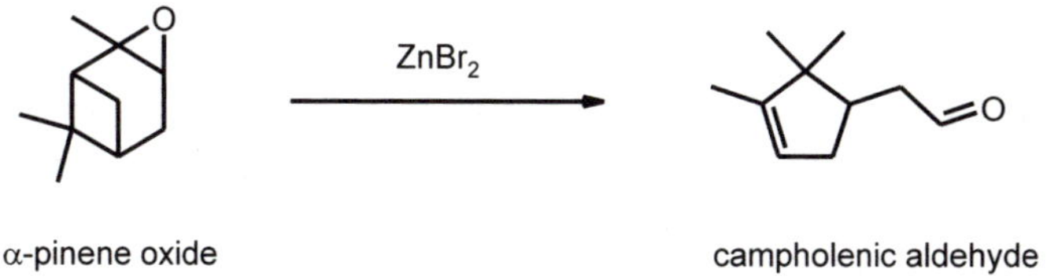

α-pinene oxide   campholenic aldehyde

Figure P14

Problem 15

As part of a synthesis of santalene sesquiterpenoids, Barrett and McKenna (1971) treated cyclopentadiene (P15.1) with the allenic acid (P15.2). This gave a mixture of two isomeric acids, (P15.3) and (P15.4). One of these, (P15.3) rearranged to the lactone (P15.5) on treatment with formic acid. The other remained unchanged. What are the structures of (P15.3) and (P15.4) and what are the mechanisms of the reactions? (Figure P15).

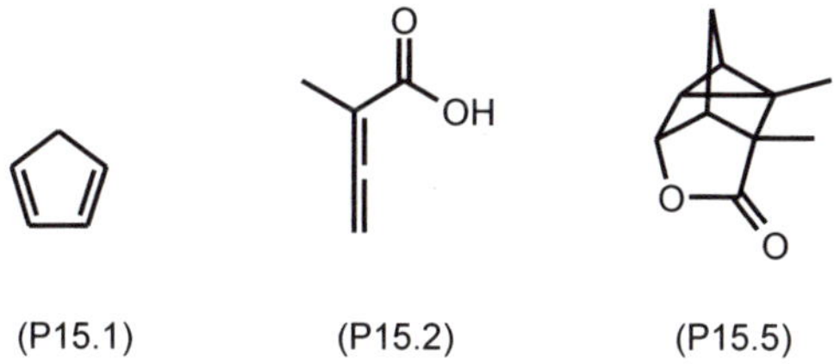

(P15.1)   (P15.2)   (P15.5)

Figure P15

Problem 16

In Chapter 6, we came across the conversion of Z-ocimenone to filifolone. Write a mechanism for this reaction which is shown in Figure P16.

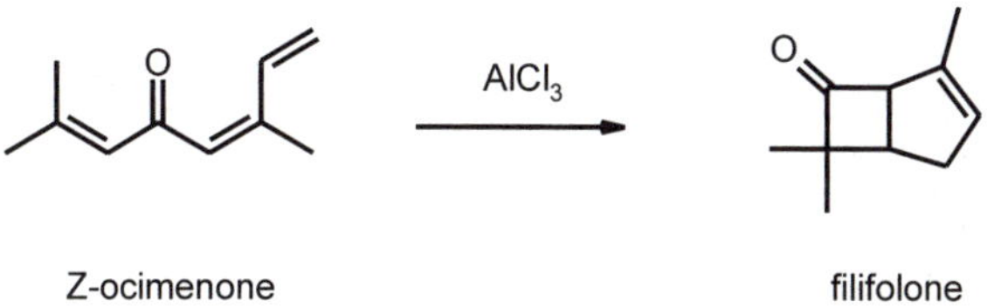

Z-ocimenone   filifolone

Figure P16

Problem 17

As seen in Chapter 6, Figure 6.33, the acid catalysed isomerisation of
thujopsene (P17.1) leads to the olefin (P17.2). Acylation of (P17.2) gives
a ketone with an excellent cedarwood odour. The relevant part of Figure
6.33 is reproduced below as Figure P17. The olefins (P17.10) and
(P17.11) are also present in the crude isomerisation mixture. Provide
mechanisms to account for their formation. Hint: one of them is formed
from carbocation (P17.4) and the other from (P17.9).

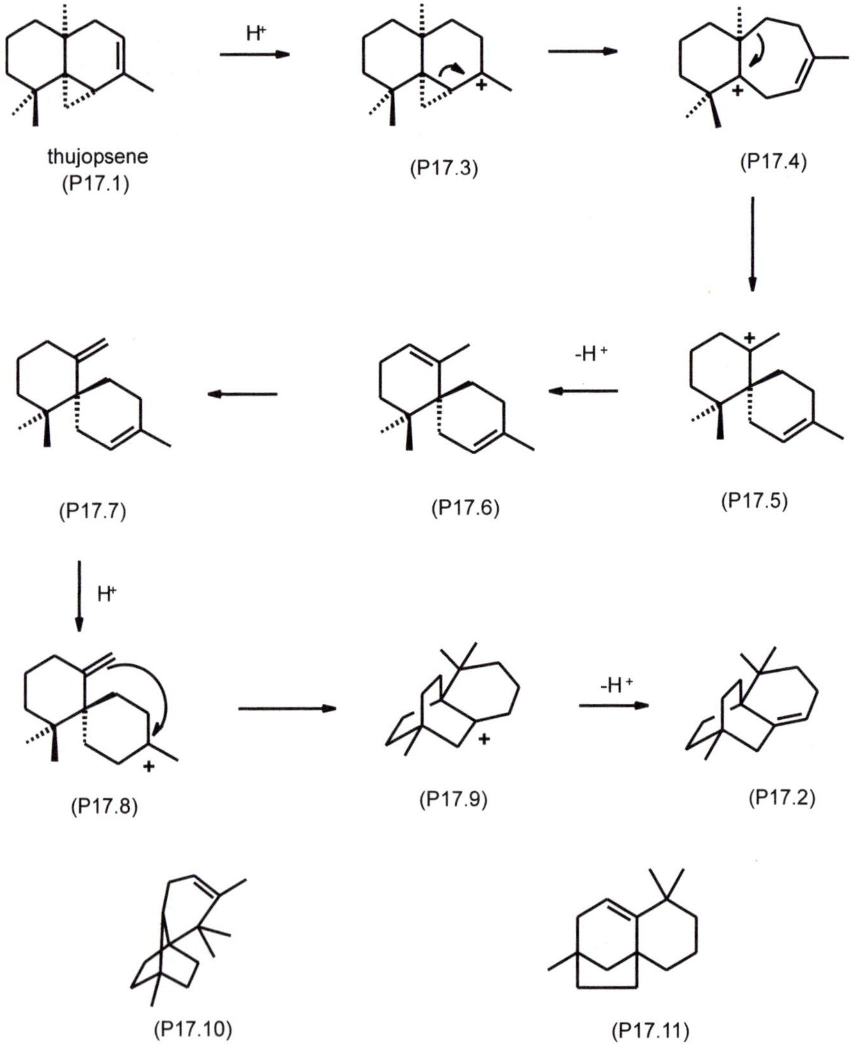

thujopsene
(P17.1)

(P17.3)

(P17.4)

(P17.7)

(P17.6)

(P17.5)

(P17.8)

(P17.9)

(P17.2)

(P17.10)

(P17.11)

Figure P17

CHAPTER 7

Problem 18

The formation of the anion (P18.2) from santonin (P18.1) by the action of hydroxide is mentioned in Chapter 7, Figure 7.9. Write a mechanism for this reaction (Figure P18).

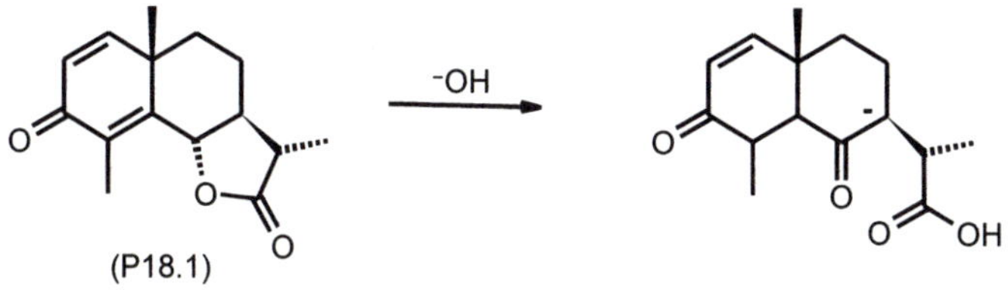

(P18.1)

santonin

(P18.2)

Figure P18

Problem 19

The starting material for Stork's synthesis of β-vetivone was shown in Chapter 7 to be prepared as shown in Figure P19. Write a mechanism for this reaction. Why is the mono-enol form of the product more stable than the diketone?

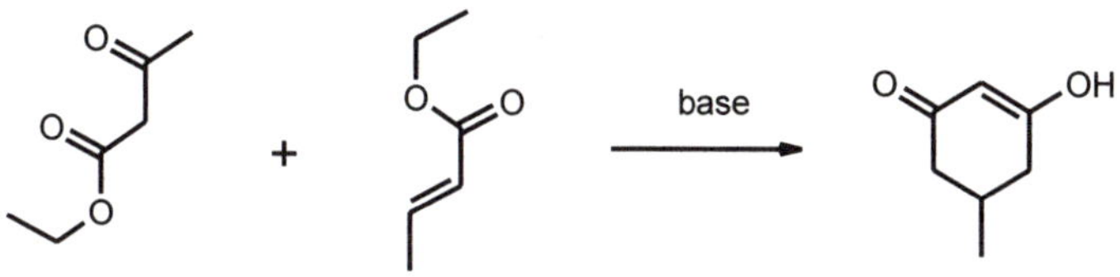

Figure P19

Problem 20

In Figure 7.18, the starting material for the synthesis of patchouli alcohol is the ketone (P20.1). Explain how this can be formed by exhaustive base catalysed alkylation of phenol (Figure P20).

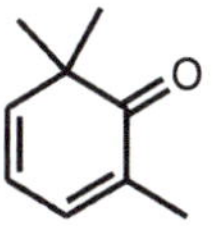

(P20.1)

Figure P20

Problem 21

Treatment of the aldehyde (P21.1) with boron trifluoride etherate at low temperature in dichloromethane, gave the alcohol (P21.2). Write a mechanism which accounts for the stereochemistry of the product (Figure P21).

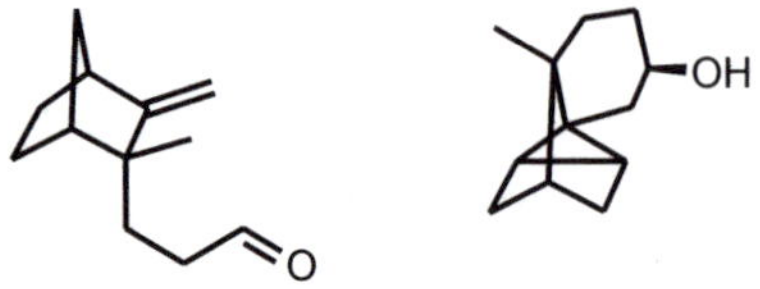

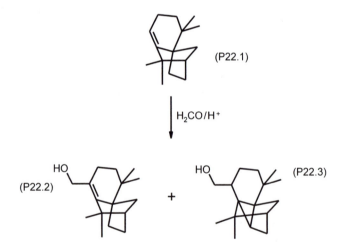

(P21.1)   (P21.2)

Figure P21

Problem 22

The Prins reaction of isolongifolene (P22.1) with formaldehyde gives two products (P22.2) and (P22.3) as shown in Figure P22. Write mechanisms to explain the formation of each.

(P22.1)

H₂CO/H⁺

(P22.2)   +   (P22.3)

Figure P22

Problem 23

Treatment of caryophyllene oxide (P23.1) with an aqueous acid leads to the formation of two products, a diol (P23.2) and an aldehyde (P23.3). Write mechanisms to account for their formation (Figure P23).

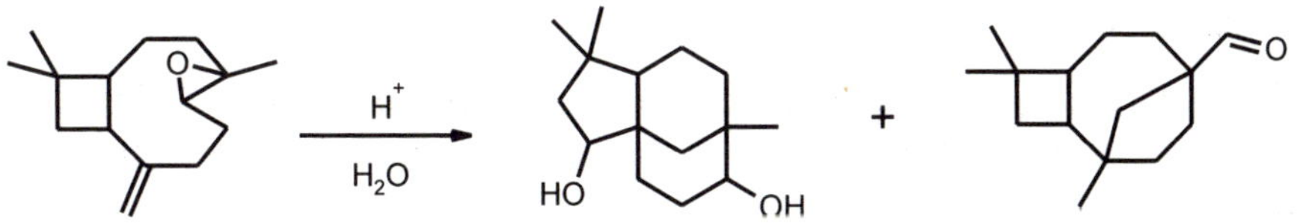

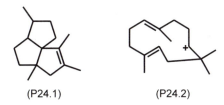

(P23.1)   (P23.2)   (P23.3)

Figure P23

Problem 24

The essential oil of *Echinops giganteus* var. *lelyi*, a culinary herb from Cameroon, contains many sesquiterpenoids. The most abundant (26.9% of the oil) is silphiperfol-6-ene, (P24.1). The biogenesis of this material from farnesyl pyrophosphate is enzyme mediated but the chemistry of the enzymatic reactions follows the basic principles of carbocation mechanisms (Figure P24). Initial cyclisation of farnesyl pyrophosphate gives the cation (P24.2). Suggest a sequence of cationic reactions which could lead from (P24.2) to (P24.1). This is a particularly difficult problem. If you would like a hint (without having to look at the solution) one will be found at the end of the problems section.

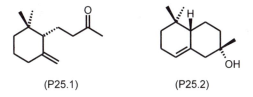

(P24.1)   (P24.2)

Figure P24

CHAPTER 8

Problem 25

Figure P25 shows two of the degradation products of ambreine. Is it possible that one of these could be easily formed from the other? Write a mechanism to justify your answer.

(P25.1)   (P25.2)

Figure P25

Problem 26

The second step of Buchi and Wuest's synthesis of Ambrox® involves the cyclisation reaction shown in Figure P26. Suggest a mechanism for this reaction.

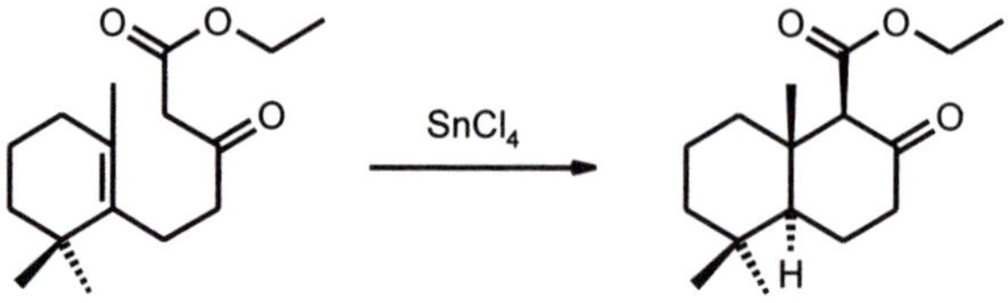

Figure P26

Problem 27

The reaction shown in Figure P27 occurs further along in Buchi and Wuest's synthesis. What is the mechanism of this reaction?

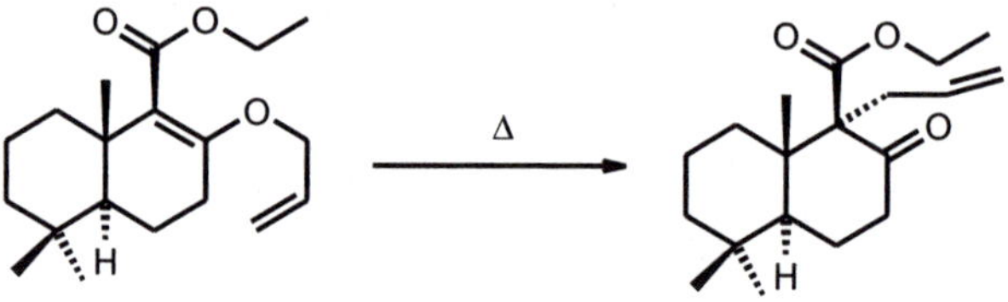

Figure P27

Problem 28

Jeger's ketal (P28.2) can be prepared by treatment of the epoxyketone (P28.1) with acid as shown in Figure P28. Suggest a mechanism to account for this.

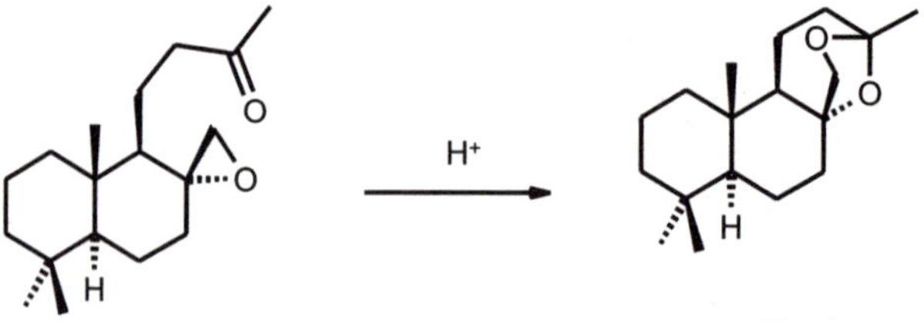

(P28.1)

(P28.2)
Jeger's ketal

Figure P28

Problem 29

The reaction shown in Figure P29 is the penultimate step of one of the syntheses of α-damascone shown in Figure 8.24. What complication might be expected in this reaction?

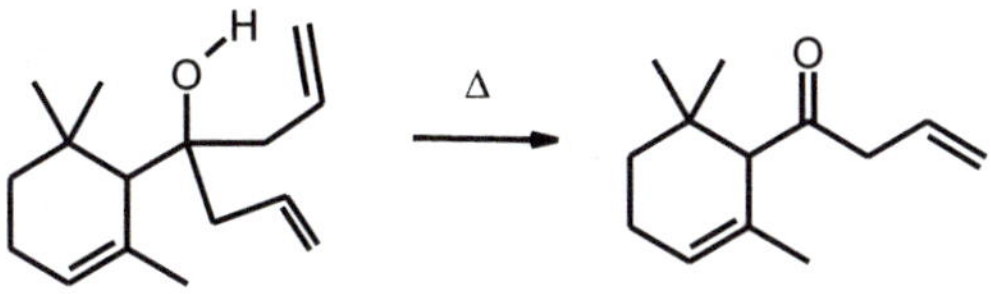

Figure P29

Problem 30

The synthesis of the theaspiranes (P30.2) from the diol (P30.1) is shown in Figure P30. Why is the selective dehydration of (P30.1) achievable? What is the mechanism of the cyclisation?

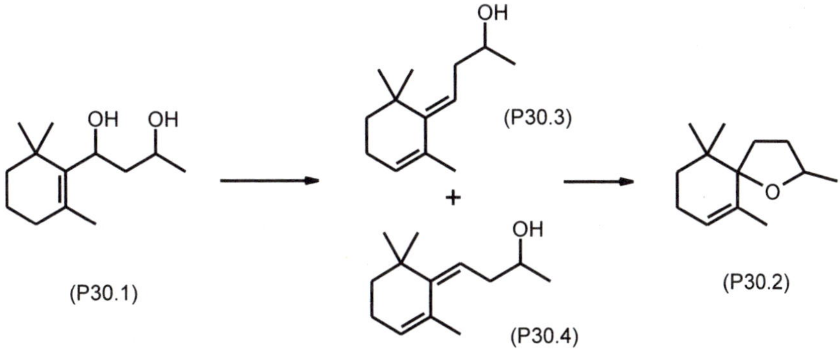

Figure P30

CHAPTER 9

Problem 31

Figure P31 shows two different routes to citral. Compare and contrast them as production processes commenting on the advantages and disadvantages of each?

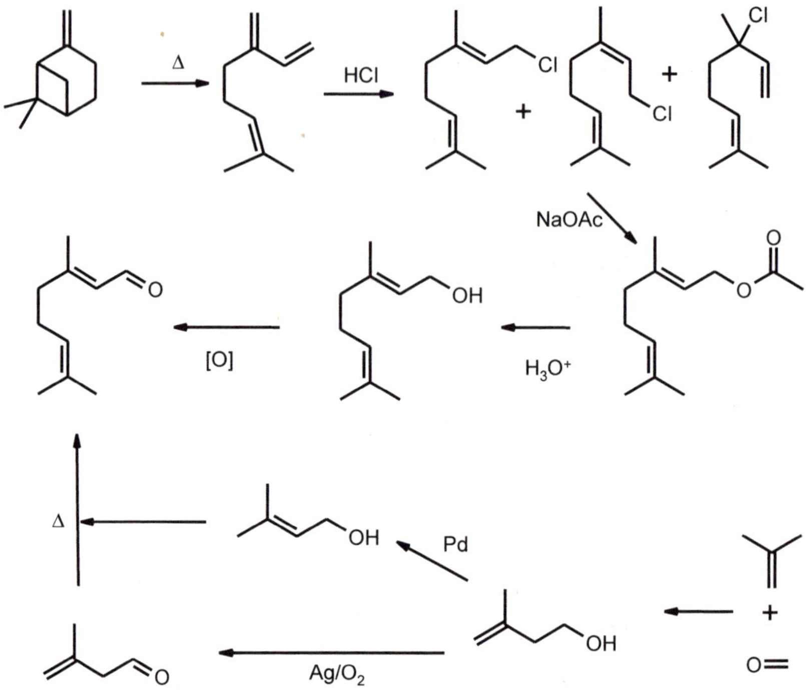

Figure P31

Problem 32

Figure P32 shows four routes to 2-methylhept-2-ene-6-one, an important synthon for terpenoid preparation. These routes are taken from the schemes shown in Figures 9.6, 9.7, 9.9 and 9.11 in Chapter 9. What is the atom efficiency of each route? (For the present purposes, consider only stoichiometric reagents and ignore solvents and catalysts.) If these routes were the result of a brainstorming exercise by your development team, what would be your priority order for testing them out in the laboratory? How would this choice be affected by the atom efficiency calculations?

CHAPTER 10

Problem 33

In Chapter 10 it is stated that Rosyrane® (P33.1) can be prepared from benzaldehyde either by Prins reaction with isoprenol (P33.2) or by a hetero-Diels–Alder reaction with isoprene (P33.3). Might one expect differences in the product from these two reactions? (Figure P33).

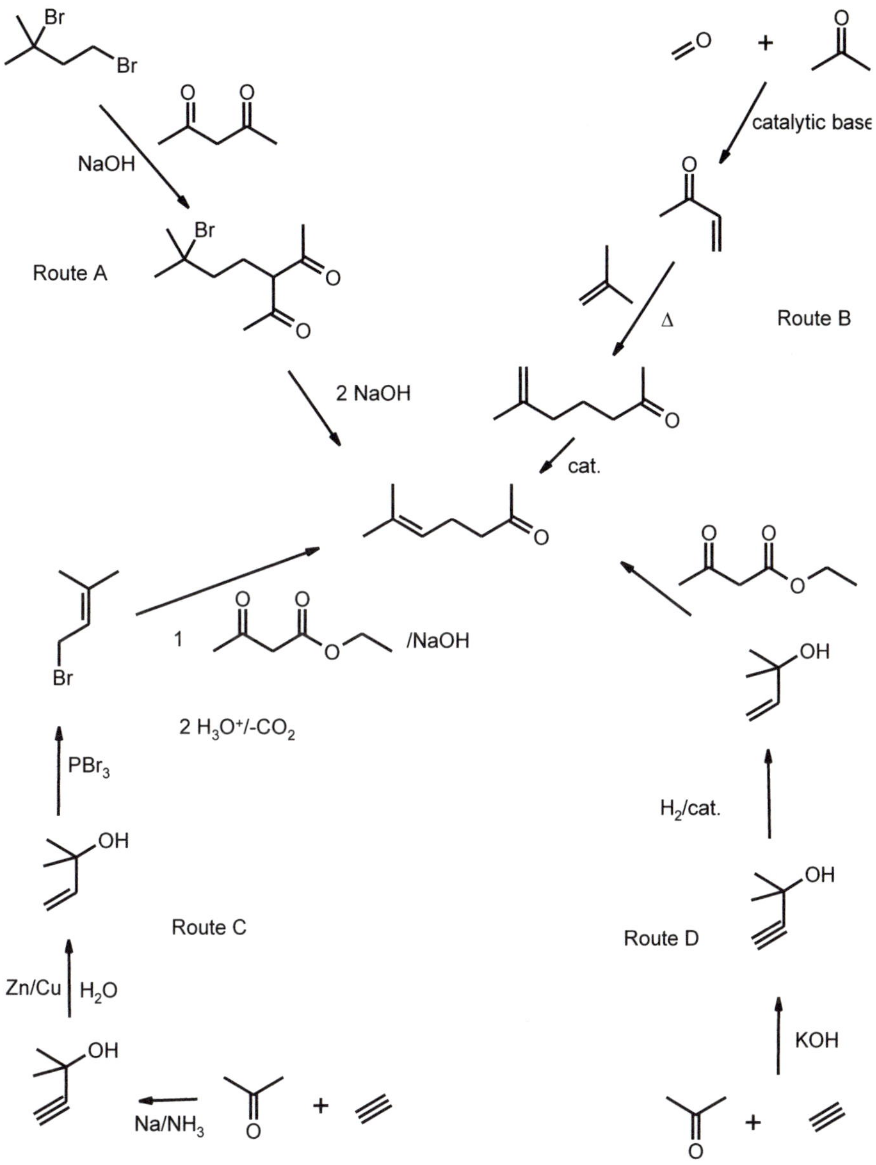

Figure P32

Problem 34

Spirambrene® (P34.1) is produced from 2-carene (P34.2). Suggest a synthetic scheme for this preparation (Figure P34).

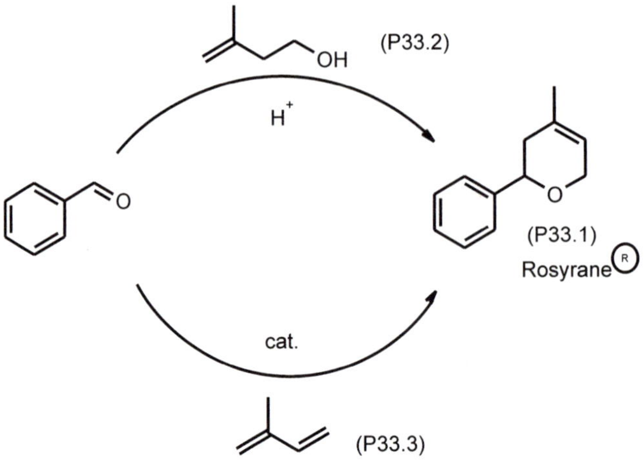

Figure P33

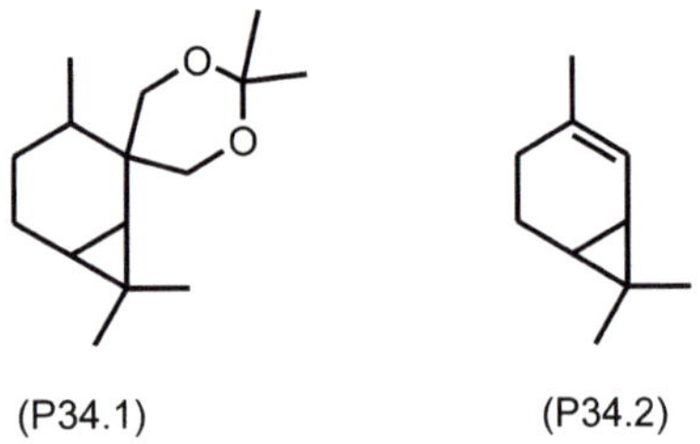

Figure P34

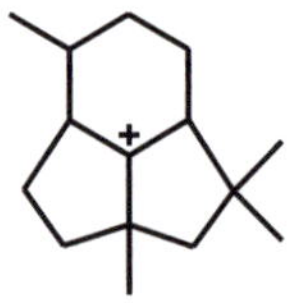

(P34.1)   (P34.2)

Figure P35

Hint for problem P24

The carbocation shown in Figure P35 is a key intermediate.

Solutions to Problems

The first four problems concern the recognition of terpenoids. There are a number of simple techniques which one can use to do this. The most obvious is to add up the total number of carbon atoms in the main skeleton of a molecule. Obviously, pendant fragments linked to the core through ester functions should be ignored as these would have been added after biogenesis of the main skeleton. If the total number of carbon atoms is a multiple of five, then the material is more likely to be a terpenoid. I say "more likely to be" rather than "is" because some non-terpenoid materials will, by chance happen to have $5n$ carbon atoms in their skeleta. Furthermore, as we saw in Chapter 8, degradation can reduce the number of carbon atoms in a terpenoid-derived material. The next technique is to look for isopropyl or gem-dimethyl fragments as these are very characteristic of terpenoid structures. Furthermore, because of the patterns in terpenoid biogenesis, these groups are often located at the point in the structure which was originally the end of the chain where the cyclisation process started. This is a very valuable point to be able to identify as it helps enormously when trying to trace out the original open chain. Long chain fragments with attached methyl groups giving a repeating pattern of $-CHMe(CH_2)_3-$ (or unsaturated equivalents thereof) are also indicative of a terpenoid. Having deduced that the material is likely to be a terpenoid, the next step is to try to identify the original open chain backbone. This can be difficult in highly cyclised structures, especially when there has been a skeletal rearrangement after the initial cyclisation steps. In these, and other difficult cases, once some isoprene fragments have been identified, they can be marked out and this will help to identify the others.

PROBLEM 1

Three of the materials are terpenoids: dehydrocostus lactone, dihydropyrocurzerenone and andrographolide. The positions of the

dehydro costus lactone   dihydropyrocurzerenone   andrographolide

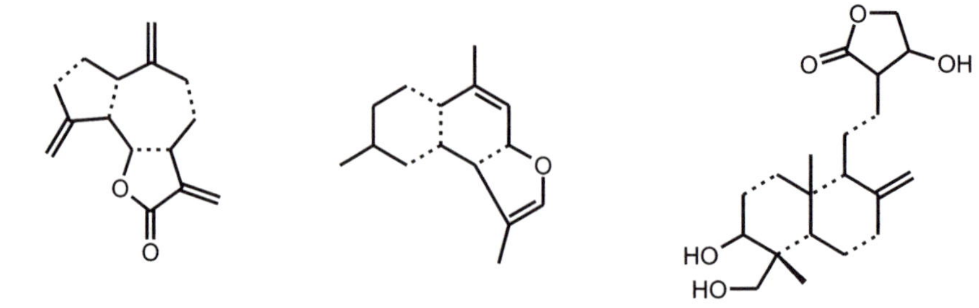

Figure S1

isoprene units in their skeleta are shown in Figure S1. Pinoresinol is a shikimic acid derivative and both prostaglandin PG-H2 and usnic acid are polyketides.

PROBLEM 2

The three terpenoids are acorone, abietic acid and betulinic acid. The positions of the isoprene units in their skeleta are shown in Figure S2. Betulinic acid is slightly unusual in that it has a tail-to-tail link in the skeleton and furthermore, one of the carbon atoms has shifted from its original position in the sequence. This is doubly confusing in this case because the rearranged isoprene unit now looks like one that has been added into the chain the wrong way round. However, the overall pattern is still clearly visible and the position of the rearrangement can easily be worked out. Hyosycamine is an alkaloid. Quercitin is a flavonoid and podophyllotoxin is a phenyltetralin lignan, both of which classes are sub-groups of the shikimic acid family.

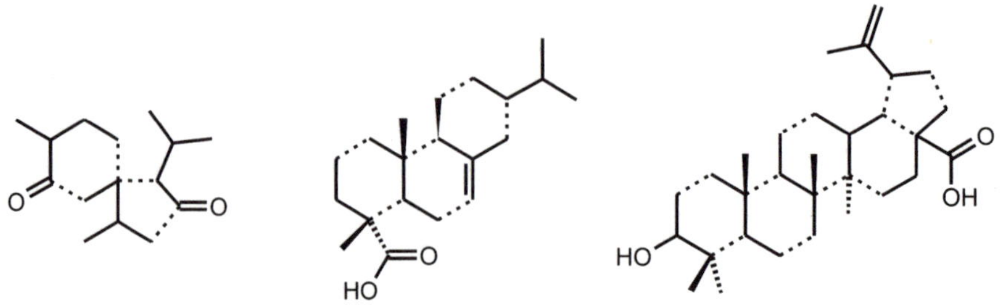

acorone   abietic acid   betulinic acid

Figure S2

PROBLEM 3

Both guaiazulene and neocembrene A are easily recognised as terpenoids. The positions of the isoprene units in their skeleta are shown in Figure S3. Pyrethrin II is partly terpenoid. Consider the ester function in the centre of the molecule. The acid portion of this ester is terpenoid, while the alcohol fragment is polyketide in origin. The isoprene units of the terpenoid acid are shown in Figure S3. In this case, the link between the terpenoid and non-terpenoid parts of the molecule occurs through an ester fragment making it easy to separate the two. It is also possible to find terpenoid and non-terpenoid fragments fused together through carbon–carbon bonds as is the case, for example, with chlorophyll and tocopherol (see Chapter 1 for structures). Griseofulvin and jasmonic acid are polyketides, while strychnine is an alkaloid.

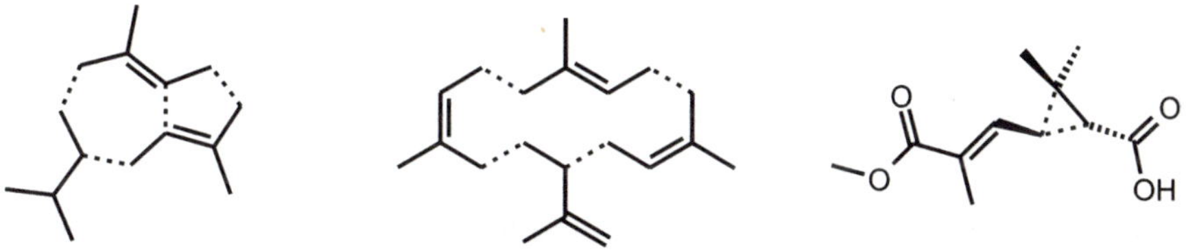

| guaiazulene | neocembrene A | acid part of pyrethrin II |

Figure S3

PROBLEM 4

Luffarin W and uvidin B are terpenoids and the positions of the isoprene units in their skeleta are shown in Figure S4. Capsaicin and capsanthin occur in chilli peppers and capsicums. Capsaicin, which is partly shikimic acid and partly polyketide derived, is responsible for the hot sensation of chillies. Capsanthin is a carotenoid, hence terpenoid, and contributes to the red colour. Since it is a carotenoid, it contains a tail-to-tail link as shown in Figure S4. The long branched chain of oravactaene may look terpenoid but do not be deceived. There is no regular isoprene pattern and it is in fact a polyketide, which has had methyl groups added to the chain. Tetrodotoxin is an alkaloid and one of the most toxic chemicals known. It occurs in the puffer fish, known to sushi eaters as fugu, and also in some species of newts.

PROBLEM 5

The formula $C_{10}H_{18}O$ is four hydrogens short of a saturated structure and therefore implies that there are either two rings or two double bonds

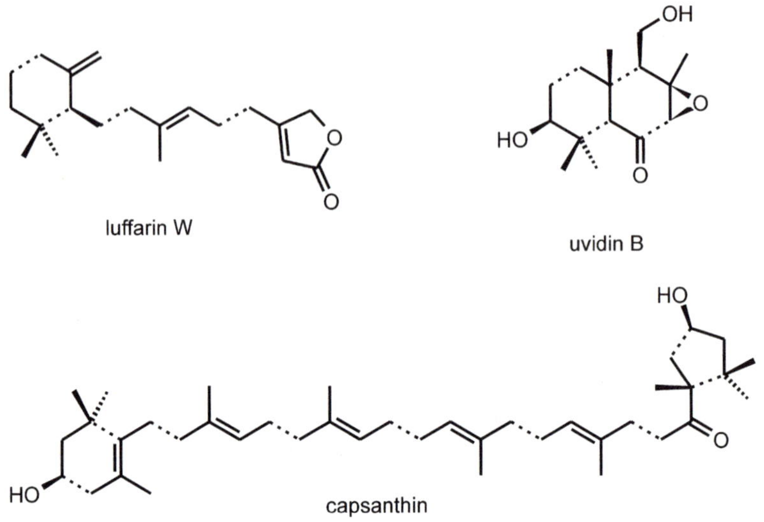

luffarin W

uvidin B

capsanthin

Figure S4

or one of each in the molecule. The uptake of four atoms of hydrogen on hydrogenation confirms that there are two double bonds. The fact that the material does not react with either hydroxylamine or 2,4-dinitrophenylhydrazine tells us that the oxygen function must be either an alcohol or an ether rather than an aldehyde or aketone. The fact that it releases hydrogen slowly when treated with sodium implies an alcohol rather than an acid and the slow rate of hydrogen evolution suggests a tertiary alcohol. Resistance to oxidation supports this proposition. Acid-catalysed rearrangement to an isomeric alcohol, which can be oxidised by manganese dioxide, suggests a tertiary allylic alcohol. Dehydration of the saturated alcohol gives the two most thermodynamically stable olefins. The ozonolysis to give 2-methylheptan-6-one tells us the bulk of the structure. The remaining two carbons must have been lost as acetaldehyde and so we can piece the structure together as shown in Figure S5. Knowledge of the isoprene rule would help in this structural determination since the empirical formula would have indicated either a 2,6-dimethyloctane skeleton or one of the monoterpenoid ring systems, depending on the number of double bonds in the molecule.

PROBLEM 6

Carvone is a double bond isomer of carvacrol. Acids can isomerise double bonds and so isomerise carvone to carvacrol. There are many

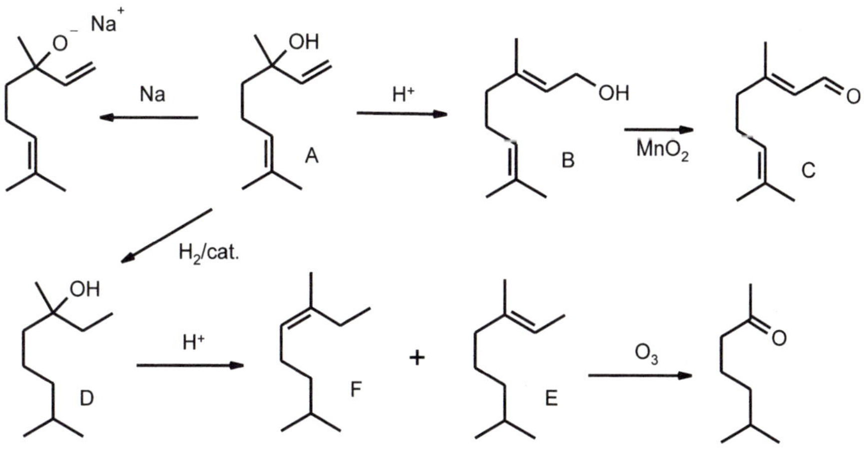

Figure S5

potential reaction sequences which can bring this about; one is shown in Figure S6. Small amounts of carvacrol would spoil the odour of carvone but more serious is the risk that extensive conversion would occur. The resonance energy released in forming an aromatic ring from three double bonds is of the order of 27 kcal mol^{-1}. On a large scale, this would represent such a liberation of heat that still would be unlikely to be able to lose it quickly enough and so there would be a serious risk of an explosion.

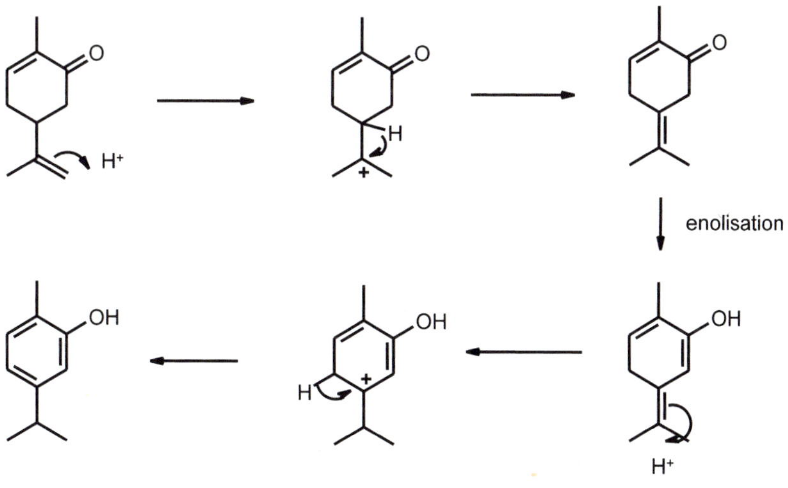

Figure S6

PROBLEM 7

The trisubstituted bond of dihydromyrcene (P7.2) is more reactive than
the one at the far end of the molecule and so, when treated with an acid,
it tends to be more easily protonated. Direct hydration of olefins is not
very favourable and so the trapping of the initial carbocation (7.4) by the
other double bond is very competitive. The regioselectivity is controlled
by the stability of the carbocation which will be produced. If the double
bond were to add at the more hindered end, a six-membered ring would
be produced but the new carbocation would be primary. Therefore, the
addition takes place the other way round, as shown in Figure S7, to give
(P7.5). This species has a slightly less favourable seven-membered ring
but a much preferred secondary carbocation. A 1,2-carbon shift leads to
a ring contraction and hence a more stable six-membered ring, thus
giving another secondary carbocation (P7.6) which is quenched by water
to give the observed product (P7.3). The problem, if we are seeking to
produce dihydromyrcenol, is that of the low nucleophilicity of water.
Increasing the proportion of water present in the system will not help
because the other double bond is always close by. Therefore, the best way
to overcome this selectivity issue is to use a better nucleophile. Acetic
acid, for instance, would lead to the formation of dihydromyrcenyl
acetate which could then be hydrolysed to the desired product. Sulfuric
acid is another possibility. In concentrated sulfuric acid the initial
carbocation (P7.4) is trapped as the sulfate ester and this is hydrolysed on
dilution of the medium.

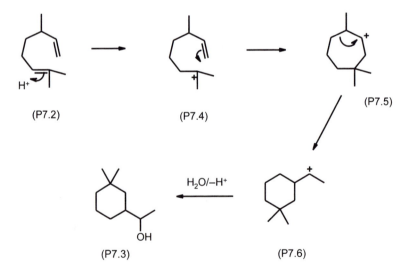

Figure S7

PROBLEM 8

The initial reaction is a simple Diels–Alder addition to give the aldehyde
(P8.2). Addition of stannic chloride to the oxygen atom of the aldehyde
function leads to a transannular addition to give the carbocation (P8.3).
A 1,2-carbon shift gives carbocation (P8.4). This carbocation is drawn in
two different ways. In the first representation, all the atoms are drawn
where they were in the starting material, only the bonds which have been
broken or made having been changed. This technique helps to avoid
making mistakes when drawing out carbocation rearrangements. Having
ensured that all the atom connectivities are correct, the structure can
then be redrawn in a way that shows the new species more clearly, as has
been done here. A quick check to verify that the connectivities of both
representations are identical is always a good idea at this point. Having
done that, we can now see that a 1,2-hydrogen shift leads to an oxygen
stabilised carbocation (P8.5) which can lose stannic chloride to give the
observed product (P8.1) (Figure S8).

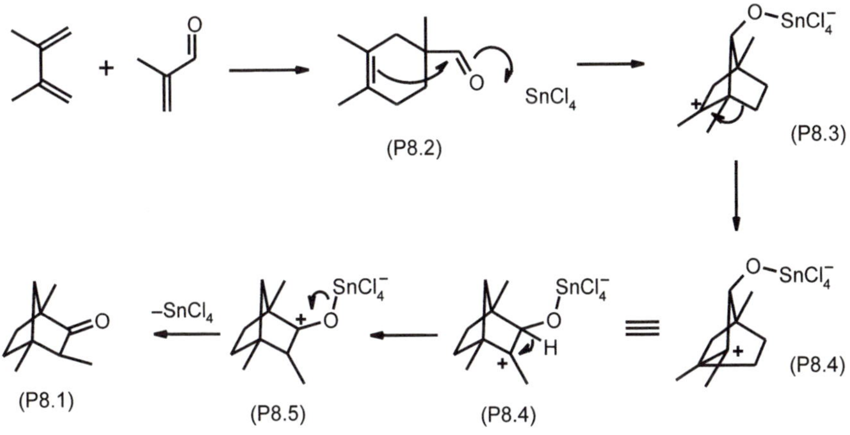

(P8.2)

(P8.3)

(P8.1)

(P8.5)

(P8.4)

(P8.4)

Figure S8

PROBLEM 9

Protonation of the cyclopropane ring of (P9.1) causes ring opening to
give the carbocation (P9.3) which can lose a proton to give the observed
product (P9.2) and the driving force is the relief of ring strain. This
example also serves to illustrate the similarities between the chemistry of
the santalanes (Chapter 6) and that of camphor and bornane type
monoterpenoids (Chapter 5) (Figure S9).

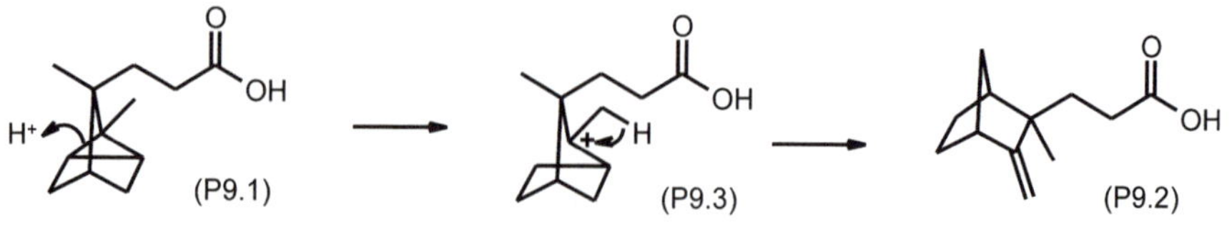

Figure S9

PROBLEM 10

The positive charge of the bisabolyl cation can be trapped by the double bond in the six-membered ring and this can happen in one of two ways. On the left of Figure S10, we see it being trapped so as to leave a tertiary

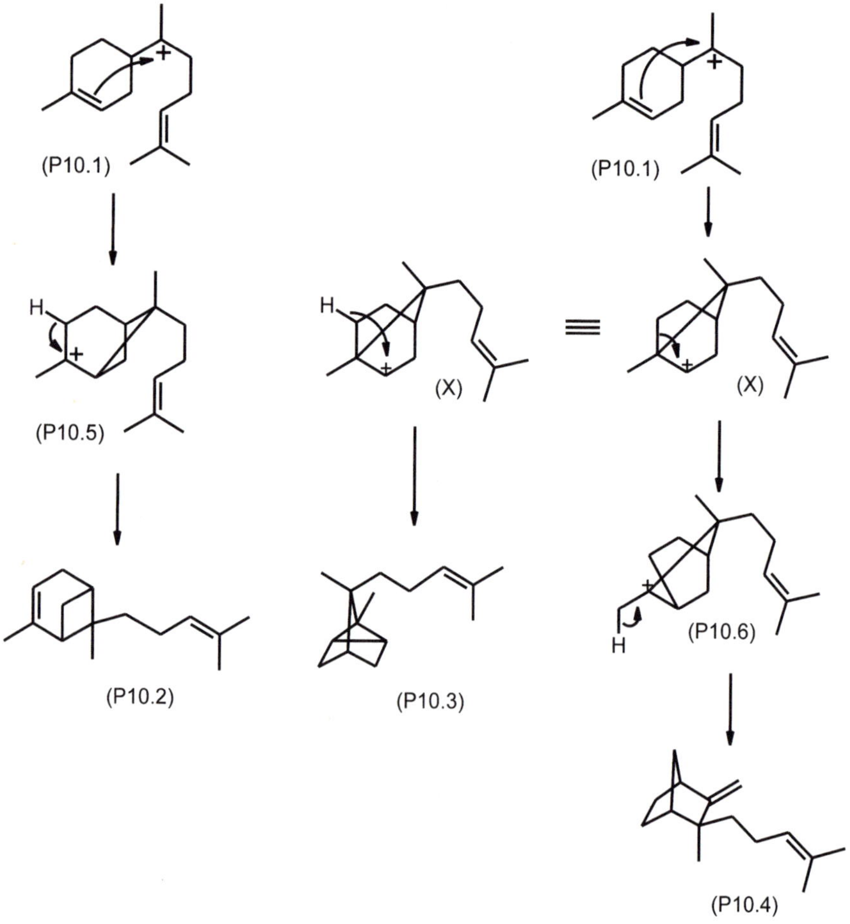

Figure S10

cation but a strained four-membered ring giving carbocation (P10.5). This carbocation can eliminate a proton to give α-bergamotene (P10.2). The alternative for (P10.1) is for the cation to add to the other end of the double bond, as shown on the right of Figure S10. This gives a less strained ring system but a less stable secondary carbocation. The resultant structure is the intermediate (X). Transannular elimination of a proton from (X) gives α-santalene (P10.3). A 1,2-carbon shift in (X) gives a new 2,2,1-bicycloheptyl carbocation (P10.6) and this can eliminate a proton to give β-santalene (P10.4).

PROBLEM 11

The transition from the bisabolane skeleton to the cuparane is not favourable energetically. Initially a secondary carbocation is formed from a tertiary, though a hydrogen shift quickly moves the positive charge back to a tertiary centre. The five-membered ring which is formed is very cluttered and, therefore, suffers from steric strain. The driving force must, therefore, come from the protein which catalyses the reaction. The 1,2-carbon shift to the chamigrane skeleton relieves some of the steric strain but this creates a spiro centre, the strain of which is relieved by the 1,2-carbon shift which creates the thujopsane skeleton.

PROBLEM 12

The cyclisation of the acorane skeleton to the cedrane skeleton creates steric and ring strain. The driving force must come from the enzyme involved. The first 1,2-carbon shift relieves much of the ring strain but still leaves steric strain but, in doing so, forms a secondary carbocation from a tertiary one. This is reversed by the second 1,2-carbon shift and the final methyl shift helps reduce steric strain.

PROBLEM 13

Figure S13 shows a possible pathway from camphene to each of the three observed carbocations. Note that one serves as an intermediate for another. Other pathways are possible also and in the reaction, probably all of them will operate to different degrees.

PROBLEM 14

Complexation of the zinc bromide (a Lewis acid) with the oxygen atom of pinene oxide leads to cleavage of the epoxide ring. This occurs in the direction which generates a tertiary carbocation. The ring strain of

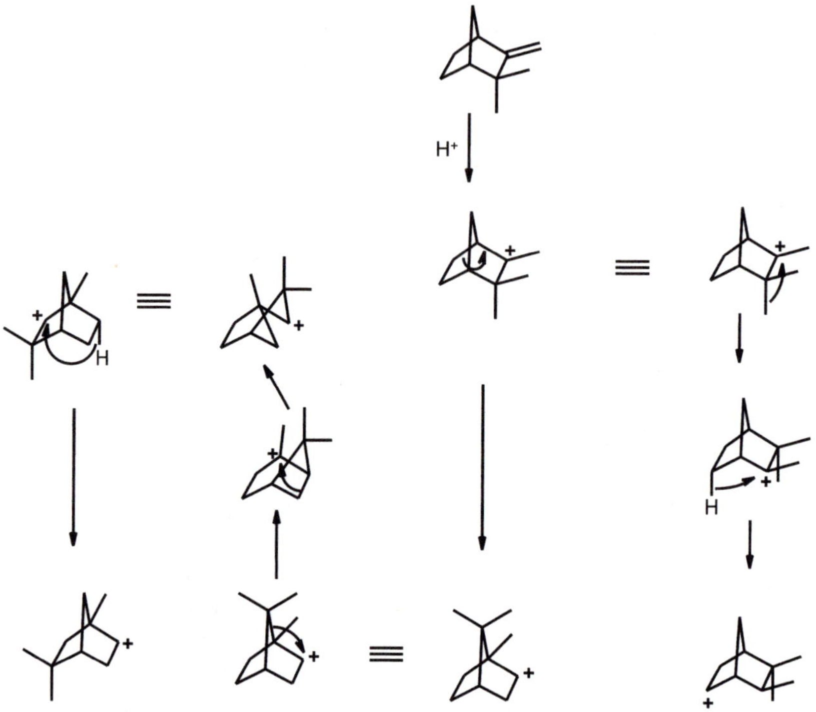

Figure S13

the four-membered ring can now be relieved and, in accordance with the rules of carbocation chemistry, it is the larger group which moves. A flow of electrons back from the zinc halide and across the molecule, quenches the positive charge and thus generates campholenic aldehyde (Figure S14).

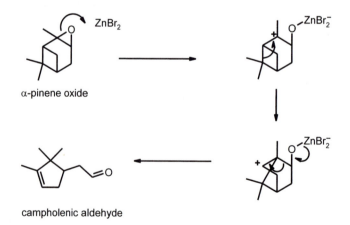

α-pinene oxide

campholenic aldehyde

Figure S14

PROBLEM 15

The initial reaction is a Diels–Alder addition to give a mixture of the two possible adducts (P15.3) and (P15.4). It can be seen that the former is perfectly set up for a very interesting double intramolecular cyclisation when treated with a Bronsted acid (Figure S15).

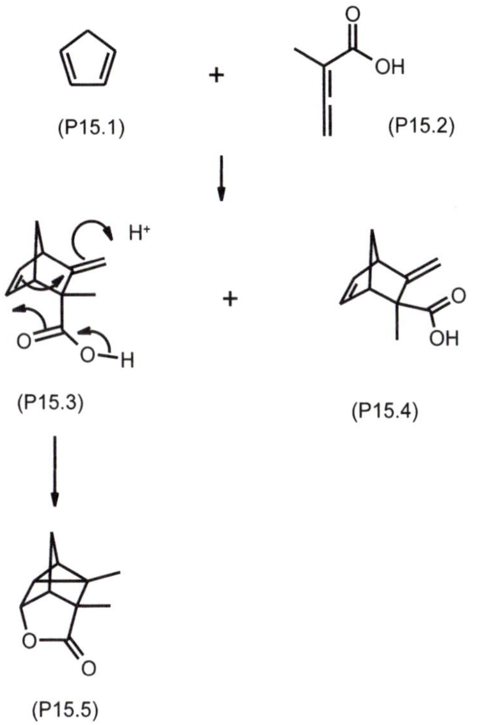

(P15.1)

(P15.2)

(P15.3)

(P15.4)

(P15.5)

Figure S15

PROBLEM 16

Complexation of the aluminium chloride with the carbonyl group of the ocimenone leads to a flow of electrons through the molecule to give a six-membered ring and a tertiary carbocation on the isopropyl substituents. A 1,2-carbon shift involving the very electron-rich carbon atom attached to the oxygen, expands the ring to a seven-membered one and produces a secondary carbocation. However, this carbocation is quickly trapped by the electrons of the very electron rich double bond opposite. This allows for charge neutralisation and forms the fused ring system of filifolone.

The reaction also takes place with *E*-ocimenone but, in that case, the initial ring formation is more difficult to visualise (and draw!) (Figure S16).

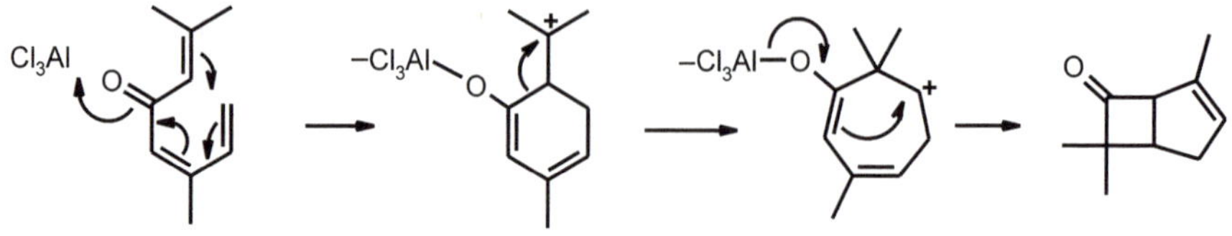

Z-ocimenone   filifolone

Figure S16

PROBLEM 17

Olefin (P17.10) is formed from (P17.4) and olefin (P17.11) from (P17.9) as shown in Figure S17.

PROBLEM 18

The first step of the process is the hydrolysis of the lactone function. This is followed by abstraction of the γ-hydrogen from the α,β-unsaturated ketone function. Reprotonation of the anion occurs at the α-position as explained in Chapter 6, giving the enol (P18.5). This can be reketonised to (P18.7) which can then be deprotonated to give the anion (P18.2). In basic media, all these forms will be present in equilibrium. However, the anion (P18.2) can undergo the intramolecular Michael addition shown in Figure 7.9 and so, eventually, all of the material will be converted through to the product (Figure S18).

PROBLEM 19

The most acidic proton in either of the starting materials is that between the ketone and ester functions of ethyl acetoacetate and this is lost to give the corresponding anion which then adds to ethyl crotonate in a Michael reaction to give, after reketonisation of the ester, the adduct (P19.1). There are three possible sites for anion formation in (P19.1) but the only anion which can react to give a six-membered, rather than a four-membered ring on intramolecular reaction is the one shown in Figure S19. This is, therefore, the one which reacts and the product is (P19.4) as shown. The ester group of (P19.4) is then hydrolysed and the resultant carboxylic acid decarboxylated to give the desired product. This

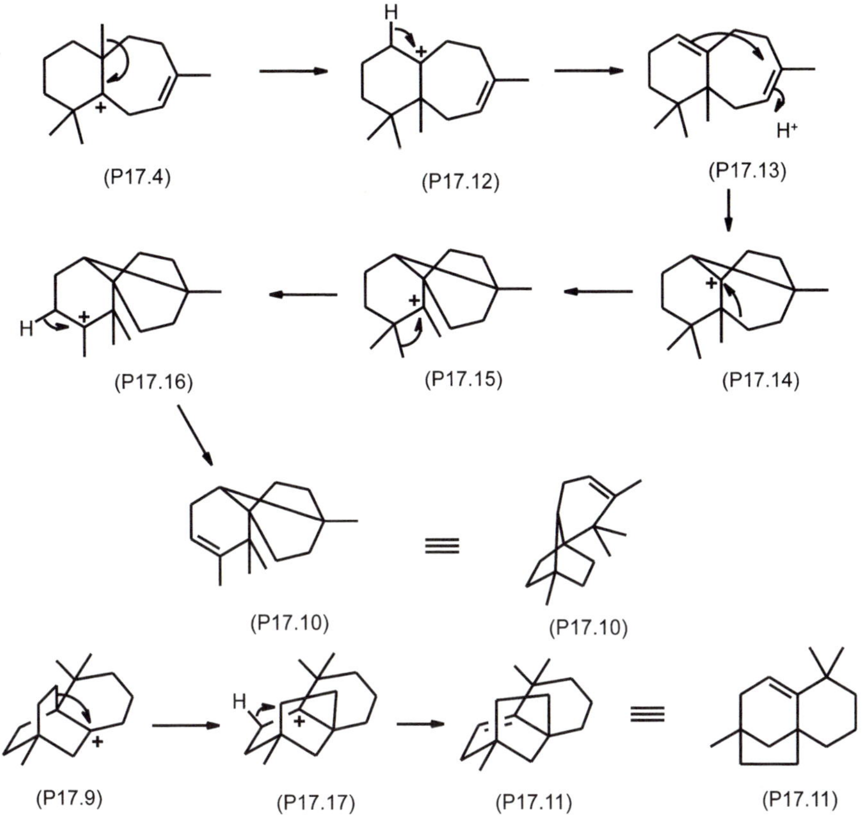

Figure S17

combination of Michael addition followed by Claisen or aldol condensation is known as the Robinson annulation reaction. The mono-enol form of the product is more stable because of the resonance energy gained by overlap between the π-electrons of the remaining ketone and those of the double bond of the enol.

PROBLEM 20

Deprotonation of phenol (P20.2) gives an anion which is equivalent to the enolate anion of cyclohexadienone (P20.3). This can be alkylated on carbon, rather than oxygen, if the reaction conditions are correct. The product ketone (P20.4) can be re-enolised to give *o*-cresol (P20.5) and the whole process repeated. After the addition of three methyl groups, the product ketone (P20.1) cannot enolise and so the alkylation process ceases (Figure S20).

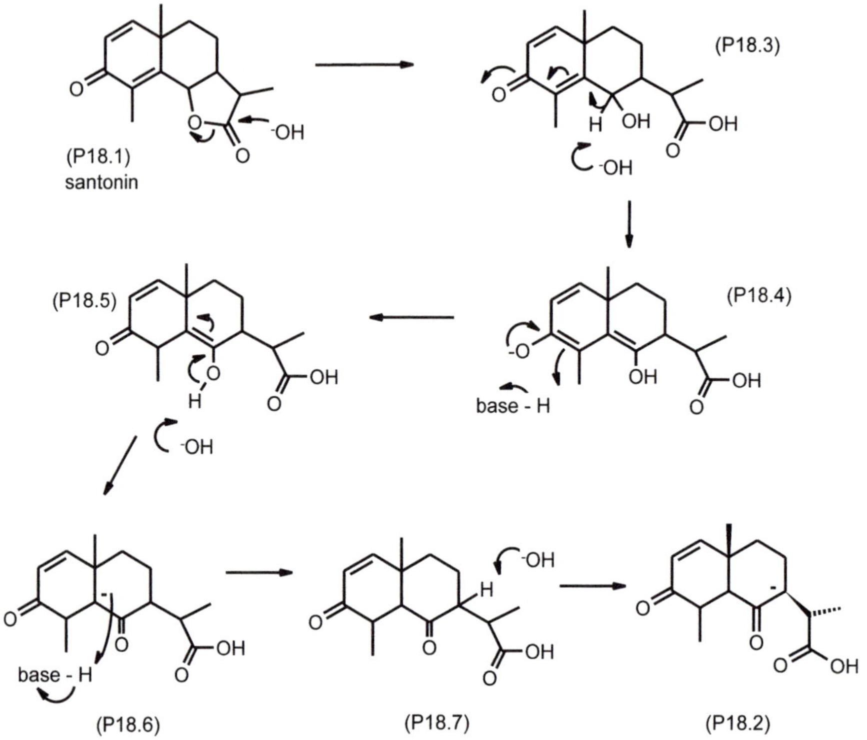

Figure S18

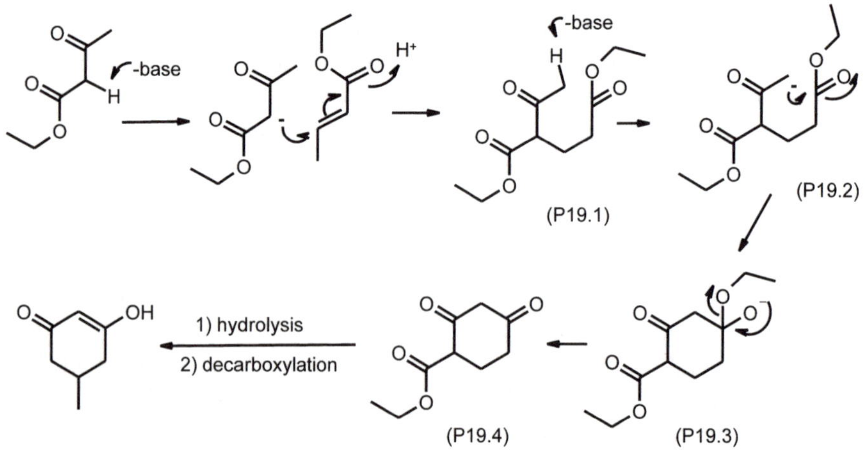

Figure S19

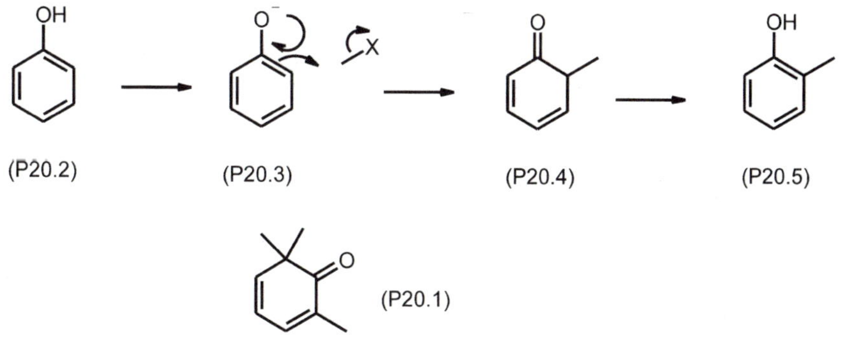

(P20.2) (P20.3) (P20.4) (P20.5)

(P20.1)

Figure S20

PROBLEM 21

The initial reaction is a Prins reaction, catalysed by the boron trifluoride complex. In order to achieve maximum overlap of the π-orbitals of the olefin and aldehyde groups, the aldehyde must approach the olefin from below, as will easily be seen using molecular models. This means that the resultant alcohol function is located on the downward side of the molecule as shown in Figure S21. A 1,2-carbon shift followed by a transannular bond formation with concomitant loss of a proton, provides the skeletal rearrangement to the product. It may not be too obvious in the figure, but an experiment with molecular models will soon

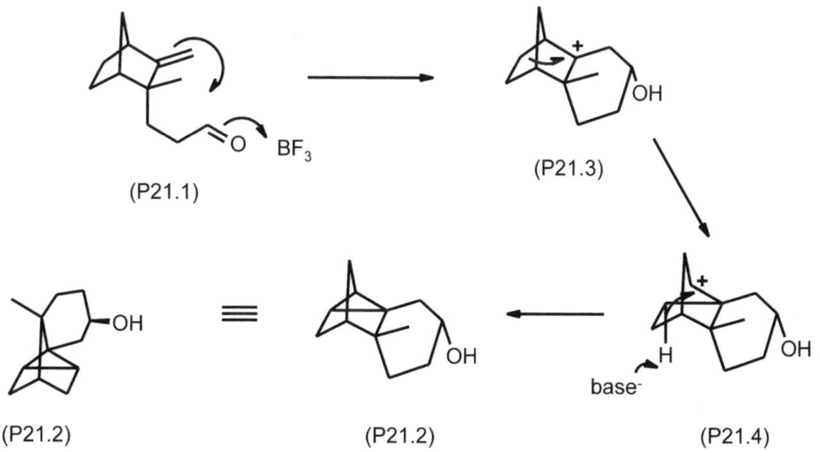

(P21.1) (P21.3)

(P21.2) ≡ (P21.2) base⁻ (P21.4)

Figure S21

confirm that the two representations of (P21.2) in Figure S21 are the same and that the stereochemistry of the alcohol function is as originally shown.

PROBLEM 22

The routes to the two products are shown in Figure S22. Both start with the addition of the formaldehyde to give the carbocation (P22.4). Elimination of the proton adjacent to the positive charge gives (P22.2). To form (P22.3) requires a transannular elimination of a proton with the formation of a new three-membered ring.

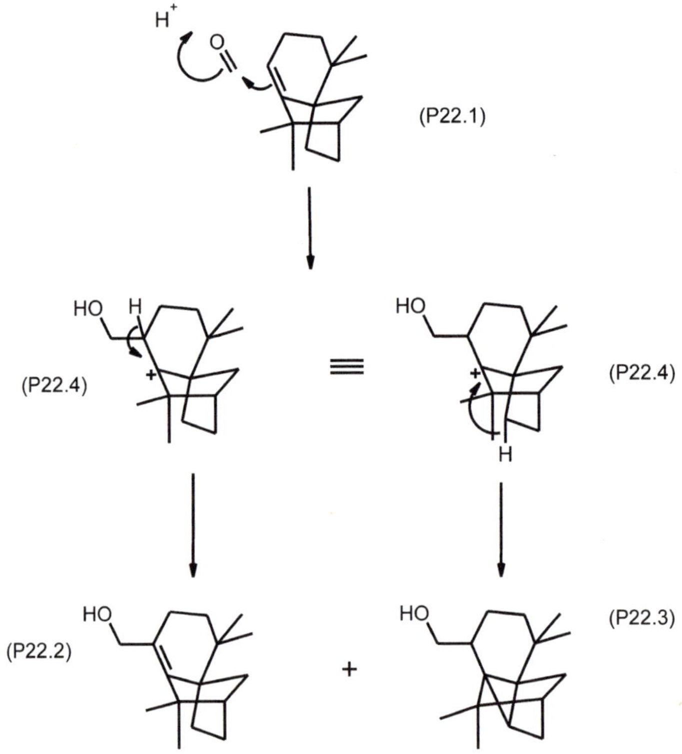

Figure S22

PROBLEM 23

As described in Chapter 7, the chemistry of caryophyllene is dominated by the drive to reduce ring strain. Forming the epoxide increases the strain because of the addition of a three-membered ring. Protonation of the epoxide allows the three-membered ring to spring open and, therefore, releases the strain that was added by its introduction. Not only that, but the strain of having a *trans* double bond in a nine-membered ring has been eliminated as the bond which was a *trans* double bond, is now a single bond. However, the carbocation (P23.4) which results still suffers from ring strain as it contains two strained rings (one four-membered and one nine) fused together. The upper part of Figure S23 shows one way in which the strain is reduced. Transannular trapping of the carbocation by the methylene double bond, serves to relieve the

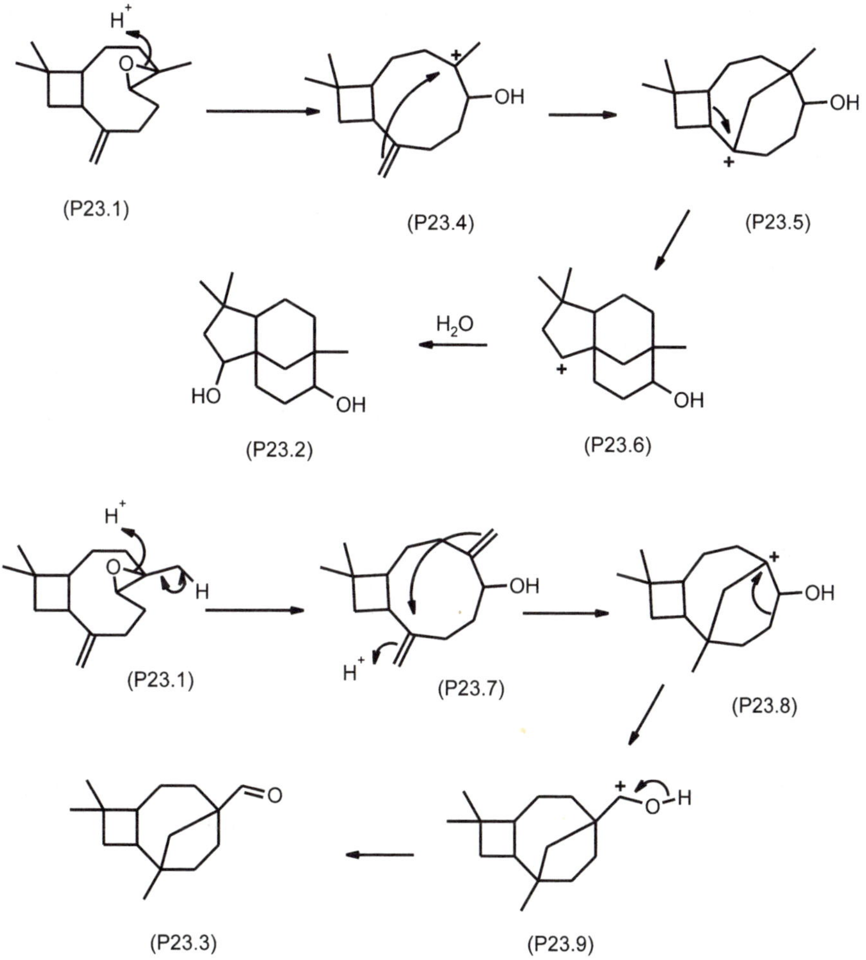
H+
O
OH
OH
+
+
(P23.1)
(P23.4)
(P23.5)
H_2O
HO
OH
OH
(P23.2)
(P23.6)
H+
O
H
OH
OH
+
H+
(P23.1)
(P23.7)
(P23.8)
O
O-H
(P23.3)
(P23.9)

Figure S23

steric strain of the nine-membered ring by formation of fused six- and seven-membered ones, giving the new carbocation (P23.5). It is now possible to release the strain of the four-membered ring by a 1,2-carbon shift to give the carbocation (P23.6) which contains only five- and six-membered rings fused together. Quenching of this last carbocation by water gives the observed diol (P23.2).

Another possible path for the initial carbocation (P23.4) is to eliminate a proton to give the dienol (P23.7) as shown in the lower part of Figure S23. Protonation of the methylene double bond next to the four-membered ring, now allows transannular closure in the opposite direction to give carbocation (P23.8). A 1,2-carbon shift reduces the strain associated with a bridgehead carbon atom. Although the cation centre has gone from tertiary to primary, it is now stabilised by the adjacent oxygen atom and loss of a proton from (P23.9) generates the aldehyde (P23.3).

PROBLEM 24

Initially, the starting carbocation (P24.2) is trapped by the nearest double bond, generating a four-membered ring (P24.3), as in the biosynthesis of caryophyllene. A 1,2-carbon shift giving (P24.4) relieves the strain of the four-membered ring. This carbocation is then trapped by the double bond across the ring relieving the strain of the nine-membered ring and producing carbocation (P24.5) which contains only five- and six-membered rings, though the way in which they are fused does create some strain. Structure (P24.5) is drawn in two ways. The first shows its relationship to its precursor (P24.4) and the second shows the ring system more clearly. Carbocation (P24.5) then undergoes a 1,3-hydrogen shift to give (P24.6) and this enables the ring strain to be relieved by a 1,2-carbon shift to give (P24.7). A 1,2-methyl shift to (P24.8) then allows elimination of a proton to give silphiperfol-6-ene (P24.1) (Figure S24).

PROBLEM 25

Ambrinol (P25.2) can be produced by a simple acid-catalysed cyclisation of dihydro-γ-ionone (P25.1) as shown in Figure S25 and so the latter is quite likely to be a precursor for the former in natural ambergris.

PROBLEM 26

The reaction takes place through the enol form of the ester function as shown in Figure S26. The electrons from the enolate add to the double bond and the resultant carbanion is quenched by a proton. The role of

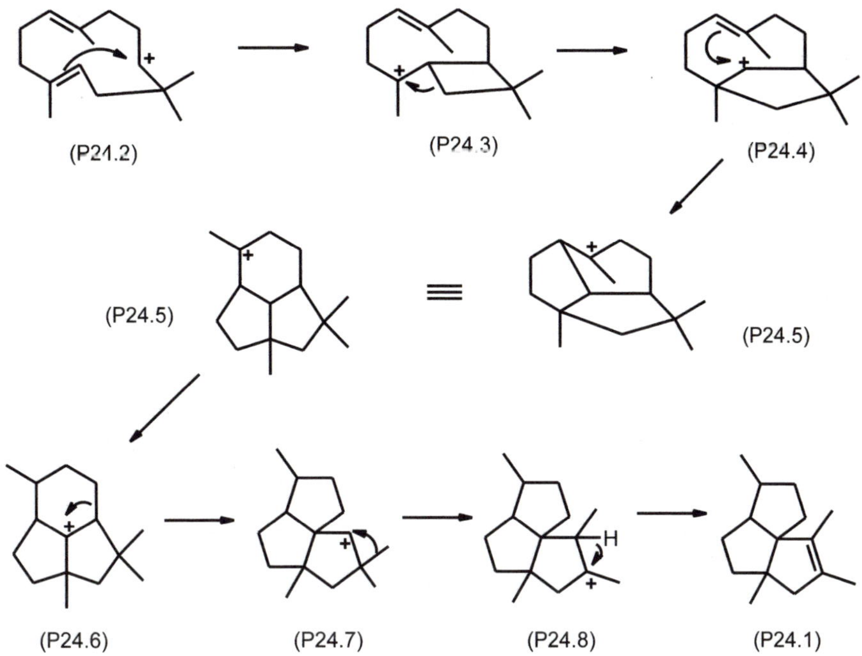

Figure S24

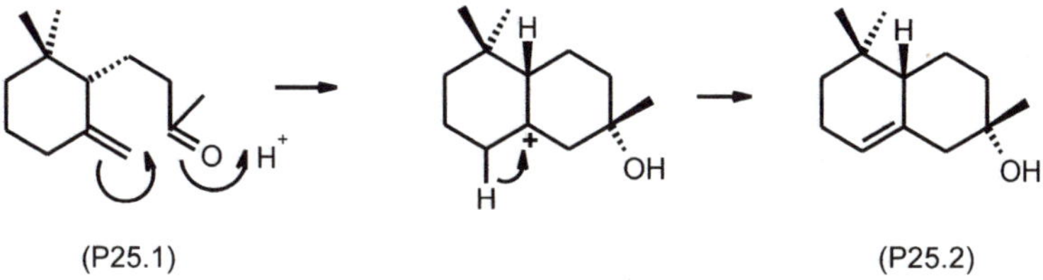

Figure S25

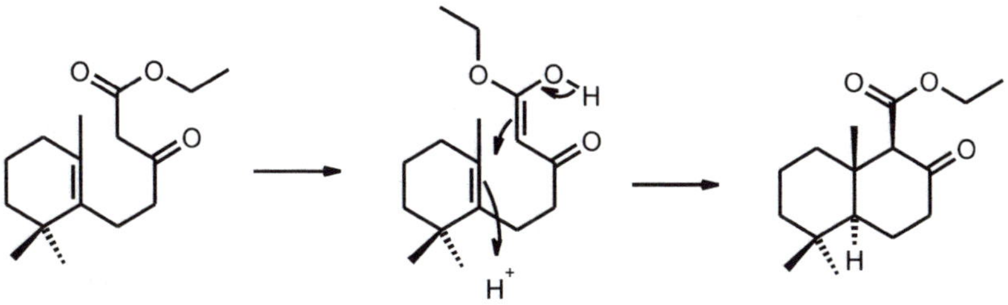

Figure S26

the stannic chloride catalyst is to activate the enol form through complexation with it. The bulky enolate complex blocks one face of the olefin and this, coupled with the *trans-anti*-periplanar effect, results in the addition of the proton from the opposite side and giving the observed *trans*-geometry of the product.

PROBLEM 27

The reaction is an electrocyclic reaction known as the Claisen Reaction. The mechanism is shown in Figure S27.

Figure S27

PROBLEM 28

Protonation of the epoxide leads to opening of it to relieve the ring strain associated with its three-membered ring. The direction of opening is such that the incipient carbocation is tertiary. However, the addition of the carbonyl group must be synchronous since the stereochemistry of the ring remains unchanged. If a free carbocation were to have been produced, then some racemisation at the oxygenated position of the decalin ring would be likely as addition to the planar cation would be possible from either side. The positive charge on (P28.3) is quenched by the electrons of the alcohol group and a proton is eliminated to give Jeger's ketal (Figure S28).

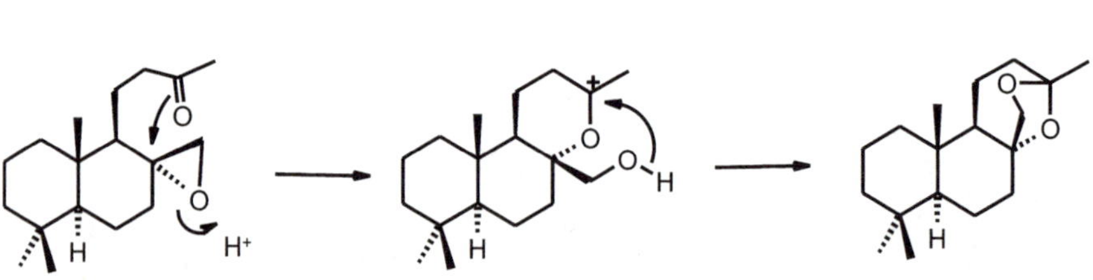

(P28.1)   (P28.3)   (P28.2)
                                    Jeger's ketal

Figure S28

PROBLEM 29

There are three double bonds each in a β,γ-position to the alcohol group. Two give identical products as shown in the top line of Figure S29 but the third gives a quite different pair of products from the same electrocyclic mechanism. Statistically, one might expect the yield of trimethylcyclohexene and hepta-1,6-diene-4-one to be 33%. In fact they only occur to the extent of 20%, indicating that there is some preference for the desired reaction pathway. The reasons for this probably lie in stereoelectronic factors.

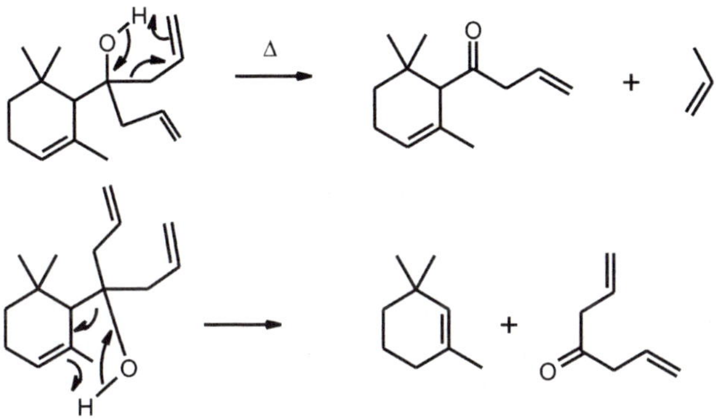

Figure S29

PROBLEM 30

The selectivity is possible because one of the alcohol groups is allylic. Protonation of this allylic alcohol gives an allylic carbocation (P30.5) in which the positive charge is stabilised by delocalisation. The cation could be trapped at this point by the adjacent alcohol group, thus inducing cyclisation. Alternatively, it could eliminate to give the dienes (P30.3) and (P30.4). In the latter case, the allylic carbocation (P30.5) is produced again as the first stage of the cyclisation. In this case it is generated by protonation of the diene function. The E/Z-stereochemistry of the double bond is lost at this stage and both isomers give the same cation. Protonation at the opposite end of the diene system would also give an allylic carbocation but in this case, intramolecular trapping by the oxygen atom would give a strained four-membered oxetane ring instead of a furan. It is possible that the alternative carbocation does form but does not react and eliminates a proton to regenerate the diene. It is also likely that the observed direction of protonation is assisted by the presence of the alcohol group at the right distance to form the furan ring as shown in Figure S30.

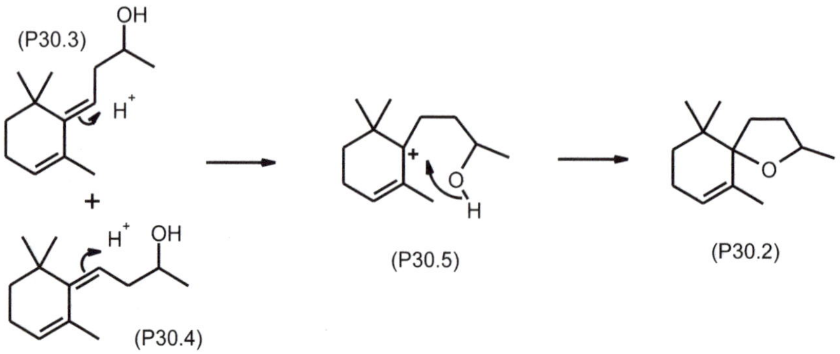

Figure S30

PROBLEM 31

Synthesis from β-pinene – Advantages

The starting material is a by-product of the paper industry, this is particularly important for companies also involved in that industry. The reactions are all very simple batch chemistry, although the first is best carried out in the gas phase.

Synthesis from β-pinene – Disadvantages

The feedstock supply is limited by the volume available from the paper industry. Introduction of chlorine into a molecule means that great care must be taken to remove it all and stringent controls put in place to ensure that the techniques employed for its total removal (below ppb levels) are effective. The use of sodium acetate as a nucleophile to displace chlorine results in the generation of sodium chloride in stoichiometric amounts as an effluent which must be disposed of. The acetic acid generated in the next step must also be either disposed of or cleaned up for recycling.

Synthesis from Isobutylene and Formaldehyde – Advantages

The starting materials are inexpensive and are available in essentially unlimited supply. The only effluent generated is water. There are only four process steps, two of which run in parallel.

Synthesis from Isobutylene and Formaldehyde – Disadvantages

The reactions are all gas phase and require careful control and skilful engineering.

Overall, the second synthesis is much better for the modern chemical industry because of its better feedstock position and vastly superior environmental performance.

PROBLEM 32

As stated in Chapter 9, the atom efficiency of a reaction sequence is defined as the molecular weight of the desired product divided by the sum of the molecular weights of all of the products of the synthesis, multiplied by 100 to give a percentage figure. Let us look at each route in turn.

In route A, the products are 2-methylhept-2-ene-6-one, two moles of sodium bromide and one of sodium acetate. The molecular weights of these are 126, 206 and 82, respectively. This gives an atom efficiency of 30%.

Route B produces only the desired ketone and one mole of water (MW = 18) and so the atom efficiency is 88%.

In addition to the desired product, route C produces hydrogen, sodium chloride and additional ammonia (from quenching of the sodium amide with ammonium chloride), zinc and copper hydroxides from the reduction (for simplicity we will assume one mole of each, and no other by-products), one third of a mole of phosphorous acid, sodium bromide, ethanol (from the ester hydrolysis) and carbon dioxide (from the decarboxylation). The molecular weights involved are therefore: 126, 1, 58.5, 17, 99, 97.5, 27.3, 103, 46 and 44, respectively. This gives an atom efficiency of 20%.

Route D generates potassium chloride, carbon dioxide and ethanol as by-products. This gives an atom efficiency of 43%.

These atom efficiencies are very useful in predicting the likelihood of a commercially attractive synthesis being produced. The very low figure for route C rules it out straight away, the low atom utilisation and the nature of the by-products indicate a serious effluent problem which will add significantly to the process costs.

Similarly, route B is by far the most attractive in terms of atom utilisation and is one of the shortest routes also, hence probably has lower process costs.

Route A is shorter than route D but has a poorer atom utilisation. Route D requires handling acetylene which is not an attractive proposition on large scale.

The conclusion would be to investigate route B first and if it proved impossible then routes A and D could be studied in competition. The environmental performance of route C is so poor that it would not warrant development work as it stands. If alternatives could be found for each stage, then it would look more attractive; in fact it would be converted into route D.

PROBLEM 33

The hetero-Diels–Alder reaction gives a single isomer (the polarisation of the two components ensure that the isoprene adds only one way round to the double bond of the benzaldehyde) whereas the Prins reaction, because it goes through a carbocationic intermediate which can eliminate a proton from any of three positions, gives a mixture of three positional isomers as shown in Figure S33.

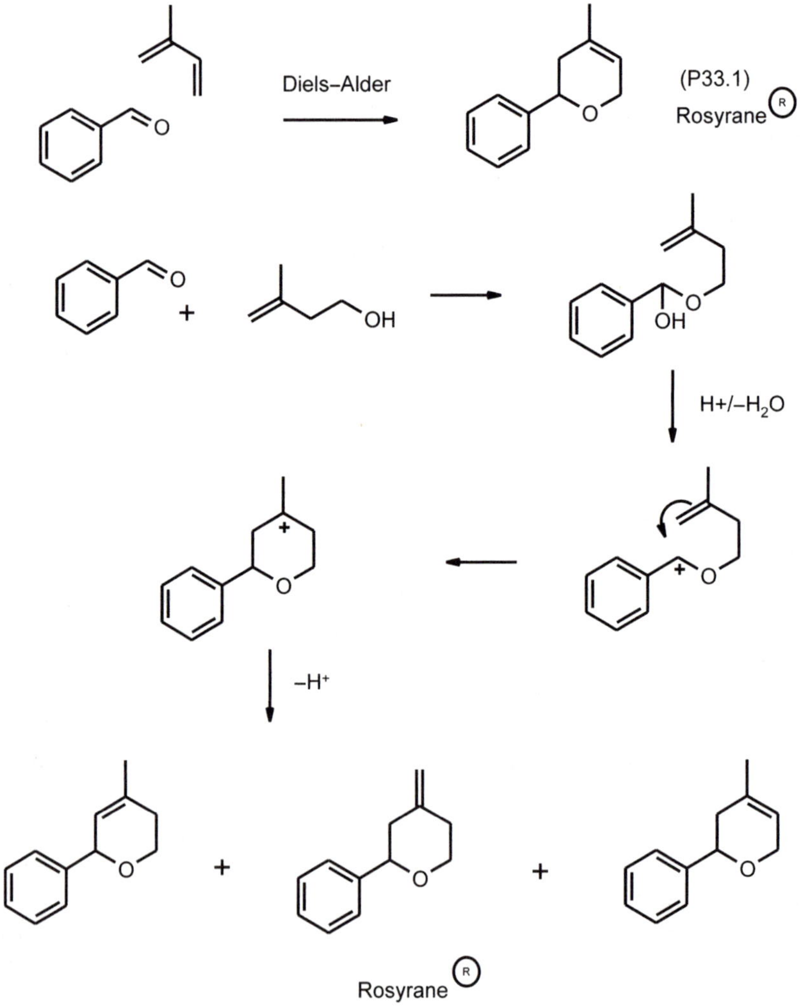

Figure S33

PROBLEM 34

The synthesis starts with hydroformylation of 2-carene (P34.2) to give formylcarane (P32.3). Hydroformylation always puts the newly introduced substituent onto the least hindered end of the double bond. Treatment of formyl carane with excess formaldehyde in the presence of base sets the Tollens reaction in motion to give the diol (P33.4). The first stage of the Tollens reaction is the aldol condensation. If there is excess formaldehyde present, every enolisable hydrogen atom will be replaced by a formyl group. In this case there is only one enolisable hydrogen in the substrate. The excess formaldehyde then reduces all the aldehyde functions in the product by means of the Canizzaro reaction. In this case there is only one carbonyl group to be reduced and the product is the diol. Ketalisation with acetone in the presence of an acid catalyst then gives Spirambrene® (P34.1) (Figure S34)

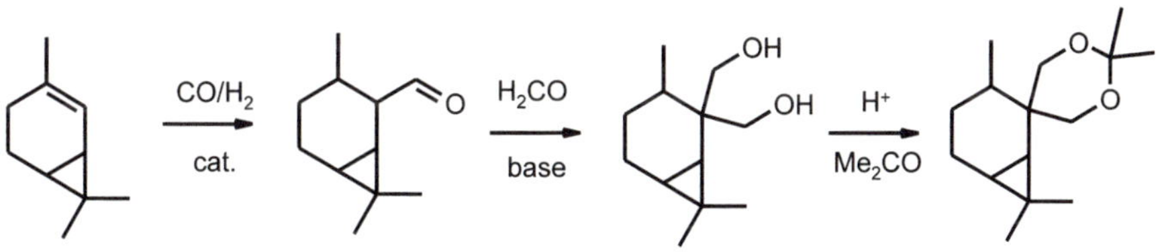

| (P34.2) | (P34.3) | (P34.4) | (P34.1) |

Figure S34

Author Index

Subject Index